African Indigenous Vegetables in Urban Agriculture

African Indigenous Vegetables in Urban Agriculture

Edited by Charlie M. Shackleton,
Margaret W. Pasquini and Axel W. Drescher

earthscan
from Routledge

First published by Earthscan in the UK and USA in 2009

For a full list of publications please contact:
Earthscan
2 Park Square, Milton Park, Abingdon, Oxon OX14 4RN
605 Third Avenue, New York, NY 10017

First issued in hardback 2016

Earthscan is an imprint of the Taylor & Francis Group, an informa business

ISBN 13: 978-1-138-14884-0 (hbk)
ISBN 13: 978-1-84407-715-1 (pbk)

Typeset by MapSet Ltd, Gateshead, UK
Cover design by Grounded Design

A catalogue record for this book is available from the British Library

Library of Congress Cataloging-in-Publication Data
African indigenous vegetables in urban agriculture / edited by C.M. Shackleton, M. Pasquini and A.W. Drescher. — 1st ed.
p. cm.
Includes bibliographical references and index.
ISBN 978-1-84407-715-1 (pbk.)
1. Vegetables—Africa. 2. Endemic plants—Africa. 3. Urban agriculture—Africa. I. Shackleton, C. M. (Charles Michael) II. Pasquini, M. (Margaret) III. Drescher, Axel W.
SB323.A35A37 2009
635.096—dc22

2008052631

Contents

List of Figures, Tables, Boxes and Plates

Figures

Tables

Boxes

Plates

The plate section is located between Chapters 6 and 7 (between pages 224 and 225)

List of Contributors

Mary O. Abukutsa-Onyango is professor of horticulture in the Department of Horticulture, Jomo Kenyatta University of Agriculture and Technology in Kenya. Professor Abukutsa-Onyango has been involved in horticultural research on African indigenous leafy vegetables since 1990, both at the academic level and with farmers, and is widely published. Selection and provision of quality seed and development of production protocols have been core focuses of her activities.

Enoch Achigan-Dako is a molecular systematist working as a research officer at the Laboratory of Plant Sciences, Faculty of Agronomic Sciences, University of Abomey Calavi, Benin. Dr Achigan-Dako is broadly interested in the conservation and utilization of plant genetic resources with an emphasis on the lineage relationships between cultivated plants and their wild relatives and how the wild potential could be used for crops breeding. Currently, Dr Enoch Achigan-Dako's work focuses on the family Cucurbitaceae, among which many species are used as vegetables in Africa.

Bianca Ambrose-Oji is an environmental sociologist working as a research fellow at CAZS Natural Resources, Bangor University, Wales. Her research interests are broadly the areas of the use and management of natural resources in Africa and Asia, with a strong focus on the contribution of non-timber forest products and wild resources to livelihoods and poverty alleviation. Much of this work has looked into the links between rural, peri-urban and urban livelihood systems. Dr Ambrose-Oji has contributed to the IndigenoVeg project activities, and is currently involved with a DARWIN-funded project on indigenous vegetables in West Africa.

Françoise Assogba-Komlan is an agronomist with a specialization in soil science. She is the leader of the Vegetable Research Programme at the National Institute of Agricultural Research in Benin (INRAB). Dr Assogba-Komlan's research focuses primarily on the regeneration and conservation of soil fertility in urban and peri-urban agricultural systems (mainly vegetable production systems). Over the past ten years, she has also been working on the sustainable production and promotion of indigenous vegetables and has led projects that inventory the indigenous species used throughout the country, the collection of seed resources, and various on-station and on-farm trials with cultivated and wild species.

Andreas de Neergaard is associate professor and director of studies at Life Sciences, University of Copenhagen, Denmark. His research centres on nutrient and carbon flows in agro-ecosystems, biological soil fertility and land use–soil quality interactions. Dr de Neergaard has worked extensively in interdisciplinary research teams, including livelihood aspects of natural resource management. He is experienced in international tertiary-level education programmes and capacity development within agricultural development.

Meissa Diouf has a long track record in horticultural research in Senegal, based at the Centre pour le Developpement de l'Horticulture (CDH), Institut Sénégalais de Recherches Agricoles (ISRA). More latterly, Dr Diouf has turned his attention to underutilized and under-studied species – in particular, African indigenous vegetables, including aspects of growth, production and seed selection and breeding. He is a plant breeder and director at ISRA-CDH.

Axel W. Drescher is a professor at the Department of Physical Geography (IPG), University of Freiburg, Germany. He is head of the section on Applied Geography of the Tropics and Subtropics (APT). His research interests are broad, but concentrated on vulnerability, risks, and food and agriculture, with a main focus on small-scale production systems, examining the integration of applied natural and social sciences.

Gudrun B. Keding is an agriculturalist specialist on tropical plant science and international nutrition, and works as a research fellow at the Institute of Nutritional Sciences, International Nutrition Unit, University of Giessen, Germany. Within the scope of her PhD study, she conducted an interdisciplinary research project on the links between vegetable diversity and nutritional health in Tanzania that was integrated within a project on indigenous vegetables (ProNIVA) at the AVRDC – World Vegetable Center (Regional Centre for Africa), Tanzania.

Christophe Kouamé is a vegetable research officer working at the AVRDC – World Vegetable Center. His research interests have focused on African indigenous vegetables, from the development of improved varieties to the promotion of sustainable production technologies in urban and peri-urban production systems. Dr Kouamé led the National Vegetable Research for Development programme in Côte d'Ivoire and coordinated multidisciplinary research projects. Currently stationed at Yaoundé, Cameroon, he coordinates the AVRDC's vegetable breeding and seed systems programmes in the humid zones of West and Central Africa.

Wakuru Magigi is an urban planner working as a lecturer at Moshi University College of Cooperative and Business Studies (MUCCoBS) in Kilimanjaro, Tanzania. His research interests are land-use planning, urban growth, environmental planning and management, housing co-operatives, and food and agriculture. Currently, Dr Magigi's work focuses on analysing urban planning practice, community interventions, livelihoods and lifestyle changes in wetland areas as a result of urbanization processes in sub-Saharan African cities, a project under the VicRes initiatives funded by the Swedish International Development Agency/Department for Research Cooperation (Sida/SAREC).

Patrick Maundu is an ethnobotanist working for the National Museums of Kenya. He is also an honorary research fellow at Bioversity International (formerly International Plant Genetic Resources Institute – IPGRI) at its Sub-Saharan Africa office in Nairobi, Kenya. He coordinates Bioversity's African Leafy Vegetable Programme for the region.

Yasuyuki Morimoto is an ethnobotanist working for Bioversity International, sub-Saharan Africa office in Nairobi, Kenya. His research interests are Africa's indigenous food crops and plants, focusing on how farmers shape crop genetic diversity to secure their livelihoods and food cultures. His work also develops participatory community protocols for identifying economic and cultural values and livelihood opportunities for local communities that maintain unique crop diversity thus contributing to their well-being and providing incentives for continued conservation.

Dr Mel O. Oluoch is Training Specialist, Horticulturist and Deputy Global Theme leader for Nutrition programmes at AVRDC – World Vegetable Center. He is based at the Regional Center for Africa in Arusha, Tanzania. His background is in vegetable crop production systems; technology and production of seed physiology; and cell biology. His key responsibilities include coordinating and conducting training programmes in Africa; carrying out research in crop production technologies, variety development and evaluation; genetic enhancement, seed physiology, production and technology; vegetable seed systems, and variety release and commercialization of exotic and indigenous vegetables. In addition, he helps to coordinate and advise on the AVRDC nutrition programme.

Margaret W. Pasquini is a geographer working as a research officer at CAZS Natural Resources, Bangor University, Wales. Dr Pasquini's research interests are directed at topics that lie at the interface between agricultural systems and environmental protection. Currently, her work focuses on the promotion, cultivation and conservation of indigenous vegetables, particularly in urban agricultural systems in sub-Saharan Africa. Dr Pasquini led the IndigenoVeg project and is also coordinating a second project funded by the UK government on indigenous vegetables in West Africa.

Germain N. Pichop is an agricultural economist and was a post-doctoral researcher at the World Vegetable Center based in Arusha, Tanzania. Dr Pichop coordinated the supply chains surveys conducted under the IndigenoVeg project. He was Deputy Global Theme Leader for Postharvest Management and Market Opportunities at AVRDC. He is currently Co-director of the Center for International Business and Chair of the Department of Business at the Oklahoma City Community College in the US.

Charlie M. Shackleton is professor and head of the Department of Environmental Science, Rhodes University, Grahamstown, South Africa. He has a diverse array of research interests, but is widely published on issues relating to the contribution of natural resource harvesting (including indigenous vegetables) to sustainable rural livelihoods in Southern Africa. This includes natural resource valuation and supply-and-demand models for key resources used by rural communities.

Drissa Silué is based at the Regional Center for Africa of the World Vegetable Center in Arusha, Tanzania. Dr Silué is a plant pathologist by training and his current research priorities focus on the diagnosis of vegetable diseases, especially those found on African indigenous vegetables, pathogen population analysis, screening of germplasm for resistance to diseases and the development of integrated pest management strategies for protecting vegetable crops against diseases. He has also been involved in multi-locational germplasm evaluation for horticultural traits and the participatory selection of crop lines by farmers.

Ineke Vorster is a plant breeder and agricultural development researcher at the Agricultural Research Council, Vegetable and Ornamental Plants Institute, Pretoria, South Africa. She has various research interests, including the documentation of indigenous knowledge associated with African indigenous vegetable use and production, and is coordinating with research and extension to help streamline research on a needs-based approach for these vegetables.

Katinka Weinberger is a socio-economist and global theme leader for Postharvest Management and Market Opportunities at the World Vegetable Center, Shanhua, Taiwan. Dr Weinberger's research interests include high-value supply chains for poor farmers and related aspects of post-harvest and food safety, nutritional health and gender. Currently, her work focuses on the identification of mechanisms that help small-scale farmers and market agents to overcome constraints related to participation in supply chains for vegetables.

Ray-Yu Yang is head of the nutrition unit of the AVRDC – World Vegetable Center head office in Taiwan, and leader of the centre's research theme on nutritional security, diet diversification and human nutrition. Dr Yang's work as theme leader emphasizes food-based approaches and the linkage of agriculture and nutrition to solve micronutrient deficiencies problems. She specializes in the nutritional and functional characterization of African and Asian indigenous vegetables.

Einir M. Young is the deputy director of CAZS – Natural Resources and head of sustainable development at the Welsh Institute of Natural Resources, Bangor University, Wales. Dr Young has extensive experience in developing countries throughout the world with respect to participatory development of low-cost, low-risk technology with resource-poor farmers in semi-arid low-income countries, as well as integrated sustainable rural development. She was also involved in strategic planning and management of the IndigenoVeg programme.

Preface

Margaret W. Pasquini and Einir M. Young

Although there are over 7000 species that can be used for food (Williams and Haq, 2002), the world today relies on just a few crops for its energy requirements. In developing countries, this situation emerged mainly as a result of the promotion of Green Revolution high-yielding varieties, which displaced many local landraces (Thies, 2000). While it is acknowledged that the Green Revolution made significant contributions to hunger and poverty alleviation in many settings in Asia and, to some extent, in Latin America, its technologies were too expensive or inappropriate for much of Africa (IFPRI, 2002). Here, and in marginal environments elsewhere that could not benefit from Green Revolution varieties (as these perform well only in irrigated or high-potential rain-fed areas), many other crops continued to play significant roles in traditional farming systems.

Over the last two decades, in a context of much stronger awareness of the manifold interactions between agriculture and environment, the limitations of the Green Revolution, concern about rapid climate change, the realization of the need for a highly diversified diet, rich in vegetables and fruit, for good health, and a shift from top–down to community-driven rural development, these neglected and underutilized species have begun to attract considerable interest for their multiple underexploited benefits in terms of nutritional and food security, income generation and medicinal value, suitability for low-input systems, and for marginal environments.

The first signs of interest from the research and development community came during the late 1980s (for example, with the establishment in 1988 of the International Centre for Underutilised Crops, or the ICUC); with increased donor funding, the 1990s saw an expansion in projects, networks and conferences, which has continued strongly during the 2000s (Jaenicke et al, 2006).

Recent noteworthy milestones have included:

- the founding of the Plant Resources of Tropical Africa (PROTA) network in 2000, with its mission to synthesize 'the dispersed information on the approximately 7000 useful plants of tropical Africa and to provide wide access to the information through web databases, books, CD-ROMs and special products' (see www.prota.org/);
- the establishment in 2002 of the Global Facilitation Unit for Underutilized Species (GFU), a multi-institutional initiative that globally seeks to create an enabling environment for stakeholders who are engaged in developing underutilized species through policy analysis and advice to decision-makers, public awareness-raising activities, and documentation, information and communication of research and other activities pertaining to underutilized crops (see www.underutilized-species.org/);
- the recognition of the need for further research by the Consultative Group on International Agricultural Research's (CGIAR's) Science Council with the new system priority of 'promoting conservation and characterization of underutilized plant genetic resources to increase the income of the poor' (CGIAR, 2005);
- the inclusion of 'indigenous vegetables for biodiversity, healthy diet and marketing opportunities' as one of the AVRDC – World Vegetable Center's strategic programme directions in its *Strategy 2010* (AVRDC, 2002).

Over the last five to ten years, in sub-Saharan Africa, growing numbers of researchers have been involved in research initiatives on one particular underutilized crop commodity group, that of indigenous or traditional vegetables.[1] In the current climate of interest for underutilized crops, this research sector is likely to continue growing strongly over the next decade.

In this context, the IndigenoVeg network[2] was established in early 2006 with the overall aim of promoting indigenous vegetables principally in urban and peri-urban areas. Sub-Saharan Africa is currently experiencing the highest urban growth rates in the world (4.58 per cent), and is projected to have over 50 per cent of its population living in urban areas by 2030 (UN-Habitat, 2007). Thus, urban consumers will be an important group to target during the promotion of indigenous vegetables.

The task of promoting indigenous vegetables to urban consumers is challenging on several fronts. First, there is the problem of image. There are many reports that indigenous vegetables are overlooked by city dwellers (particularly young people) because they are perceived to be 'old fashioned' and 'poor man's food'. Second, there is a question of convenience. As a result of urbanization and modernization forces, as well as busier lifestyles, it is possible that urban consumers are increasingly turning to 'fast' foods, consuming less fresh produce, especially if it requires a lot of preparation effort (as is the case with many indigenous vegetables). And lastly, there is the influence of wider economic forces, resulting, for example, from globalization, which has increased economic opportunities for a restricted number of commodity groups, leading to the marginalization of local agro-biodiversity.

Nevertheless, there are opportunities for promoting indigenous vegetables in urban and peri-urban areas. For example, urban consumers, particularly in East Africa, are starting to be concerned about the safety of their food, and with their lower requirements for external inputs, indigenous vegetables could tap into this market. Although their popularity is threatened, knowledge of their uses and preparation is still current; thus, promotional campaigns could be developed to reach a large audience in urban areas that have better infrastructure and wider coverage by different communication media than rural areas.

The IndigenoVeg network has recognized significant lacunae in our understanding of the current and potential place and role of indigenous vegetables in urban and peri-urban environments,[3] particularly with respect to their production. Thus, the network has drawn together specialists from the two fields of indigenous vegetables and urban and peri-urban agriculture, with the specific objectives of coordinating existing research programmes and identifying the major research gaps to help guide future research and development efforts that aim to promote indigenous vegetables in urban and peri-urban areas.

Over the years, various definitions of urban and peri-urban agriculture have been used. One recent comprehensive definition is given by Mougeot (2000, p10), who states that:

> *Urban agriculture is an industry located within (intra-urban) or on the fringe (peri-urban) of a town, a city or a metropolis, which grows and raises, processes and distributes a diversity of food and non-food products, (re-)using largely human and material resources, products and services found in and around that urban area, and in turn supplying human and material resources, products and services largely to that urban area.*

Numerous scientific studies have highlighted significant benefits of urban and peri-urban agriculture, including improved food and nutritional security for urban dwellers, employment and income-generation opportunities.[4] Various authors point to the potential environmental benefits – for example, if managed efficiently, large quantities of waste produced in cities could be turned into compost for urban and peri-urban agriculture, recycling valuable soil nutrients, enhancing soil properties and alleviating the waste disposal problem, which is very serious in many developing countries (Cofie et al, 2006). While recognizing that there are many problems affecting urban and peri-urban production systems in low-income countries[5] that need addressing, the IndigenoVeg network has taken the position that agriculture does have, and should continue to have, a vital place in an urban environment, and should therefore be explicitly incorporated within urban planning processes.

Urban and peri-urban agriculture supplies significant amounts of the vegetables consumed in many sub-Saharan African cities today (see the review by Tixier and de Bon, 2006), and could be potentially transformed into a significant source of indigenous vegetables for urban areas. The starting point for this to happen will be for researchers to investigate which indigenous vegetables already belong to this system, the extent to which they are cultivated and the reasons why these particular vegetables are cultivated. Developing a strategy for increasing their production in urban and peri-urban areas will require a thorough understanding of their agronomy, cultural acceptability and marketing structures.

Of course, it is also vital to consider what impact stimulating demand for indigenous vegetables in urban and peri-urban areas could have upon rural areas. Even if urban and peri-urban agriculture provides significant amounts of vegetables for the city, a proportion of the demand will continue to be met from rural sources, particularly in the case of wild-collected varieties. Increased popularity of these varieties in urban areas could have beneficial economic impacts for rural areas, but could also lead to over-harvesting and loss of biodiversity. Promoting indigenous vegetables in urban and peri-urban environments could also result in increased demand by the rural population. A recent survey on the uses of indigenous vegetables in Mali provides an example of how new eating habits can spread from urban to rural areas (Pasquini and Ambrose, 2007). *Corchorus* species are widely consumed in villages across the country and have a good market value. Key informants explained that this is a relatively new development: while *Corchorus* spp. have always been known in the wild, they were not traditionally consumed until young village women migrated to the capital city for work, learned how to prepare the vegetable and transmitted the information back to their families in the village.

Drawing extensively from (but not restricted to) the discussions and research of the IndigenoVeg network, this book synthesizes existing knowledge and new information, bringing together the fields of indigenous vegetables and urban and peri-urban agriculture on the African continent. In doing so, it highlights the potential synergies of the two fields with benefits for sustainable livelihoods, food security and biodiversity conservation, illustrated with new case studies from across sub-Saharan Africa. It concludes by reflecting critically on what is understood about indigenous vegetables and urban agriculture, and what areas require further research, identifying the areas for policy intervention to bring the two fields together.

Notes

1 These terms are used interchangeably to describe indigenous and naturalized species that formerly have not been the object of scientific breeding work.

2 IndigenoVeg is the acronym for a European Union (EU) Framework Six-funded project entitled Networking to Promote the Sustainable Production and Marketing

of Indigenous Vegetables through Urban and Peri-Urban Agriculture in sub-Saharan Africa (see www.indigenoveg.org/).

3 There has been much debate about what constitutes peri-urban areas – an interesting contribution is given by Iaquinta and Drescher (2000), who have developed a peri-urban typology tool.

4 Mougeot (2005) and Cofie et al (2003) offer useful compilations of the information from various studies on the contributions from urban food production to urban food supply and urban employment, income and food expense savings.

5 The incorrect use of agrochemicals poses considerable health and environmental risks; farmers may use contaminated sewage or grey water for irrigation, or apply waste that has not been treated properly, contaminating crops; there could also be increased competition for water between farmers and residents. See the review by Cofie et al (2003) for details on some of the problems.

References

AVRDC (Asian Vegetable Research and Development Centre) (2002) *Empowering Small-Scale Farmers for Knowledge-Based Agriculture: AVRDC Strategy 2010*, AVRDC, Shanhua, Taiwan

CGIAR (Consultative Group on International Agricultural Research) Science Council (2005) *System Priorities for CGIAR Research 2005–2015*, CGIAR, Rome, Italy

Cofie, O. O., van Veenhuizen, R. and Drechsel, P. (2003) 'Contribution of urban and peri-urban agriculture to food security in sub-Saharan Africa', Paper presented at the Africa session of the 3rd WWF Symposium, Kyoto, 17 March 2003

Cofie, O., Bradford, A. A. and Drechsel, P. (2006) 'Recycling of urban organic waste for urban agriculture', in van Veenhuizen, R. (ed) *Cities Farming for the Future: Urban Agriculture for Green and Productive Cities*, RUAF Foundation, IDRC and IIRR, the Philippines, pp209–242

Iaquinta, D. L. and Drescher, A. W. (2000) 'Defining periurban: Understanding rural–urban linkages and their connection to institutional contexts', Paper presented at the Tenth World Congress, IRSA, Rio de Janeiro, 1 August 2000

IFPRI (International Food Policy Research Institute) (2002) *Green Revolution – Curse or Blessing?*, IFPRI Brief, Washington, DC

Jaenicke, H., Hoschle-Zeledon, I. and Manning, N. (eds) (2006) *Proceedings of Regional Consultation Workshop: Strategies for Research and Development of Underutilised Plant Species in Africa*, 24–26 May 2006, Nairobi, Kenya, International Centre for Underutilised Crops (ICUC), Colombo Sri Lanka and Global Facilitation Unit for Underutilized Species (GFU), Rome, Italy

Mougeot, L. J. A. (2000) 'Urban agriculture: Definition, presence, potential and risks', in Bakker, N., Dubbeling, M., Gundel, S., Sabel-Koschella, U. and de Zeeuw, H. (eds) (2000) *Growing Cities, Growing Food: Urban Agriculture on the Policy Agenda. A Reader on Urban Agriculture*, German Foundation for International Development (DSE), Feldafing, Germany, pp1–42

Mougeot, L. J. A. (2005) *AGROPOLIS: The Social, Political, and Environmental Dimensions of Urban Agriculture*, Earthscan/IDRC, London

Pasquini, M. W. and Ambrose-Oji, B. A. (2007) 'Conservation of vegetable biodiversity in Mali and Benin: Threats and existing management systems', Paper presented to the Darwin Initiative, Ref 15/003, Mid-Term Review Workshop, 17–19 July 2007, Bamako

Thies, E. (2000) *Promising and Underutilized Species Crops and Breeds*, Deutsche Gesellschaft für Technische Zusammenarbeit (GTZ) GmbH, Eschborn, Germany

Tixier, P. and de Bon, H. (2006) 'Urban horticulture', in van Veenhuizen, R. (ed) *Cities Farming for the Future: Urban Agriculture for Green and Productive Cities*, RUAF Foundation, IDRC and IIRR, the Philippines, pp315–347

UN-Habitat (2007) 'Urbanization: A turning point in history', Secretary General's visit to Kibera, Nairobi, 30–31 January 2007, www.unhabitat.org/downloads/docs/Press_SG_visit_Kibera07/SG%205.pdf

Williams, J. T. and Haq, N. (2002) *Global Research on Underutilised Crops: An Assessment of Current Activities and Proposals for Enhanced Cooperation*, ICUC, Southampton, UK

List of Acronyms and Abbreviations

AIDS	acquired immune deficiency syndrome
AIV	African indigenous vegetable
ALV	African leafy vegetable
ANOVA	analysis of variance
APT	Applied Geography of the Tropics and Subtropics (University of Freiburg, Germany)
AVRDC	World Vegetable Center (*formerly* Asian Vegetable Research and Development Center)
Bt	*Bacillus thuringiensis*
°C	degrees Celsius
CBO	community-based organization
CDH	Centre pour le Developpement de l'Horticulture
CGIAR	Consultative Group on International Agricultural Research
cm	centimetre
CO_2	carbon dioxide
DMRT	Duncan's Multiple Range Test
EPU	extensive peri-urban
EU	European Union
FANRPAN	Food, Agriculture and Natural Resources Policy Analysis Network
FAO	United Nations Food and Agriculture Organization
FAO-SAFR	United Nations Food and Agriculture Organization Regional Office for Southern Africa
g	gram
GDP	gross domestic product
GFU	Global Facilitation Unit for Underutilized Species
ha	hectare
HIV	human immunodeficiency virus
IAV	indigenous African vegetable
IBPGR	International Board for Plant Genetic Resources
ICUC	International Centre for Underutilised Crops
IDA	iron-deficiency anaemia

IDRC	International Development Research Centre
IFPRI	International Food Policy Research Institute
ILV	indigenous leafy vegetable
INRAB	National Institute of Agricultural Research (Benin)
IPG	Department of Physical Geography (University of Freiburg, Germany)
IPGRI	Biodiversity International (formerly International Plant Genetic Resources Institute)
IPM	integrated pest management
ISRA	Institut Sénégalais de Recherches Agricoles
IU	intensive urban
IUCN	World Conservation Union
IV	indigenous vegetable
kg	kilogram
km	kilometre
LER	land equivalent ratio
m	metre
m^2	square metre
MDG	Millennium Development Goal
mg	milligram
MIS	market information systems
MND	micronutrient deficiency
MUCCoBS	Moshi University College of Cooperative and Business Studies
n	total sample population size
NARES	National Agricultural Research and Extension Systems
NGO	non-governmental organization
NHSDP	National Human Settlement Development Policy
NO_x	nitrogen oxide
NPK	nitrogen, phosphorus and potassium
NTFP	non-timber forest product
O_3	ozone
OECD	Organisation for Economic Co-operation and Development
PEM	protein-energy malnutrition
PROTA	Plant Resources of Tropical Africa
RCA	Regional Centre for Africa
RCBD	randomized complete block design
RDA	recommended daily allowance
RUAF	Resource Centres on Urban Agriculture and Food Security
RUC	rural–urban continuum
SADC	Southern African Development Community
Sida/SAREC	Swedish International Development Agency/Department for Research Cooperation
SIPU	semi-intensive peri-urban
SO_2	sulphur dioxide
SSA	sub-Saharan African

t	tonne
TAV	traditional African vegetable
TALV	traditional African leafy vegetable
TLV	traditional African leafy vegetable
UAP	Urban Agriculture Programme
UDASEDA	Ubungo Darajani Settlement Development Association
UK	United Kingdom
UNDP	United Nations Development Programme
UNICEF	United Nations Children's Fund
UPA	urban and peri-urban agriculture
US	United States
USDA	US Department of Agriculture
VAD	vitamin A deficiency

CTA

partageons les connaissances au profit des communautés rurales

sharing knowledge, improving rural livelihoods

The Technical Centre for Agricultural and Rural Cooperation (CTA) was established in 1983 under the Lomé Convention between the ACP (African, Caribbean and Pacific) Group of States and the European Union Member States. Since 2000, it has operated within the framework of the ACP-EU Cotonou Agreement.

CTA's tasks are to develop and provide services that improve access to information for agricultural and rural development, and to strengthen the capacity of ACP countries to acquire, process, produce and disseminate information in this area.

CTA is financed by the European Union.

CTA
Postbus 380
6700 AJ Wageningen
The Netherlands
Website: www.cta.int

1

Urban Food Systems and African Indigenous Vegetables: Defining the Spaces and Places for African Indigenous Vegetables in Urban and Peri-Urban Agriculture

Bianca Ambrose-Oji

Urban food systems and trends in vegetable production in urban and peri-urban areas

This chapter presents a broad picture of the role of urban and peri-urban agriculture for urban food security, and how the production or collection of indigenous vegetables along the rural–urban continuum (RUC) contributes to this is in sub-Saharan African (SSA) cities and peri-urban areas. Drawing upon a wide literature from a number of different disciplines (including geography, agriculture, ethnobotany and rural sociology), it aims to review the importance of vegetable production, particularly indigenous vegetable production, in urban and peri-urban systems. In doing so, it provides definitions of urban agriculture and the spaces and places in which indigenous vegetables are produced and collected, and ends with a typology to understand the complexity of the spaces for indigenous vegetable production and harvest.

Globally, the growth of cities and urbanized areas continues at an exponential rate, with the fastest and most expansive growth being experienced in developing areas of the world. The urban population of the world is estimated to increase from 2.86 billion in 2000 to 4.98 billion by 2030 (UN,

2002; World Bank, 2002). The mean rate of urban growth in less developed regions (i.e. non-Organisation for Economic Co-operation and Development (OECD) countries) between 2000 and 2005 was just under 3 per cent per annum. The corresponding figure for rural population growth was only 0.5 per cent (UN, 2006). Examining the differences by world regions, the proportion of Africans currently living in urban areas is the lowest in the world at around 30 per cent; but the figures for SSA show urbanization running at the highest levels worldwide, at approximately 4.3 per cent per annum. Projections vary; but there is a suggestion that sometime between 2020 and 2030, 50 to 60 per cent of Africa's population will be living in urban areas, compared to just 15 per cent in 1950 and 34 per cent in 1994 (Gordon et al, 2000; UN, 2006). As urbanization takes place, another important trend is revealed – namely, the locus of poverty in Africa is slowly shifting from rural to urban areas. The World Resources Institute, for example, estimated that more than 56 per cent of the world's absolute or chronic poor are concentrated in cities and urban areas (WRI, 1996, cited in Drescher and Mackel, 2000).

The question of food supply and urban food systems is consequently becoming an issue of some significance for international agencies as well as urban politicians and policy-makers, along with the urban administrations and populations themselves (Argenti, 2000). One of the important reasons for this is that compared with rural areas, urban livelihood strategies tend to rely much more heavily on the cash economy and the potential for the sale of labour to secure cash income (Drakakis-Smith, 2000). The ability to earn income in the first place and the real purchasing power of that cash therefore become crucial to household well-being and to urban economic activity. As much as 60 to 80 per cent of the income of the urban poor is spent on the purchase of food (Maxwell et al, 2000). So, for most urban households, the amount of cash available to spend on food, set against the price of food available in urban markets, has a significant impact upon household food security overall and well-being. If the price of food is high, the food security of many urban households, particularly of the poorest sections of society, is undermined. In these circumstances, the quantity of food consumed reduces, and the diversity and nutritional quality of the food purchased also tend to decline.

As urban areas grow and poverty within them increases, understanding how urban food systems work and finding ways to ensure that they remain sustainable is a mounting preoccupation. The globalization of trade in agricultural products and foodstuffs, and the industrialization and commercialization of global and regional food systems have changed not only the way in which urban centres are supplied with food products, but the types of food that are made available and the nutritional character of those foods. This is true not only in the developed North, but also in the fast-changing urban centres of SSA. While some researchers put forward a case that these emerging food systems provide cheaper food products and greater nutritional diversity and calorific value which benefit the majority of urban households, others argue that poorer sections of society are adversely affected and at greater risk from under-nutrition as foods become increasingly processed and nutritionally homogenized, more Westernized, with high levels of fat and sugar, as well as increasingly expensive and beyond the reach of lower income groups (see, for example, Kimbrell, 2002). In short there is a 'nutritional transformation' centred on towns and cities, away from traditional diets to increased consumption of dairy products and meat, less complex carbohydrates, and a general decline in dietary diversity (Dixon et al, 2007). There are also perceived disadvantages on the supply side, where increasing integration within global food systems means that international and national regulations maintaining food standards as well as seed supply often work against the interests and capacity of local producers. Local products and packaging may not conform to hygiene regulations, they may not be standardized in terms of size and shape, and local varieties and landraces or wild products may not be recognized by seed certification or food product standards. The result is a continuation of the erosion of agro-biodiversity and nutritional and dietary diversity, and the development of a 'dual' urban food supply system with a formal and informal sector servicing different sections of urban society.

What is clear is that urban food systems are becoming increasingly complex and are drawing upon larger spatial networks and systems. The food supply systems and delivery networks include a wide range of actors, ranging from producers to providers of services such as transport and credit; there is a correspondingly broad set of functions from field-level production, processing, packaging, wholesaling and retailing; and there is an equally complex set of laws, regulations, facilities and infrastructure to ensure that food arrives in urban centres from a variety of local, regional and international supply centres. This increased level of complexity relies at the same time on systems and processes that may be far from the control of either municipal authorities or national-level governments (see, for example, Fold and Pritchard, 2005). Contemporary systems are more often in the hands of a limited number of transnational companies (e.g. in 2002, the ten largest food companies controlled 24 per cent of all global processed food sales; Fold and Pritchard, 2005) and international conventions and governance frameworks, than metropolitan authorities or consumers. The reliability of urban food systems is therefore exposed to global market and economic conditions, and to socio-political or climatic events in ever more distant areas of the globe. This brings with it a level of uncertainty about the security of urban food supply. In the current context of unpredictable oil prices and weakness in global financial and economic systems, and one in which climate change is expected to have an ever increasing impact on agricultural production (Millennium Ecosystem Assessment, 2005), policy-makers and planners increasingly need to find adaptive capacity within urban food systems.

This is slowly forcing a shift in thinking away from seeing food supply as merely a household-level issue to one that is perceived as a systems and process issue manifest at a landscape and urban unit planning level. In many instances, this means re-evaluating the trade-off between land for urban food supply as opposed to other urban and peri-urban uses, and the role of agriculture in building urban food system resilience along a continuum from the urban centres, the peri-urban 'buffer zones' and the rural hinterlands. With the increasing rate of urbanization, the use of space in urban areas and the peri-urban fringe becomes ever more intensive, accommodating competing land uses such as housing, urban infrastructure, storm water drains, industry, and industrial waste management systems. The challenge for planners and policy-makers is to recognize how agriculture can fit into this complexity and provide some of the social and economic resilience that food systems generally, and urban ones in particular, frequently lack.

It is important to understand that despite these significant pressures, agriculture continues to be practised in most SSA cities. Indeed, the United Nations Development Programme (UNDP, 1996) estimated that more than 800 million people are engaged in some form of urban agriculture worldwide, with the prevalence of urban farming in SSA cities ranging between 30 and 80 per cent of all urban households (UNDP, 1996).

There are important household-level determinants that account for this persistence of agriculture along the rural–urban continuum. Livelihood analyses have demonstrated that if the range of assets available for a household to make a living is broadened, different and additional livelihood options appear. This is particularly true if households gain access to natural capital assets such as land, which can significantly alter the interaction between the use of labour and the generation of income in urban settings. If household members have the chance to engage in urban or peri-urban agriculture, labour can be used to grow food for household consumption, rather than generate the cash to purchase food, and labour can also be used to grow crops as an additional source of income. Since many of those people practising agriculture in urban contexts are some of the poorest people in society, urban agricultural systems can make a significant contribution to mediating livelihood vulnerability through offsetting the high costs of food (e.g. Zeeuw et al, 2000; van Averbeke, 2007). For those newly arrived in city areas, urban agriculture may play a key part in the vulnerable transition from farm to non-farm livelihoods, providing a route for socio-economic change and household adjustment to the urbanization process. In addition to this, the surplus generated by households for sale into urban markets often provides that important informal supply which continues to meet the demands of the poorer sections of urban society. So, in many SSA cities, just as the city boundaries push further forward into rural hinterlands, the number of households engaging in urban and peri-urban agriculture is increasing. In Dar es Salaam, for example, the number of families engaged in agricultural production has increased from 18 per cent in 1967 to 67 per cent in 1991, making urban agriculture the second largest employer after petty trade (Ratta and Nasr, 1996, cited in Bryld, 2003). The number of households involved in urban farming in Ouagadougou (Burkina Faso) is 36 per cent; in Maputo (Mozambique) it is 35 per cent; and in Lusaka (Zambia) it reaches 45 per cent. In some cities, such as Nairobi (Kenya), as much as 50 per cent of food consumed by low-income groups comes from urban agriculture (Argenti and Marocchino, 2005).

Small household-level producers (see Plate 1.1) are not the only actors providing products to the urban food system. There are larger-scale, more commercial actors active in peri-urban areas (see Plate 1.2) supplying the more formal sectors with fresh produce and products for processing. There are also producers, both large and small, supplying food items from rural areas served by urban transport corridors. Indeed, Mougeot (2000) argues that the total food supply contribution of urban and peri-urban agriculture should not be downplayed, and estimates that as much as 15 per cent or more of all food consumed in urban areas is provided by urban agriculture. Mougeot (2000) goes on to predict that this 15 per cent share is likely to increase by 100 per cent over the next two decades.

Numerous studies in SSA cities (e.g. Lachance, 1993; Drakakis-Smith, 1994; Guyer, 1997; Diouf et al, 1999; Drescher, 1999; Dongus, 2001; Asomani-Boateng, 2002; van den Eynden et al, 2003; Weinberger and Msuya, 2004; Kasambula et al, 2007; Vorster et al, 2007a), demonstrate that a wide range of crops are grown in urban agricultural systems, including staples such as maize (*Zea mays*), cassava (*Manihot esculenta*), potatoes (*Solanum tuberosum*), bananas/plantains (*Musa* spp.), and sorghum (*Sorghum* spp.). The horticultural sector in urban and peri-urban agriculture supplying fruit and, particularly, vegetables, occupies a significant niche in food systems, providing often highly perishable produce to urban consumers. This is where the economic advantage of urban agriculture traditionally lies since the transport of vegetables over long distances is often not an economically viable proposition to urban populations with low purchasing power. Reporting on global trends, Weinberger and Lumpkin (2005) highlighted significant increases in the production and supply of horticultural crops (fruit and vegetables), including those produced in urban and peri-urban areas, where the per capita supply of vegetables outpaced fruit, growing at a rate of 1.8 per cent between 1990 and 2000. However, they point out, too, that although the per capita global supply of fruits and vegetables had risen to 173kg (i.e. 112kg of vegetables and 61kg of fruit) in 2002, changes are regionally differentiated. Per capita supplies rose most in China and Latin America. But in SSA, the average annual growth in per capita supply of horticultural produce was, in fact, negative between 1971 and 2000, and the per capita supply sits at 106kg, significantly less than other regions (Weinberger and Lumpkin, 2005). What this demonstrates is that the range of crops provided by urban areas is diverse, and the supply of vegetables is significant, providing an important nutritional contribution to urban diets. Where the supply of horticultural produce to urban areas is not maintained, there is an important public health issue which urban planners and policy-makers will need to attend to.

Despite the potential of urban and peri-urban agriculture to provide a significant contribution to urban food security now and in the future, urban areas continue to suffer from a significant crisis of legitimacy with strong historic roots. Looking back to the earliest records of urbanization and city-building, colonial administrations in most African countries expressly forbade urban agriculture as an activity within city limits (Drakakis-Smith et al, 1995; Ashebir et al, 2007), particularly the growing of indigenous crops. Although there were a few exceptions in some cities where vegetable production was encouraged as it was needed to feed the European expatriate populations, agriculture in the city was otherwise perceived as 'backwards', 'untidy' and an activity belonging to rural hinterlands rather than the seat of new governments and 'civilization'. In many African cities, peri-urban agriculture has also been used as a tool of racial segregation, providing a thin green line or buffer zone between expatriate or wealthy, and 'native' or poor sections of the metropolitan community (e.g. Freeman, 1991; Toulmin and Quan, 2000). However, there have been times when urban food supplies have been disrupted, which resulted in a tolerance of urban agriculture in some African countries – for example, the government-endorsed development of urban agriculture in Ghana during the economic crisis of 1972 to 1976 (Asomani-Boateng, 2002); the national drive for increased home production in Tanzania in response to the oil crisis and drought, also during the 1970s (Jacobi et al, 2000); responses to the 'economic war of liberation' in Kampala (Uganda) during the mid 1970s and 1980s, followed in the late 1980s by the structural adjustments programmes of the World Bank (Jamal, 1985; Pinstrup-Andersen, 1989; Bigsten and Kayizzi-Mugerwa, 1992); and the politically expedient tolerance and quiet administrative support for urban agriculture to cushion policies of government retrenchment in Cameroon (Page, 2002).

So, the case for urban agriculture providing an element of resilience to urban food systems and a safety net to urban households seems already to have been proven. But urban agriculture, in particular, and peri-urban agriculture, more generally, continue to be perceived as illegal, temporary and transitory land-use activities in an unstoppable process of urbanization that is focused on the residential and industrial development of urban land rather than on the preservation of open space and the development of sustainable cities and livelihoods. In many city areas it is an activity that may just be tolerated, but where land tenure continues to be insecure, and where metropolitan administrations may initiate 'crop slashing', 'clearance' and 'clean-up' programmes and the prosecution of urban farmers at any time.

In summary, urban and peri-urban agriculture, and the production of horticultural crops within this, presents an important topic not only for academic research examining poverty and livelihoods, urbanization, agro-biodiversity, food security and food supply chains, but also for those involved with urban and peri-urban planning and policy. The subject has seen increasing attention: for many SSA cities today, urban agriculture sits within a policy lacuna, falling between the dichotomous realms of urban and rural development strategies. While rural development policy is predicated on increasing the efficiency and profitability of farming and natural resource use, as well as protecting the natural resource base, urban policy focuses on the protection and expansion of income-earning employment opportunities that are rarely predicated on natural resource use, and on the use of land and space for building and the provision of services and service infrastructure. Maxwell (1995) went so far as to claim that urban policy and programme design that was cognizant of urban agriculture was often based on speculation and anecdote rather than on evidence-based and integrated land- and resource-use planning.

While, on the one hand, the persisting colonial legacy and restrictive urban planning promulgate laws and by-laws forbidding or restricting urban and peri-urban agriculture, metropolitan authorities and government agencies are, on the other hand, being pushed (by external donors and global policy instruments such as the Millennium Development Goals) to reconsider urban agriculture not as a problem to be addressed by spatial policy, but rather by cross-sectoral policies spanning social, economic and environmental spheres (Cabannes, 2004; van Veenhuizen, 2006). The social and economic policy dimensions focus attention on maintaining access to food for the poor, as well as on increasing the nutritional status of urban populations, and continuing to provide income for those engaged in commercial production. Environmental policy concerns the maintenance of green spaces in city areas as part of urban greening and the management of environmental quality. The challenge for SSA city authorities is to put these new policy imperatives into action and to secure the place of urban horticulture and vegetable production within them.

Defining urban agriculture and characterizing productive spaces for vegetables along the rural–urban continuum

The introductory section of this chapter has used the term urban agriculture without providing a clear definition. There are a variety of definitions; but at its simplest it can be described as agriculture located within or on the fringe of a town or city. Depending upon the local context, in some city areas it may be labelled as 'urban gardening' rather than 'farming' or 'agriculture'. Urban agriculture is not limited to the production of agricultural and horticultural crops, but may also include forestry, floriculture, aquaculture and livestock production. As suggested in the introduction above, the horticultural component of urban agricultural systems is that part which includes the production of fruit and vegetables, including indigenous African vegetables.

One of the most often quoted and now widely accepted definitions of urban agriculture comes from Mougeot (2000, p10), who adds more complexity to his conceptualization by saying that:

> *Urban agriculture is an industry located within (intra-urban) or on the fringe (peri-urban) of a town, a city or a metropolis, which grows or raises, processes and distributes a diversity of food and non-food products, (re-)using largely human and material resources, products and services found in and around that urban*

> *area, and in turn supplying human and material resources, products and services largely to that urban area.*

In both these formulations urban agriculture includes agriculture practised in the inner city as well as towards the outer edge of urban areas where the distinction between city and countryside is more difficult to characterize. There is a clear recognition that urban agriculture exists along a spatial continuum. It is this variation in the characteristics of a production system that is found in different locations, performing different functions that make a clear formulation of urban agriculture so problematic.

Mougeot's (2000) definition concentrates on two key ideas: first, that urban agriculture is an industry (i.e. commercialized and differentiated into steps and multiple actors) concerned mainly with the production of crops; and, second, that there is a cycling of resources (human, environmental and economic) and agricultural products largely within the urban area. The dynamics of urban agricultural systems within these limits are becoming increasingly well known and documented. The production of high-value perishable products such as dairy items (milk, yoghurt and cheese) and green leafy vegetables is a characteristic feature of urban agriculture. Producers make the most of the valued added afforded by close proximity to markets and large populations. Indeed, some studies suggest that as much as 20 per cent of SSA urban food needs are met from production in urban (i.e. intra-urban and peri-urban) agricultural plots rather than being brought in from rural areas (Armar-Klemesu, 2000). There are, of course, other definitions of urban agriculture. The Resource Centres on Urban Agriculture and Food Security (RUAF) takes a systems-based approach and states:

> *... the most striking feature of urban agriculture, which distinguishes it from rural agriculture, is that it is integrated into the urban economic and ecological system: urban agriculture is embedded in – and interacting with – the urban ecosystem, influenced by urban policies and plans; it also has positive and negative effects on urban ecologies.* (RUAF, www.ruaf.org)

For Maxwell (1995), urban agriculture has more to do with household livelihoods, and he defines it as the deliberate use of land and labour for the production of subsistence crops to reduce dependence upon financial or exchange entitlements in the provision of food.

As the introduction to this chapter demonstrated, there are elements of each of these definitions that hold true, particularly in the dualism between formal and informal urban food systems. However, both Mougeot's (2000) and Maxwell's (1995) definitions belie some of the increasing complexity found in urban and peri-urban agriculture across SSA. Urban agriculture may not be commercialized; urban and peri-urban agricultural products are not necessarily sold to the market as they may be used for household consumption or even in barter exchange; and wild and semi-wild plants may be gathered and collected from agricultural and other spaces rather than 'produced'. Most importantly, perhaps, is the need to stress the 'porous' boundaries surrounding urban areas and the increasing recognition that cities and towns are not closed systems. There are strong rural–urban linkages within urban food systems that may promote or facilitate the movement and exchange of resources of, for example, labour (members of rural households may work in urban and peri-urban agriculture on a seasonal or temporary basis), capital and agricultural inputs (e.g. seeds, fodder and manure) that contribute to urban agriculture and urban livelihood outcomes as much as those in the countryside (e.g. Tacoli, 1998; Bah et al, 2003; Bryld, 2003). In addition, the wide range of tenurial and property rights regimes found in urban and peri-urban areas (from common property, to private land, to no formal rights at all) will affect the form and function of plots of land and the resources on them along a spatial continuum between urban and more rural areas. In short, urban agriculture is complex, and it is difficult to disentangle the way in which it is integrated with and serves urban food systems.

African indigenous vegetables and exotic crops in urban agriculture

The beginning of this chapter also introduced the importance of vegetable production in urban and peri-urban agriculture to urban food systems. However, disaggregated figures for the sector are hard to find. Chapters 6 and 7 in this book explore the extent of African indigenous vegetables (AIVs) in urban agriculture further; but it is worth noting here that some studies estimating the value of markets for vegetables in African cities reinforce the importance of traded vegetables, including indigenous vegetables. In 1996 in Cameroon, for example, the value of urban markets for indigenous vegetables was estimated as US$22 million, a situation that was expected to continue as newly urbanizing populations were based on first- and second-generation immigrants with strong preferences for traditional food plants (Gockowski et al, 2003). Other research shows that the supply of vegetables to cities is almost entirely predicated on urban and peri-urban agriculture – for example, in Bangui (Central African Republic), the entire city's vegetable supply is from urban and peri-urban agriculture (Tixier and de Bon, 2006); in Bissau (Guinea-Bissau), Antananarivo (Madagascar) and Dar es Salaam (Tanzania), it is approximately 90 per cent of supplies (Jacobi and Amend, 1997; Moustier and David, 1997; Moustier, 2000); in Accra (Ghana), between 80 and 90 per cent comes from peri-urban production (CENCOSAD, 1994; Maxwell et al, 2000); and in Zambia, between 40 and 50 per cent of urban and peri-urban farmers in Lusaka supply vegetables to the city's markets (Drescher, 1999).

There are two main classes of vegetables in SSA urban and peri-urban agriculture. One group is exotic vegetables that originate from outside of the continent, and the second group comprises indigenous or traditional African vegetables. Providing a single and widely accepted definition of a traditional African vegetable is fraught with difficulty, and is open to as much, if not more, debate than that surrounding definitions of urban agriculture. There are a host of terms describing traditional African vegetables, including indigenous African vegetable (IAV); indigenous leafy vegetable (ILV); African leafy vegetable (ALV); traditional African vegetable (TAV); traditional African leafy vegetable (TALV or TLV) – and all are subject to contested meanings. According to the United Nations Food and Agriculture Organization (FAO, 1988), traditional vegetables are all categories of plants whose leaves, fruits or roots are acceptable and used as vegetables by urban and rural communities through custom, habit and tradition. Before the introduction of exotic crops and associated weeds, traditional vegetables would have been found in the wild or were semi-domesticated varieties of the indigenous flora. For some academics and practitioners, this means that 'traditional African vegetables' are defined as 'wild' plants, or semi-domesticated species that are part of traditional diets and may often be relied on as foods during periods of crop failure or famine. Gockowski et al (2003) define traditional leafy vegetables as those leafy green vegetables that have been originally domesticated or cultivated in Africa for the last several centuries. As time has passed, however, those vegetables which are now used 'according to custom and tradition' may include introduced species, so that for some people the term African traditional vegetables goes as far as including exotic produce such as tomatoes which are now customarily used by African populations. The distinction continues to be made with separate green leafy vegetables, casually referred to as 'African spinaches', as a particular group with stronger ties to the indigenous flora and with specific nutritional characteristics.

In this introductory chapter we define African indigenous vegetables or traditional African vegetables as names that refer to those plants which originate on the continent, or those which have such a long history of cultivation and domestication to African conditions and use that they have become 'indigenized', e.g. sweet potato (*Ipomoea batatas*) and pumpkin (*Cucurbita* spp.). To be specific, the predominant exotic vegetables found in SSA metropolitan areas are carrot (*Daucus carota* var.), tomato (*Lycopersicon esculentum* var.), green beans (*Phaseolus vulgaris*), onions (*Allium* spp.), cabbage (*Brassica oleracea*), lettuce (*Lactuca sativa*), and chard (*Beta vulgaris* var.). Important indigenous or traditional African vegetables include okra (*Abelmoschus esculentus*), sweet potato (*Ipomoea batatas*), cowpea (*Vigna unguiculata*), yams (*Discorea* spp.), and taro tubers (*Colocasia esculenta* and *Xanthasoma* spp.) (see Chapters 3 and 6 for more details on specific species and their relative use). In terms of the leafy species alone, a range of species from several major families of plants are used, with the genera *Amaranthus*, *Agathosma*, *Bidens*, *Cleome*, *Chenopodium*, *Corchorus*, *Crotalaria*, *Cucurbita*, *Ipomoea*, *Solanum*, *Vernonia* and *Vigna* being the most conspicuous (Coetzee et al, 1999; Shackleton, 2003; Pichop, 2007). There is variation in the dominant plant families and species utilized by region and country according to the interplay of ecology and cultural preference. In West and Central African cities, the most common and popular leafy vegetables are sweet potato leaves (*Ipomoea* spp.), pumpkin (*Cucurbita* spp.) and wild spinach (*Amaranthus* spp.); in East and Southern Africa the African nightshades (*Solanum* spp.), wild spinach (*Amaranthus* spp.), spider plant (*Cleome* spp.) and pumpkins (*Cucurbita* spp.) predominate (Pichop, 2007). It is also worth noting the increasing significance of Ethiopian kale (*Brassica carinata*), which is being promoted in urban agriculture across the continent as a more nutritious indigenous alternative to exotic cabbage.

Despite this variety of African indigenous vegetables, the focus of contemporary urban and peri-urban vegetable production in most SSA cities has turned to the production of exotic crops and varieties. A real tension exists between the two groups of vegetables, and there are many surveys now that support the contention that in absolute and relative terms, there is a general decrease in the production of indigenous vegetables (Weinberger and Msuya, 2004). This is likely to be related to an apparent reduction in consumer demand. The declining demand is linked to changes in contemporary tastes and preferences for less bitter vegetables (indigenous species and varieties often have a stronger, more bitter taste), for vegetables that are quicker to cook and prepare, and for vegetables that are perceived as more symbolic of a modern and sophisticated lifestyle than those species associated with a rural past. Indeed, in some SSA countries, the association between indigenous leafy vegetables and 'hunger' and 'famine' foods, or rural poverty, is such that it can be regarded as humiliating to eat them. In Tanzania, the value of indigenous vegetable consumption has fallen from previous reported levels of 20 per cent to approximately 11 per cent of the value of all food consumption, with the figure being even lower at just 2 per cent for the wealthiest households.

However, there are examples of exceptions to these general patterns. Gockowski et al (2003) report that in contrast to other urban markets such as Nairobi, Accra and Harare, where *Brassica* species have replaced traditional vegetables, traditional leafy vegetables appear to be maintaining their importance in Cameroon, even though people from the higher-income categories were less likely to continue purchasing certain ones, such as *Gnetum* spp. and cassava leaf (Gockowski et al, 2003). A surprising result from the formal urban food supply networks in East Africa has been the increased commercialization of indigenous vegetable production and contract supply to supermarkets. In the main urban centres in Kenya (e.g. Maseno and Nairobi), increased promotion and marketing of indigenous vegetables has been matched by increased demand by consumers for some species of leafy vegetables such as *Amaranthus* spp., with a concomitant increase in the urban and peri-urban cultivated area put down to this crop (Abukutsa-Onyango et al, 2007) (see also Chapters 6 and 7). This pattern of change is reflected in Kampala (Uganda) as formal markets begin supplying traditional vegetables in wet and processed forms (Kasambula et al, 2007).

One of the characteristic features of traditional African vegetables is that they often contain higher levels of essential minerals and micronutrients than exotic vegetables, as well as being high in protein. Chapter 4 discusses the nutritional value of traditional vegetables in greater depth; but it is worth noting here that African leafy vegetables can make significant contributions to the diets of urban populations in the provision of elements, particularly iron, zinc and vitamin B complex and vitamins C and A, which have been noted as deficient in many parts of SSA (e.g. Guarino, 1997; United Nations University, 2005). In the case of zinc, for example, the staple crops common in SSA countries (maize, sorghum, rice and beans) do not absorb zinc as growing crops and therefore cannot provide zinc in the diet. The traditional combination of starchy staple with a green vegetable sauce means that indigenous vegetables play an important role in balancing dietary quality (Laker, 2005). So the case has been made that traditional vegetables can contribute to urban food security to an even greater extent than exotic crops. The fact that urban agriculture has an impact upon nutritional status has been tested and found to have a positive association with nutritional outcomes for 'producing households' (Maxwell et al, 1999). The trend towards increased production of exotic species in urban agriculture is therefore of concern since many of these vegetables, notably the brassicas, have a far lower nutritional value than traditional vegetables (see Chapter 4).

There is evidence to suggest that there are differences in the nutritional status of individuals and households at different points along the rural–urban continuum. There has been a well-propagated notion that nutrition, particularly child nutrition, is better in urban areas than in rural areas (von Braun et al, 1993; Ruel et al, 1998). The main causal factors attributed to this are better access to potable water and to primary health services, metropolitan campaigns promoting health and sanitation, and an increased variety in the available urban foodstuffs which contribute to balanced diets. However, Fotso (2007) and Fotso and Barthelemy (2005) show that the real situation is more complex than this and, looking at changes over time, show that in some city areas such as Kampala there is a measurable improvement of child nutrition, but in other cities undergoing rapid urbanization (e.g. Nairobi), the nutritional status of urban populations remains low and is moving back closer to that of rural populations. Fotso (2007), Fotso and Barthelemy (2005) and Smith et al (2004) all show in their research that it was socio-economic status (e.g. income, education, household size and residential area) that had an effect on nutritional outcomes rather than location along the continuum *per se*, even though they acknowledge interplay between social status and community social status in terms of the infrastructure and other resources available in particular areas along the rural–urban continuum. So even within city boundaries one would expect differences in the nutritional outcomes of poorer and richer households with spatial inferences according to where they tend to reside, linked with the foods supplied there through urban agriculture sources or more globalized food supply systems.

In summary, access to African indigenous vegetables through urban agriculture can improve the nutritional outcome of poorer groups in urban and peri-urban areas. At the same time, there are also implications for middle-income groups: the supply of African leafy vegetables to supermarkets and retail outlets that they are likely to use is an important outcome of urban and peri-urban agriculture for the middle income group's own nutritional needs.

Spatial variation in urban and peri-urban vegetable production

Chapter 8 of this book takes a closer look at the extent to which indigenous African vegetables are integrated within urban agriculture. However, in our exploration of the definition of urban agriculture and of AIVs, it is still important to map out where vegetables are produced as a way of understanding the sometimes contradictory spatial patterns and trends associated with this particular segment of the urban food system. Variation exists in terms of the systems employed, the types of vegetable being grown, and whether the objective of production is for home consumption or for the market. Erenstein et al (2006) established differences in geospatial production patterns around West African cities according to the value of crops with respect to other livelihood activities, to the biophysical conditions and ability to support productive cropping, and, finally, to access rights to land. Although the influencing variables are not described, some studies indicate a partition between those crops grown closer to cities (i.e. green leafy vegetables), compared to those grown further away (e.g. fruits, legumes and root crops), as is the case in Nairobi (Abukutsa-Onyango et al, 2007) and for Dar es Salaam, where leafy vegetable production is found within city limits and non-leafy vegetables are produced in the peri-urban zone (Stevenson et al, 1996). Yet other research shows that the production of vegetables is concentrated in the peri-urban rather than the urban zone, where the urban zone tends instead to be home to mixed farming plots largely for household subsistence. This is the case in cities such as Accra (Ghana), Yaoundé (Cameroon), Kampala (Uganda), Abidjan (Côte d'Ivoire) and Cotonou (Benin) (e.g. Gockowski et al, 2003; Assogba Komlan et al, 2007; Assouma et al, 2007; Kasambula et al, 2007). There may also be a seasonal or temporal dimension to the spatial distribution of vegetables in urban and peri-urban agriculture. For example, Kasambula et al (2007), working in Uganda, note that in the smaller plots used by small-scale vegetable producers in the Mbale urban area, vegetable cropping tends to be in pure stands. This contrasts with the larger plots used in the peri-urban zone of Kampala where many medium-sized producers have commercial stands of leafy vegetables that are grown in mixed associations of crops to incorporate temporal differentiation in growing conditions and harvesting times. In contrast to the situation in Uganda, it is the larger medium-scale commercial producers in and around Nairobi who produce in largely mono-cropped systems; small-scale production tends to be mixed cropping and associated with subsistence systems (Abukutsa-Onyango et al, 2007).

The kind of plots in which people grow vegetables varies very substantially, and the production spaces include intensive irrigated plots and fields on the urban fringe; plots in open spaces and unscheduled land such as roadsides, airport buffer zones, drainage systems and valley bottoms, steep slopes, and empty building plots in both the urban and peri-urban zones; home gardens within and beside residential plots; community and allotment gardens in communal or government assigned areas; and rooftop and container gardens on and within buildings.

All of these different spaces have varying property rights regimes and tenurial arrangements. Only 20 per cent of the urban agricultural cultivation in the city areas of developing countries happens on land owned by the cultivator (Bryld, 2003). The diverse tenurial arrangements provide access for different types of people and, consequently, to different production functions and associated possibilities. Irrigated plots, fields, home gardens and some community and allotment gardens often have recognized *de facto* and *de jure* tenure systems[1] that tend to be associated with investment in longer-term agricultural system improvements and inputs such as the development of irrigation systems, building of bunds, improved soil management and the use of agricultural inputs. Urban agriculture practised in open spaces and unscheduled land, in valley bottom roadsides, and in other areas where ownership and use rights might be disputed, semi-legal or even illegal may not be subject to the same investment in agricultural development or labour intensity, and such spaces may be more likely to hold fast-maturing short-cycle crops such as leafy vegetables and staples such as maize.

It would be as well to point out here that immigrants to a city area may represent medium-sized commercial producers with a significant impact upon the horticultural sector, rather than the more easily imagined marginalized immigrant farmer supplying only small quantities of vegetables to the market. In Côte d'Ivoire, for example, the production of indigenous African vegetables for Abidjan is almost all carried out by large numbers of non-Ivoirians, many from Burkino Faso (Assouma et al, 2007).

In terms of the numbers of people engaged in urban and peri-urban agriculture, the importance of smaller producers should not be underestimated since more smallholders may be cultivating indigenous vegetables than commercial producers. For example, in Tanzania, as many as 40 per cent of farmers who cultivate small plots of land are engaged in the cultivation of indigenous vegetables, while only 25 per cent of relatively large-scale farmers are engaged in the cultivation of indigenous vegetables (Weinberger and Msuya, 2004).

Maxwell (1995) made important distinctions between the reasons that different groups of people might have for practising urban agriculture. He identifies four categories of urban farmers:

1 commercial farmers producing for the urban market;
2 those interested in self-sufficiency where the primary motivation is the production of food for the household;
3 those providing a measure of food security through the production of a surplus or reserve supply of food on their urban plots; and lastly
4 those with no other means, trying to produce the food they need but often forced to sell goods to meet other household expenses.

These different objectives for practising urban and peri-urban agriculture combine with tenure issues to affect the choice of crops and the production systems employed. In terms of the medium-scale specialized commercial production of vegetables, it is the larger irrigated plots with secure tenure that tend to be used. However, there are many accounts of small-scale market-oriented production of vegetables by those who have little recourse but to turn to urban agriculture as an essential activity, including the poorest younger men and new immigrants to city areas, in the less secure lands to which they can effectively gain access (Ambrose-Oji, 2004; Gündel, 2006). Households working to produce for subsistence may invest significant amounts of labour and other agricultural inputs even where their tenure is not secure since urban agricultural production is their livelihood imperative.

The spatial distribution of exotic and indigenous or traditional African vegetables (leafy and non-leafy) in urban and peri-urban areas is complex and there are no clear trends. Commercial producers may produce either exotic or indigenous/traditional vegetables, and so may smallholders and those interested only in the production of vegetables for home consumption where a mix of traditional and exotic crops may be preferred. There is evidence to suggest that indigenous and traditional African vegetables and leafy vegetables, in particular, have a significant role to play for those producing in the least secure areas. For such groups of people in such plots, the investments demanded by exotic species (e.g. the high costs of seed, the need to irrigate and the need for fertilizers and pesticides) may not match the quicker returns from indigenous vegetable crops which may be faster maturing, less needy of pesticides and more tolerant of marginal soil and nutrient conditions (Jansen van Rensburg et al, 2007; Vorster et al, 2007b). While some of these general characterizations of the interplay between security of tenure, the investment of labour and resources into urban and peri-urban agriculture, and the systems that exist as a consequence may hold true in many cases, Slater and Twyman (2003) remind us that we should be wary of generalizations. As they point out, if people are willing to construct houses on land without formal tenure, then they may also be willing to invest in the medium- and longer-term production of natural resources: 'it is not clear, therefore, what the precise differences are between rural and urban areas with respect to tenure and the sustainability of livelihoods' (Slater and Twyman, 2003, p18).

There are emerging trends of a move in urban agriculture in some areas away from smaller plots owned or used by poorer groups of people for subsistence or mixed subsistence/market purposes towards the consolidation of larger plots owned by the more wealthy entrepreneurial class in whose plots the poor now labour. In some cities the rich concentrate their production on the highest-value commodities offering the best potential for commercialization, which tend to be fruits and exotic vegetable species (Binns and Lynch, 1998; Benson, 2006). However, in some countries (e.g. Kenya and Uganda), where there is an expanding middle class and a growth in urban market demand, the commercial production of indigenous vegetables such as *Solanum nigrum*, *Cleome gynandra* and *Amaranthus* spp. is also increasing through the efforts of commercial producers and through direct links to supermarkets who are beginning to offer contracts to supply indigenous vegetables to them (e.g. Abukutsa-Onyango et al, 2007).

In addition to those indigenous and traditional African vegetables that are cultivated, wild-harvested vegetables continue to be significant in some SSA city areas both in the peri-urban as well as the urban zones. Alternative characterizations for the species included here are wild resources, underutilized crops, minor crops, wild foods and famine foods, weeds, minor forest products, and even non-timber forest products (NTFPs). There is now a very comprehensive literature on the use of wild food plants and of wild vegetables, although these tend to concentrate on use by rural African populations (Getahun, 1974; Fleuret, 1979; Peters et al, 1984; Becker, 1986; Campbell, 1986; Scoones et al, 1992; Zemede, 1997; Shackleton et al, 1998; Chweya and Eyzaguirre, 1999; Cavendish, 2000; Engle and Altoveros, 2000; Grivetti and Ogle, 2000; Zemede and Mesfin, 2001; Harris and Mohammed, 2003; Letšela et al, 2003; Shackleton, 2003; Lado, 2004; Ambrose-Oji, 2005; Bedigian, 2005; Balemie and Kebebew, 2006; Cocks, 2006; Delang, 2006; Dovie et al, 2007). Much less has been documented on the use of wild foods or vegetables by peri-urban or urban populations, although there is certainty that wild vegetables are being collected for consumption and for sale by urban and peri-urban households, and that for some groups of people this may represent an important input into their livelihood systems (Shackleton et al, 2007). For example, Cocks (2006) reported that 42 per cent of urban households collected wild AIVs in and around their homestead plot, consuming a mean of 31kg per year. While some of these wild vegetables may be used as the primary vegetable component of a meal, others might be added as just one ingredient of a sauce or a stew, and yet others are recognized as having nutraceutical[2] properties and are actively sought out for direct consumption or to add medicinal or therapeutic qualities to a meal.

It is clear that the boundaries between 'wild' and domesticated 'crops', or between what is or is not part of an urban and peri-urban agricultural system, are both difficult to distinguish and hard to maintain. The peculiar features of urban and peri-urban agriculture disrupts the general perception that wild spaces and agricultural spaces are separate, and that cultivated species are planted in agricultural plots while wild species[3] grow spontaneously, mainly in non-cultivated and rural areas. In Soshanguve (South Africa), for example, peri-urban and urban farmers collect wild species from their plots for home consumption. Occasionally these might be spontaneous wildings; at other times the farmers themselves might scatter seed from the wildings, intentionally encouraging their spread over the plot. At yet other times they might 'weed out' those same wild plants. The species that farmers in Soshanguve might choose to utilize in this way include, for example, *Amaranthus* spp., *Cleome gynandra*, *Bidens* spp. and *Chenopodium* spp. The distinction between wild and cultivated or domesticated is hard to clarify here since these are the same species that might in other contexts be actively cultivated as a crop (IndigenoVeg Project, 2007a; Vorster et al, 2007b). There may be no distinct spatial separation between 'wild' and 'cultivated' agricultural plots and spaces. Using South Africa as an example again, there are complex relationships between urban and peri-urban vegetable or crop production and the harvesting of wild species. Urban farmers may contract out the right, or charge a fee, for collectors to gather wild species such as *Cleome gynandra* and *Amaranthus* spp. from their horticultural or staple-producing fields and plots. For the farmer, an important agricultural task is being undertaken (i.e. weeding), and for the collectors, the wild resource is used as a consumption or income-generating livelihood input (IndigenoVeg Project, 2007a; Vorster et al, 2007b). Indeed, Dovie et al (2007), working with South African households, record larger quantities of wild species being collected from arable plots than from 'wild' rangelands. Drescher and Mackel (2000), researching Lusaka (Zambia), Shackleton (2003) and Shackleton et al (2007) in Durban and Limpopo Province (South Africa), and Kasambula et al (2007) in Mbale District (Uganda) show how wild harvested species continue to provide important inputs to urban and peri-urban households. In Lusaka, nearly 40 per cent of urban and 76 per cent of peri-urban respondents said that they gathered wild vegetables for additional food or income; in Durban, nearly 100 per cent of urban and peri-urban farmers said that they collected species from the wild; and in the small city of Mbale (Uganda), 45 per cent of the farmers and retailers interviewed reported collecting wild *Solanum* spp. intended for urban sale.

Matrix of spaces and places in urban and peri-urban agriculture for the production of African indigenous vegetables

The evidence presented in this chapter shows that there is spatial differentiation in vegetable production and agricultural systems along the rural–urban continuum. However, because of the complexity of vegetable production and collection in urban and peri-urban agricultural systems, it is difficult to synthesize a coherent characterization of the places and spaces for this in African city areas. It is not only the dynamics of household or commercial decision-making within city areas that vary to produce significantly different agricultural outcomes. There are differences within and between cities that influence the form and function of agricultural activity. Within cities there are differences between areas that are formalized and organized and those that are not. The more formalized areas might provide water for irrigation and institutions for community farming in contrast with less organized zones where water supplies, community organizations and the possibilities for cooperative action are fewer.

There are differences between larger and smaller cities in terms of their resource endowments and the intensity of development within and around them. These factors affect the opportunities to access resources such as land and water for urban and peri-urban agriculture, or to find and access areas in which to collect and harvest wild plants. For example, there are distinct differences between Kampala (a large city) and Mbale (a small city) in Uganda in terms of the production systems, the availability of wild species within plots and city boundaries, and differences in the tastes and preferences of the urban populations. The socio-historical context of urbanization is also important. Some cities have a history of attachment to agricultural markets as the historic centres of trade for primary commodities (e.g. Douala in Cameroon or Dar es Salaam in Tanzania) and some an historic link to the use of land for agriculture, which persists to alter the attitudes and possibilities for urban and peri-urban agriculture within metropolitan governance systems (see, for example, Lynch, 1994; Page, 2002; Slater and Twyman, 2003).

What other spatial models of urban agriculture have been described that could be useful in unravelling this complexity? Using agricultural production as the basis for their categorization, Gockowski et al (2003) proposed three distinct spaces along the rural–urban continuum:

- intensive urban (IU);
- semi-intensive peri-urban (SIPU); and
- extensive peri-urban (EPU).

While even this categorization may present a simplified picture of production spaces, it does provide some notion of change along a rural–urban continuum and suggests that there is a difference in system features, such as the labour and production inputs and realized yields. We need to remember here that although this classification is production based, there exists a continuum associated with spaces and species which are 'domestic' (i.e. cultivated areas and cultivated species and varieties), through to semi-domestic, through to wild. Interestingly, Gockowski et al (2003) showed that agriculture in the EPU was a female domain, in contrast to the SIPU where men predominate. There is additional evidence to support the idea of gender differentiation of production through the urban–rural continuum. Synthesizing the results from a survey of urban and peri-urban production of traditional and indigenous vegetables in urban and peri-urban agriculture in seven different African countries (IndigenoVeg Project, 2007b; see Chapter 7) showed that the production for urban markets is increasingly carried out by men in urban areas (except in South Africa), and more often by women further along the continuum towards rural areas. It is the specific production for the market, where men perceive vegetable production as a cash-generating business, where these patterns hold true. Otherwise, where the primary motivation for urban and peri-urban agriculture is either survival or the production of food for home consumption, it is still largely women who undertake the agricultural activities (Bryld, 2003). It is also worth mentioning here that there may be differentiation in the spatial distribution of different age groups engaged in urban and peri-urban indigenous African vegetable production. Where there is market demand, then the age profile of producers for the market appears to be younger (e.g. Maseno and Nairobi in Kenya), compared to urban and peri-urban areas where there is either low market demand or there are poor or weak marketing systems (e.g. Pretoria in South Africa).

Dongus (2001) uses the 1998 *Dar es Salaam Spatial Development Plan* to put forward a classification of urban open spaces in which urban and peri-urban agriculture might be located. These different categories include:

- productive open spaces that are confined to peri-urban areas and include the peri-urban agricultural zone;
- utility open spaces that include resource lands such as forest reserves and lakes, urban resources such as reservoirs and landfills or municipal waste areas, and flood control and drainage areas;
- green open spaces that include urban parks, recreational areas or shopping malls, traffic islands, protected areas, wilderness areas and national parks (it may seem strange to include national parks within a classification of urban space; but in SSA cities a situation does exist where national parks occur within or adjoining metropolitan boundaries – for example, Nairobi National Park is located within the city's boundaries close to residential areas and just a few minutes from the city centre; on the slopes of Mount Meru in Tanzania lies Arusha National Park, which abuts the peri-urban zone of Arusha, the country's second city; Table Mountain in the heart of in Cape Town in South Africa is a national park); and
- corridor open spaces that include rights of way, car parking areas and docks.

It is useful to classify urban and peri-urban open spaces in this way since it reflects the conceptualizations of urban and economic development planning bodies concerned with managing the natural resource endowments within metropolitan areas. Formalized systems of urban governance, of which planners form an important stakeholder group, may not accord priority to urban agriculture in their categorizations; but they do recognize the need for a reservoir of natural resources to maintain the ecological systems and functions required by urban populations (e.g. hydrological functions in the supply of water and discharge of waste). The challenge remains the need to recognize formally the multifunctional nature that many of these spaces play, and the fact that some open space may have an agricultural food production function as well as a utility or green space function.

Developing the characterization based on resource availability and use further, Slater and Twyman (2003) propose a typology linked to the livelihoods framework where the natural resources available in a city area support livelihoods in different ways either as commodities, inputs or services directly or indirectly through primary use (i.e. direct use of the natural resources, such as land and wild plants) to support livelihoods (through consumption or sale); secondary use where the natural resources form the basis of production, such as livestock rearing or crop cultivation; or tertiary use where the natural resources provide the basis of service provision or the creation of employment opportunities (e.g. labouring in commercially viable plots). There is overlap here with the typology of household production suggested by Maxwell (1995) mentioned earlier in this chapter. For many households growing and collecting indigenous African vegetables, their subsistence, surplus, commercial and essential use is a primary use of natural resources. For those individuals engaged in the production or collection of indigenous African vegetables on other people's commercial plots, the tertiary use of natural resources forms the fundamental basis of their livelihood systems. While Slater and Twyman's (2003) characterization focuses on the use of urban and peri-urban natural resources, it is important to remember that urban and peri-urban agriculture contributes to livelihoods in other important ways. As well as providing important physical, natural and financial capital inputs, urban and peri-urban agriculture also adds to human, social and political capital dimensions of livelihoods. Van Averbeke (2007), for example, shows that with regard to immigrants into a city area, urban and peri-urban agriculture can recreate elements of rural environments that may be particularly important to women seeking to maintain their cultural identity as the providers and producers of traditional food products for the household. Other features of urban and peri-urban agriculture mentioned as important were the ability to build social networks through farming, the self-worth, enjoyment and mental well-being provided by agricultural activity, and the chance to use knowledge and skills (van Averbeke 2007).

We have already seen that urban and peri-urban agriculture can provide food, income and other forms of individual and household well-being. Additionally, we should also emphasize the vulnerability context surrounding livelihoods, and the role that urban and peri-urban agriculture (particularly the provision of AIVs within these spaces) can have in mediating livelihood vulnerability. There are important seasonal differences in the availability of vegetables as cultivated crops, and the availability of wild species and resources, and these seasonal dimensions are also played out in the use of urban and peri-urban spaces. In many environments, the combination of cultivated and wild AIVs can provide a year-round supply of sustenance that can avoid the nutritional deficits that might otherwise occur in agricultural cycles (e.g. the hunger gap at the end of the main harvesting season and the onset of the dry season). The interplay of spatial variety in collection and production spaces for both cultivated and wild collected species means that the use of

Table 1.1 *The spaces and places for indigenous African vegetable production along the urban agriculture continuum*

Urban agriculture locations and harvesting sites	*Within city (high density development)*	*The urban–rural continuum* *Peri-urban (lower density development)*	*Peri-urban (more rural)*
Non-open space	Container gardening Rooftop gardening Largely indigenous, but some exotic vegetables to provide nutritional supplements/surplus	Less commonly reported	Less commonly reported
Home gardens	Mixed systems; may include wild species – mainly women for subsistence or as supplements/surplus Intensive vegetable production, exotics and indigenous – mainly men for commercial market sales	As within the city	As within the city
Farm plots	Less common Commercial vegetable production; largely exotic but some indigenous – mainly men for very intensive commercial market sales May be poorer sections of the community working for richer land owners Labourers here may be either men or women for income generation	May be more extensive and farm plots Commercial vegetable production; largely exotic but some indigenous, mainly by men for commercial market sales Wild-harvested vegetables; may be collected from farm plots – often the domain of women for subsistence or for market sale	Often traditional gender-based differentiation of agricultural production with women growing vegetables for subsistence, supplements/surplus and for income May be some commercialized production by men Wild-harvested vegetables from farm plots common; may be by women for subsistence and supplements/surplus; may be by men for commercial market sale
Scheduled land (utility and corridor spaces)	Mixed systems may include wild species – may be men or women for subsistence, nutritional supplements/surplus and for income generation May be the site of production for immigrants and new arrivals to the city (often men), as well as the poorer sections of society as intensive system for cash sales or subsistence as 'only means' production	Community-based production and group gardens may be present Mixed systems mainly for subsistence and surplus but some community-based systems also supply to the market Wild-harvested vegetables may be collected by men or women for income or consumption	As within lower-density peri-urban areas Wild-harvested vegetables may be collected by men or women for income or consumption
Unscheduled land	Often by men who are newly arrived immigrants; may be mixed or intensive as an 'only means' activity for subsistence or for sale Wild-harvested vegetables may be collected by men or women for income or consumption	As above	Wild-harvested vegetables may be collected by men or women for consumption
'Wild lands' (common lands, utility and green open space)	Less commonly reported in urban areas Green open spaces and utility lands may be the site of wild-harvested leafy vegetables	As above	Wild-harvested vegetables may be collected by men or women for income or consumption

valley bottoms, river-side areas and the location of plots close to grey water sources in urban utility, wild and corridor spaces can extend the growing season to provide a year-round supply of vegetables to households and markets.

Using a combination of the characterizations of urban and peri-urban production, planning-oriented and livelihoods-oriented spaces presented in this chapter, the matrix presented in Table 1.1 shows where all these different spaces and places for the production and collection of African indigenous leafy vegetables are situated, and what their main contribution to urban and peri-urban households might be.

Urban and peri-urban agriculture in the process of urbanization and urban planning

This chapter began by sketching how and why urban and peri-urban agriculture has suffered a crisis of legitimacy over time. Urban economic and development planners may well have an appreciation of the need for environmental resources to contribute to the well-being of urban populations; but the demarcation of productive spaces, green places and areas for the protection and collection of wild resources are still seen as minor considerations rather than major concerns within the urban arena. Examining the specific issues that mitigate against the production of traditional African vegetables and leafy vegetables, in particular, there are many fears that occupy development and planning professionals which largely stem from the defining characteristic of urban agriculture that it utilizes and recycles urban resources (e.g. organic waste and wastewater).

The health implications of urban and peri-urban vegetable production are probably the most significant. Chapters 2 and 8 of this volume explore these issues in greater depth; but planners and policy-makers tend to believe that vegetable production in urban areas presents a health risk because of the specific use of water in production systems. Standing water in irrigation channels is perceived as providing breeding grounds for mosquitoes and therefore an important vector in the transmission of malaria (Afrane et al, 2004; Klinkenberg et al, 2008). The combination of standing water and leafy vegetation is also perceived to be an excellent environment for pest species injurious to human health, including rats, other rodents and snakes. The perceptions and beliefs around the use of wastewater from urban ditches and streams is that this represents a significant health issue because of the increased coliform load and other pathogenic organisms being applied to the large surface areas of leafy vegetables. In addition, the rapid industrialization and proliferation of new larger- and smaller-scale industries in most peri-urban areas has led to the pollution of local water sources with a range of harmful contaminants. Many AIVs, especially leafy species, are known to be super-accumulators of anti-nutritive elements, such as some of the heavy metals present in industrially contaminated water, as well as being super-accumulators of nitrogen and potassium, which are excessive and harmful to human health. Just as there have been historic incidents that have led to the toleration of urban and peri-urban agriculture by metropolitan authorities, there have been others that have caused 'health scares', which have seen the restriction, if not prohibition, of urban and peri-urban agriculture. For example, in Dar as Salaam during March 2007, a city-wide outbreak of cholera caused a rapid reaction from city authorities that banned the continued production of crops in urban areas, including Sinza, Kinondoni Municipality, where urban farmers were displaced to the outskirts of the city (Pichop, pers com, 2007). Whether or not urban agriculture actually contributed to the outbreak and spread of cholera was never tested; it was assumed that the connection with standing water and wastewater presented an acute hazard to the transmission of the disease.

There is, in fact, a dearth of research examining the health-related aspects of urban vegetable production. One of the few studies to test the effects of wastewater use in India showed that although there were measurable levels of pathogens on vegetables irrigated with grey water, these were usually below levels that posed a risk to human health, although farmers standing in wastewater certainly were suffering from dermatological problems (Bradford et al, 2003). It is not only the association with wastewater that worries planners. Also of concern is the practice of using composted waste to produce horticultural (leafy) crops. Depending upon local circumstances, this composted waste may be made up of household refuse, human waste or urban waste contaminated with a whole range of undesirable elements (e.g. hospital refuse, partially composted human waste and the by-products of industrial processing). Household-level composting may present less of a hazard than mixed urban waste, with the nuisance effects of flies and other pests, and the potential nutritive impacts limited to the household concerned. However, the use of urban waste presents a range of important issues to urban planners. In some city areas, there are organized systems of sorting, composting and auctioning urban waste. Under these often regulated and organized systems, some of the problems of managing mixed waste are limited. However, in other cities such as Bamako (Mali), urban waste is collected and distributed without sorting and composting, and applied directly to urban and peri-urban fields with all the associated problems that this may imply (though as yet unexamined). Large areas of open space in the urban and peri-urban zone are consequently covered in waste that is foul smelling, contains a very high fraction of plastic waste which is subsequently distributed by the wind, and has a fraction of metal and hospital waste that presents a significant threat to people working in or passing through areas fertilized in this way.

There are further concerns about the use of groundwater by urban and peri-urban vegetable farmers, competing against the needs of urban populations for potable water, particularly in SSA cities located in arid and semi-arid areas. There are some studies which show that the quality of groundwater in urban areas is not sufficient for drinking water; so these competitive effects are not necessarily supported by credible evidence (e.g. Dongus, 2001). There is, however, a firmer basis for the contention that urban and peri-urban production, particularly of vegetables, is responsible for the contamination of groundwater with pesticides, herbicides and nitrogen. It is information like this that has helped to construct policy narratives that perpetuate the antagonistic responses of policy-makers to urban and peri-urban agriculture.

There is also a political and governance dimension to urban and peri-urban agriculture that affects the actions and strategies promoted by decision-makers (Municipal Development Programme, 2001). Urban developers of residential and industrial estates and complexes have better political capital than urban and peri-urban communities, especially where this is supported by formal urban development plans. Consequently, the property or usufruct rights of small-scale farmers and households are constantly under pressure in areas experiencing growth and demand for land. The pressure for land development usually prompts a decline in the security of land tenure and other property rights even where these are supported by formal title. In some peri-urban zones, this planning speculation (or planning blight, depending upon the point of view) has fostered a process of land abandonment, which (while providing no security of tenure) provides space for temporary agriculture of short-rotation quick-maturing crops such as African indigenous vegetables. In other city areas, the development of a formal land market can displace previous systems of customary land allocation and tenure so that formal purchase and titling is often concentrated amongst the more wealthy and elite sections of the community.

The institutional complexity in the peri-urban zone, where it is often unclear whether rural or metropolitan development plans have effective priority, adds further complications to the decision-making environment and the room for agriculture and vegetable production within it. Those areas of unscheduled and wild lands, as well as green spaces where individuals and households might collect wild species and resources, are also coming under increasing pressure. As cities spread out further into the peri-urban zone, there is a concomitant loss of these open areas. The same planning pressures apply to lands within city boundaries that are reserved as zones for environmental security, such as those known to be 'hazard areas' (e.g. prone to flooding and therefore zoned as reserved areas) or other spaces classified for urban utility. There are examples of these lands being developed for industrial and residential purposes in many if not all SSA city areas. As urbanization increases, the areas available for wild collection and the continued presence of a seed bank of AIV species in the urban milieu decline.

As well as the range of threats to including urban and peri-urban agriculture and the continued production and collection of African indigenous vegetables, there has been a change in attitude in some areas of the world towards the integration of urban and peri-urban agriculture within some cities. As van Veenhuizen (2006) reports, Latin America appears to be leading the movement for change in this area with Cuba, Argentina and Brazil all examples of countries where there are proactive urban agriculture policies.

There is change on the African continent as well, and well-known examples here include Benin, Botswana and Zambia, where the development of thematic programmes and strategies such as the national food security policies, poverty reduction strategies and sustainable city development policies include elements that support, or at least are sympathetic with, urban agriculture. There is more proactive planning in Kampala, Dar es Salaam, Bulawayo (Zimbabwe) and Port Elizabeth (South Africa), where local governments are formulating specific urban agriculture policies.

So, in Kampala, for example, the structure plan was revised to include urban and peri-urban agriculture as a legitimate land use, and an urban agriculture unit was set up under the Kampala City Council administration to ensure effective management of the urban agriculture strategy. In some other areas, it is the relative value of urban and peri-urban agricultural production that has changed the dynamics of urban and peri-urban land use so that the continued production of vegetables in these areas no longer seems precarious. For example, in Kenya, there has been a concerted move away from a national economy based on international primary commodity exchange towards the growth and supply of national and regional urban markets with domestic production of higher-value commodities such as vegetables. As consumer demand increases for these products and new markets are established in neighbouring countries, the value of vegetable production in urban and peri-urban areas increases such that its loss would represent a significant impact upon local city economies.

Conclusions: Future prospects

It is clear that urban and peri-urban agriculture is an important land use that has much to contribute to the livelihoods of urban populations. These contributions may be through the provision of income, increased access to food, household subsistence and individual well-being through engagement in agricultural activity for all sections of urban society, poorer and richer, men and women, children and the elderly, long-term residents and newly arrived migrants. The importance of the production and collection of wild and cultivated indigenous African vegetables within these spaces is something which requires particular attention. African indigenous vegetables contribute to, and have further potential to contribute to, a certain 'value added' within urban and peri-urban food systems. They provide significant nutritional and health benefits beyond those provided by exotic vegetables; they also contribute to agro-biodiversity, nutritional diversity and the diversity of food cultures in national as well as strictly urban arenas.

There are promising prospects for the future. The ecological model of planning used in Dar es Salaam (see Chapter 8) and Lusaka is finding increasing purchase in African planning circles. Pushed forward by the concepts of city ecological footprints, and a recognition that urban populations require as many product and service functions from their environments and natural resource endowments as rural communities, the ecological planning model emphasizes the city system as a biological system. Planning uses open and green spaces as an important tool to manage environmental quality, public health and public well-being. The position of urban agriculture in this model is as a tool for managing nutrient and waste recycling, the use of open space to provide a sink for rainwater and to change the temperature dynamics of city centres, a positive use of otherwise neglected or derelict open spaces, a way to meet nutritional needs, and a method of keeping cities green and more aesthetically interesting. For planners adopting such a positive model of urban and peri-urban agriculture, the problems of health and pollution associated with vegetable production have solutions that can be managed by city authorities, such as the regulation of inputs and use of wastewater and the promotion of proper washing and preparation of urban vegetables. The latest phase of the RUAF-supported Cities Farming for the Future programme has also found ways to find safe methods of using urban resources and, importantly, has pushed forward on developing governance frameworks that have allowed more positive policy planning supporting urban agriculture initiatives. Success in 47 partner cities has included Bulawayo (Zimbabwe), Ndola (Zambia), Porto Novo (Benin) and Ibadan (Nigeria), where there has been action planning in support of urban agriculture, the integration of urban agriculture through the appointment of appropriate staff in metropolitan councils and planning departments, the development of by-laws allowing urban agriculture a place in urban and peri-urban spaces, and the involvement of cross-departmental and multiple stakeholder urban agriculture forums.

In short, the many and multiple uses of green 'growing spaces' and 'breathing places' within the concrete and brick confines of African cities, and the potential for urban and peri-urban agriculture to provide greater resilience to urban food systems, seems to hold some promise for the continued existence of urban agriculture to continue to produce, and provide the opportunity to collect, African indigenous leafy vegetables into the future.

Notes

1 *De facto* tenure is ownership or a use right that is recognized in customary law or informal systems of property rights, whereas *de jure* tenure is associated with formal and legal title and use rights upheld by legal systems and associated documentations and proofs of title. In some African countries *de facto* and *de jure* systems may coexist and even complement each other; but by and large *de jure* systems are regarded as more certain and reliable claims to land and resource ownership or usufruct.

2 The term nutraceutical was coined during the 1990s by Dr Stephen DeFelice and is loosely defined as 'food, or parts of food, that provide medical or health benefits, including the prevention and treatment of disease'. However, the term is a contested one without strict biophysical or medical definitions and is often a more general description of the perceived health and well-being characteristics than tested 'fact'.

3 The definition of 'wild' or of 'wild resources' (plants or animals) is another arena of continuing scientific debate. The World Conservation Union (IUCN) have

developed the notion of 'wild resources' rather than wild species and is promoting a code of practice with regard to the harvest of wild resources that defines wild resources in broadly practical terms (see www.iucn.org/themes/ssc/susg/bgrnd/supoleng.html). The IUCN has also given its support to a definition of wild resources expressed by Murray and Simcox (2002) as 'whole organisms living in the wild (and the parts or derivatives of such organisms), or in a wild state ... It excludes organisms that are captive reared (for at least part of their life history) before release or which are cultivated, planted or otherwise caused to grow by man in the wild' (see www.iucn-uk.org/pdf/iucnWLRresources.pdf). Wild species, as opposed to wild resources, might be characterized slightly differently as those plant and animal taxa that have not been subject to a process of domestication and therefore retain the genotypic characteristics of wild rather than cultivated or anthropogenically managed populations.

References

Abukutsa-Onyango, M., Mwai, G. N., Otiato, D. A. and Onyango, J. (2007) *Urban and Peri-Urban Indigenous Vegetables Production and Marketing in Kenya*, Report presented to the 6th Thematic Meeting of the IndigenoVeg Project, Maseno University, Maseno, Kenya

Afrane, Y. A., Klinkenberg, E., Drechsel, P., Owusu-Daaku, K., Garms, R. and Kruppa, T. (2004) 'Does irrigated urban agriculture influence the transmission of malaria in the city of Kumasi, Ghana?', *Acta Tropica*, vol 89, pp125–134

Ambrose-Oji, B. (2004) *Livelihoods Synthesis Study: Key Determinants of Poor People's Livelihood Strategies and Natural Resources-Related Opportunities*, NSRP, DFID, London, UK

Ambrose-Oji, B. (2005) *Participatory Biodiversity Conservation and Livelihoods: Socio-Environmental Dynamics in the Forest Zone of Southwest Cameroon*, Environment Department, Imperial College at Wye, University of London, London, UK

Argenti, O. (2000) *Food for the Cities: Food Supply and Distribution Policies to Reduce Urban Food Insecurity*, FAO, Rome, Italy

Argenti, O. and Marocchino, C. (2005) *Urban Food Supply and Distribution in Developing Countries and Countries in Transition: A Guide for Planners*, FAO, Rome, Italy

Armar-Klemesu, M. (2000) 'Urban agriculture and food security, nutrition and health', in Bakker, N. (ed) *Growing Cities, Growing Food: Urban Agriculture on the Policy Agenda*, Deutsche Stiftung für Internationale Entwicklung, Feldafing, Germany

Ashebir, D., Pasquini, M. W. and Bihon, W. (2007) 'Urban agriculture in Mekelle, Tigray State, Ethiopia: Principal characteristics, opportunities and constraints for further research and development', *Cities*, vol 24, pp218–228

Asomani-Boateng, R. (2002) 'Urban cultivation in Accra: An examination of the nature, practices, problems, potentials and urban planning implications', *Habitat International*, vol 26, pp591–607

Assogba Komlan, F., Akplogan, F. M. and Adegbola, Y. P. (2007) *Indigenous Vegetable Market Chains in Urban and Peri-Urban Areas of Cotonou and Lokossa, Benin*, Report presented to the 6th Thematic Meeting of the IndigenoVeg Project, INRAB, Cotonou, Benin

Assouma, Y., Djidji, H., Fiacre Alla, J. and Kouame, C. (2007) *Market Chain Analysis of Indigenous Vegetables in the Districts of Abidjan and Yamoussoukro, Côte D'Ivoire*, Report presented to the 6th Thematic Meeting of the IndigenoVeg Project, INRAB, Cotonou, Benin

Bah, M., Salmana, C., Diyamett, B., Lerise, F., Okali, D., Okpara, E., Olawoye, J. and Tacoli, C. (2003) 'Changing rural–urban linkages in Mali, Nigeria and Tanzania', *Environment and Urbanization*, vol 15, pp13–28

Balemie, K. and Kebebew, F. (2006) 'Ethnobotanical study of wild edible plants in Derashe and Kucha districts, south Ethiopia', *Journal of Ethnobiology and Ethnomedicine*, vol 2, pp53–60

Becker, B. (1986) 'Wild plants for human nutrition in the Sahelian zone', *Journal of Arid Environments*, vol 11, pp61–64

Bedigian, D. (2005) 'Slimy leaves and oily seeds: Distribution and use of wild relatives of sesame in Africa', *Economic Botany*, vol 58: ppS3–S33

Benson, A. (2006) 'The constraints and opportunities for commercial horticulture in the urban and peri-urban areas of Bamako, Mali: A gendered perspective', unpublished MSc Thesis, University of Wales Bangor, Bangor, Wales

Bigsten, A. and Kayizzi-Mugerwa, S. (1992) 'Adaption and distress in the urban economy: A study of Kampala households', *World Development*, vol 20, pp1423–1441

Binns, T. and Lynch, K. (1998) 'Feeding Africa's growing cities: The potential of urban agriculture', *Journal of International Development*, vol 10, pp777–793

Bradford, A., Brook, R. and Hunshal, C. S. (2003) 'Wastewater irrigation in Hubli-Dharwad, India: Implications for health and livelihoods', *Environment and Urbanization*, vol 15, pp157–170

Bryld, E. (2003) 'Potentials, problems, and policy implications for urban agriculture in developing countries', *Agriculture and Human Values*, vol 20, pp79–86

Cabannes, Y. (2004) 'Public financing and investments for urban agriculture', Paper prepared for the WUF 2004, IPES/UMP–LAC/IDRC, Barcelona, Spain

Campbell, B. M. (1986) 'The importance of wild fruits for peasant households in Zimbabwe', *Food and Nutrition*, vol 12, pp38–44

Cavendish, W. (2000) 'Empirical regularities in the poverty–environment relationship of rural households: Evidence from Zimbabwe', *World Development*, vol 28, pp1979–2003

CENCOSAD (1994) *Urban Market Gardens in Accra*, Centre for Community Studies, Action and Development and the Mega Cities Project, Accra, Ghana

Chweya, J. A. and Eyzaguirre, P. B. (1999) *The Biodiversity of Traditional Leafy Vegetables*, International Plant Genetics Resources Institute, Rome, Italy

Cocks, M. (2006) *Wild Resources and Cultural Practices in Rural and Urban Households in South Africa: Implications for Bio-Cultural Diversity Conservation*, Wageningen Universiteit, Wageningen, the Netherlands

Coetzee, C., Jefthas, E. and Reinten, E. (1999) 'Indigenous plant genetic resources of South Africa', in Janick, J. (ed) *Perspectives on New Crops and New Uses*, ASHS Press, Alexandria, Egypt

Delang, C. O. (2006) 'Not just minor forest products: The economic rationale for the consumption of wild food plants by subsistence farmers', *Ecological Economics*, vol 59, pp64–73

Diouf, M., Diop, M., Lô, C., Dramé, K. A., Sène, E., Bâ, C. O., Guèye, M. and Faye, B. (1999) 'Prospection de légumes feuilles traditionnels de type africain au Sénégal', in Chweya, J. A. and Eyzaguire, P. (eds) *Biodiversity of Traditional Leafy Vegetables in Africa*, International Plant Genetic Resources Institute (IPGRI), Rome, Italy

Dixon, J., Omwega, A. M., Friel, S., Burns, C., Donati, K. and Carlisle, R. K. (2007) 'The health equity dimensions of urban food systems', *Journal of Urban Health*, vol 84, pp118–128

Dongus, S. (2001) 'Urban vegetable production in Dar es Salaam (Tanzania) – GIS-supported analysis of spatial changes from 1992 to 1999', *Applied Physiogeography of the Tropics and Subtropics (APT) 12*, Institute of Physical Geography, University of Freiburg, Freiburg, Germany

Dovie, D. B. K., Shackleton, C. M. and Witkowski, E. T. F. (2007) 'Conceptualizing the human use of wild edible herbs for conservation in South African communal areas', *Journal of Environmental Management*, vol 84, pp146–156

Drakakis-Smith, D. (1994) 'Food systems and the poor in Harare under conditions of structural adjustment', *Geografiska-Annaler Series B*, vol 76B, pp3–20

Drakakis-Smith, D. (2000) *Third World Cities*, 2nd edition, Routledge, London, UK

Drakakis-Smith, D., Boywer-Bower, T. and Tevera, D. (1995) 'Urban poverty and urban agriculture: An overview of the linkages in Harare', *Habitat International*, vol 19, pp183–193

Drescher, A. W. (1999) 'Urban agriculture in the seasonal tropics: The case of Lusaka, Zambia', in Koc, M., MacRae, R., Mougeot, L. J. and Welsh. J. (eds) *For Hunger-Proof Cities: Sustainable Urban Food Systems*, International Development Research Centre, Ottawa, Canada

Drescher, A. W. and Mackel, R. (2000) 'Urban and peri-urban food security: The example of Lusaka (Zambia)', APT–Reports 11, *Urbanization, Vulnerability and Resource Management in Developing Countries*, Freiberg, Germany, pp56–79

Engle, L. M. and Altoveros, N. (2000) *Collection, Conservation and Utilization of Indigenous Vegetables*, Proceedings of a Workshop 16–18 August 1999, AVRDC, Tainan, Taiwan

Erenstein, O., Oswald, A. and Mahaman, M. (2006) 'Determinants of lowland use close to urban markets along an agro–ecological gradient in West Africa', *Agriculture, Ecosystems and Environment*, vol 117, pp205–217

FAO (United Nations Food and Agriculture Organization) (1998) *Traditional Food Plants*, Food and Nutrition Paper 42, FAO, Rome, Italy

Fleuret, A. (1979) 'The role of wild foliage plants in the diet: A case study from Lushoto, Tanzania', *Ecology of Food and Nutrition*, vol 8, pp87–93

Fold, N. and Pritchard, B. (2005) *Cross-Continental Food Chains*, Routledge, London, UK

Fotso, J.-C. (2007) 'Urban–rural differentials in child malnutrition: Trends and socio-economic correlates in sub-Saharan Africa', *Health and Place*, vol 13, pp205–223

Fotso, J.-C. and Barthelemy, K.-D. (2005) 'Socioeconomic inequalities in early childhood malnutrition and morbidity: Modification of the household-level effects by the community SES', *Health and Place*, vol 11, pp205–225

Freeman, D. (1991) *A City of Farmers: Informal Urban Agriculture in the Open Spaces of Nairobi, Kenya*, Queen's University Press, McGill, US

Getahun, A. (1974) 'The role of wild plants in the native diet in Ethiopia', *Agro-Ecosystems*, vol 1, pp45–56

Gockowski, J., Mbazo'o, J., Mbah, G. and Fouda Moulende, T. (2003) 'African traditional leafy vegetables and the urban and peri-urban poor', *Food Policy*, vol 28, pp221–235

Gordon, A., Davis, J., Long, A. and Meadows, K. (2000) *The Role of Natural Resources in the Livelihoods of the Urban Poor*, Natural Resources Institute Policy Series 9, Chatham, UK

Grivetti, L. E. and Ogle, B. M. (2000) 'Value of traditional foods in meeting macro- and micronutrient needs: The wild plant connection', *Nutrition Research Reviews*, vol 13, pp31–46

Guarino, L. (ed) (1997) *Traditional African Vegetables: Promoting the Conservation and Use of Underutilized and Neglected Crops*, IPGRI, Rome, Italy
Gündel, S. (2006) *Final Report of a Synthesis of Information Produced on Urban and Peri-Urban Agriculture Research by the RNRRS Programme, 1995–2006*, Report to DFID, RNRRS Programme, Edinburgh, Scotland
Guyer, J. (1997) *An African Niche Economy: Farming to Feed Ibadan 1968–1998*, Edinburgh University Press, Edinburgh, Scotland
Harris, F. M. A. and Mohammed, S. (2003) 'Relying on nature: Wild foods in northern Nigeria', *Ambio*, vol 32, pp24–29
IndigenoVeg Project (2007a) *Synthesis Report of the 5th Thematic Meeting – Livelihoods and IVs*, Pretoria, www.indigenoveg.org
IndigenoVeg Project (2007b) *Synthesis Report of the 6th Thematic Meeting – Marketing IVs*, Arusha, www.indigenoveg.org
Jacobi, P. and Amend, J. (1997) 'Vegetable farming in Dar es Salaam: An important source of income, and not only for the poor', *Agriculture and Rural Development*, vol 4, no 2/97, pp52–54
Jacobi, P., Amend, J. and Kiango, S. (2000) 'Urban agriculture in Dar es Salaam: Providing an indispensable part of the diet', in *Growing Cities, Growing Food: Urban Agriculture on the Policy Agenda*, Deutsche Stiftung für internationale Entwicklung (DSE), Feldafing, Germany
Jamal, V. (1985) *Structural Adjustment and Food Security in Uganda*, International Labour Organization, Geneva, Switzerland
Jansen van Rensburg, W., van Averbeke, W., Slabbert, R., Faber, M., van Jaarsveld, P., van Heerden, I., Wenhold, F. and Oelofse, A. (2007) 'African leafy vegetables in South Africa', *Water SA*, vol 33, pp317–326
Kasambula, P., Pichop, G., Maro, F. and Marealle, R. (2007) *Indigenous Vegetables Supply Chain in Uganda: Significance, Constraints and Opportunities – Case of Kampala and Mbale Districts*, Report presented to the 6th Thematic Meeting of the IndigenoVeg Project, NARO, Kampala, Uganda
Kimbrell, A, (2002) *The Fatal Harvest Reader: The Tragedy of Industrial Agriculture*, Island Press, Washington, DC
Klinkenberg, E., McCall, P. J., Wison, M. D., Amerasinghe, F. P. and Donnelly, M. J. (2008) 'Impact of urban agriculture on malaria vectors in Accra, Ghana', *Malaria Journal*, vol 7, pp151–162
Lachance, A. (1993) *A Plot of One's Own in West African Cities*, IDRC Research Reports, vol 21, no 3, Ottawa, Canada
Lado, C. (2004) 'Sustainable environmental resource utilisation: A case study of farmers' ethnobotanical knowledge and rural change in Bungoma district, Kenya', *Applied Geography*, vol 24, pp281–302
Laker, M. C. (2005) 'The global impact of zinc micronutrient deficiencies', in *Combined FSSA and SASRI Symposium on Micronutrients in Agriculture*, 25 February 2005, Mount Edgecombe, South Africa, CD-ROM
Letšela, T., Witkowski, E. T. F. and Balkwill, K. (2003) 'Plant resources used for subsistence in Tsehlanyane and Bokong in Lesotho', *Economic Botany*, vol 57, pp619–639
Lynch, K. (1994) 'Urban fruit and vegetable supply in Dar es Salaam', *The Geographical Journal*, vol 160, pp307–318
Maxwell, D. G. (1995) 'Alternative food security strategy: A household analysis of urban agriculture in Kampala', *World Development*, vol 23, pp1669–1681
Maxwell, D. G., Levin, C. and Csete, J. (1999) 'Does urban agriculture help prevent malnutrition? Evidence from Kampala', *Food Policy*, vol 23, pp411–424

Maxwell, D. G., Levin, C., Armar-Klemesu, M., Ruel, M., Morris, S. and Ahiadeke, C. (2000) *Urban Livelihoods and Food and Nutrition Security in Greater Accra, Ghana*, IFPRI Research Report 112, IFPRI, Washington, DC

Millennium Ecosystem Assessment (2005) *Ecosystems and Human Well-Being – Biodiversity Synthesis*, World Resources Institute, Washington, DC

Mougeot, L. J. A. (2000) *Achieving Urban Food and Nutrition Security in the Developing World: The Hidden Significance of Urban Agriculture, A 2020 Vision for Food, Agriculture, and the Environment*, Focus 3, Brief 6 of 10, International Food Policy Research Institute (IFPRI), Washington, DC

Moustier, P. (2000) 'Urban and peri-urban agriculture in West and Central Africa: An overview', Provisional paper (30/10/00) for SIUPA stakeholder meeting and strategic workshop, sub-Saharan region, 1–4 November 2000, Nairobi, Kenya

Moustier, P. and David, O. (1997) *Étude de Cas de la Dynamique du Maraîchage Périurbain en Afrique Sub-Saharienne, FAO N-DT/02/96, Projet: Approvisionnement et Distribution Alimentaires des Villes d'Afrique Francophone*, FAO, Rome, Italy

Municipal Development Programme (2001) *The Political Economy of Urban Agriculture: Report of a Preparatory Workshop*, Unpublished report, Municipal Development Programme, Harare, Zimbabwe

Murray, M. and Simcox, H. (2002) 'Sustainable use of wild-living resources in the UK: A review'. Unpublished report to the IUCN UK Committee, London

Page, B. (2002) 'Urban agriculture in Cameroon: An anti-politics machine in the making?', *Geoforum*, vol 33, pp41–54

Peters, C. R., O'Brien, E. M. and Box, E. O. (1984) 'Plant types and seasonality of wild-plant foods, Tanzania to southwestern Africa: Resources for models of the natural environment', *Journal of Human Evolution*, vol 13, pp397–414

Pichop, G. (2007) *Baseline Production and Marketing Supply Chain Survey: Synthesis Report*, Report presented to the 6th Thematic Meeting of the IndigenoVeg Project, AVRDC, Arusha, Tanzania

Pinstrup-Andersen, P. (1989) 'The impact of macroeconomic adjustment: Food security and nutrition', in Commander, S. (ed) *Structural Adjustment and Agriculture: Theory and Practice in Africa and Latin America*, Overseas Development Institute, London, UK

Ruel, M. T., Garrett, J. L., Morris, S. S., Maxwell, D., Oshaug, O., Engle, P., Menon, P., Slack, A. and Haddad, L. (1998) *Urban Challenges to Food and Nutrition Security: A Review of Food Security, Health, and Caregiving in the Cities*, Food Consumption and Nutrition Division Discussion Paper 51, International Food Policy Research Institute, Washington, DC

Scoones, I. C., Melnyk, M. and Pretty, J. (eds) (1992) *The Hidden Harvest: Wild Foods and Agricultural Systems – A Literature Review and Annotated Bibliography*, Overseas Development Institute, London, UK

Shackleton, C. M. (2003) 'The prevalence of use and value of wild edible herbs in South Africa', *South African Journal of Science*, vol 99, pp23–25

Shackleton, S. E., Shackleton, C. M., Dzerefos, C. M. and Mathabela, F. R. (1998) 'Use and trading of wild edible herbs in the central lowveld savanna region, South Africa', *Economic Botany*, vol 52, pp251–259

Shackleton, C. M., Ernst, L., Mthembu, T., Paumagarten, F. and Auerbach, R. (2007) *The Growing and Marketing of African Indigenous Vegetables in the Durban Metropolitan Area, South Africa*, Report presented to the 6th Thematic Meeting of the IndigenoVeg Project, Rhodes University, Grahamstown, South Africa

Slater, R. and Twyman, C. (2003) *Hidden Livelihoods? Natural Resource-Dependent Livelihoods and Urban Development Policy*, Overseas Development Institute, London, UK

Smith, L., Ruel, M. and Ndiaye, A. (2004) *Why is Child Malnutrition Lower in Urban than Rural Areas? Evidence from 36 Developing Countries*, IFPRI, Washington, DC

Stevenson, C., Xavery, P. and Wendeline, A. (1996) *Market Production of Fruits and Vegetables in the Peri-Urban Area of Dar es Salaam*, Urban Vegetable Promotion Project, Dar es Salaam, Tanzania

Tacoli, C. (1998) *Bridging the Divide: Rural–Urban Interactions and Livelihood Strategies*, IIED, London, UK

Tixier, P. and de Bon, H. (2006) 'Urban horticulture', in van Veenhuizen, R. (ed) *Cities Farming for the Future: Urban Agriculture for Green and Productive Cities*, International Institute of Rural Reconstruction and ETC Urban Agriculture, Silang, the Philippines

Toulmin, C. and Quan, J. (eds) (2000) *Evolving Land Rights, Policy and Tenure in Africa*, DFID/IIED/NRI, London, UK

UN (United Nations) (2002) *World Urbanization Prospects: The 2001 Revision*, United Nations, New York, NY

UN (2006) *World Urbanization Prospects: The 2005 Revised Population Database*, United Nations Population Division, New York, NY

UNDP (United Nations Development Programme) (1996) *Urban Agriculture: Food, Jobs and Sustainable Cities*, UNDP, Publication Series for Habitat II, vol 1, New York, NY

United Nations University (2005) 'Special issue on micronutrient deficiencies in food and nutrition in developing countries', *Food and Nutrition Bulletin*, vol 25, no 1, special issue

van Averbeke, W. (2007) 'Urban farming in the informal settlements of Atteridgeville, Pretoria, South Africa', *Water SA*, vol 33, pp337–342

van den Eynden, V., Cueva, E. and Cabrera, O. (2003) 'Wild foods from southern Ecuador', *Economic Botany*, vol 57, pp576–603

van Veenhuizen, R. (2006) 'Cities farming for the future', in van Veenhuizen, R. (ed) *Cities Farming for the Future: Urban Agriculture for Green and Productive Cities*, ETC – Urban Agriculture, Leusden, the Netherlands

von Braun, J., McComb, J., Fred-Mensah, B. and Pandya-Lorch, R. (1993) *Urban Food Insecurity and Malnutrition in Developing Countries: Trends, Policies, and Research Implications*, International Food Policy Research Institute, Washington, DC

Vorster, I., Pichop, G., Maro, F. and Marealle, R. (2007a) *Indigenous Supply Chain: Case of Soshanguve (Tshwane Metropolitan Area) South Africa*, Report presented to the 6th Thematic Meeting of the IndigenoVeg Project, ARC, Roodeplat, South Africa

Vorster, H. J., Jansen van Rensburg, W. S., Venter, S. L. and van Zijl, J. J. B. (2007b) 'The importance of traditional leafy vegetables in South Africa', *African Journal of Food, Agriculture, Nutrition and Development*, vol 7, no 4, pp1–13

Weinberger, K. and Lumpkin, T. A. (2005) *Horticulture for Poverty Alleviation*, AVRDC – World Vegetable Center, Shanhua, Taiwan

Weinberger, K. and Msuya, J. (2004) *Indigenous Vegetables in Tanzania: Significance and Prospects*, AVRDC – World Vegetable Center, Shanhua, Taiwan

World Bank (2002) *World Development Report 2002: Building Institutions for Markets*, Oxford University Press for the World Bank, New York, NY

Zeeuw, H., Guendel, S. and Waibel, H. (2000) 'The integration of agriculture in urban policies', in Bakker, N., Dubelling, M., Gundel, S., Sabel-Koschella, V. and Zeeuw, H. (eds) *Growing Cities, Growing Food: Urban Agriculture on the Policy Agenda*, Food and Agriculture Development Centre (ZEL), Feldafing, Germany

Zemede, A. (1997) *Indigenous African Food Crops and Useful Plants: Survey of Indigenous Food Crops, Their Preparations and Home Gardens*, United Nation University Institute for Natural Resources in Africa, Nairobi, Kenya

Zemede, A. and Mesfin, T. (2001) 'Prospects for sustainable use and development of wild food plants in Ethiopia', *Economic Botany*, vol 55, pp47–62

2
Urban and Peri-Urban Agriculture in African Cities

Andreas de Neergaard, Axel W. Drescher and Christophe Kouamé

Introduction

Urban agriculture can be widely defined as 'any agricultural activity within the administrative boundary of an urban centre' (Foeken and Owuor, 2000, cited in Brock and Foeken, 2006). In terms of agricultural production systems, this ranges from the smallest imaginable single plants in pots, to large farming systems covering several hectares. It includes production strictly for home consumption, as well as entirely commercialized enterprises. Production may take place on private land with secure tenure, on rented or borrowed land, or even as squatter farming with no tenure security at all. The production system may be a low- or zero-input system of haphazard nature, ranging over more or less consciously designed urban agro-ecosystems with recycling of organic wastes or manures, to input-intensive horticultural systems with large inputs of fertilizers and agrochemicals. Finally, the choice of crops will reflect local preferences, commercial opportunities, bioclimatic conditions and availability of seeds or germplasm. In short, the variety of production systems is as diverse as the inhabitants of African cities; hence, any review of urban agricultural typologies is bound to balance between overgeneralizations and context-specific case studies. In this chapter we have attempted an overview of farming systems in African cities, the people involved in the farming, and the role of urban agriculture in urban food production and nutrient flows. Some attention is devoted to recycling of organic wastes – seen from an agronomic perspective. The chapter also includes a discussion of the role of land tenure bordering on that found in Chapter 8, which more thoroughly discusses agriculture in the urban geography.

The chapter is predominantly focused on urban crop production, although urban livestock rearing is widely practised and, in some cases (namely, fish, poultry and even pig production in Asia), may contribute more to the local supply than urban crops. Reference to livestock systems is included when they interact with the crop production system. This priority reflects the focus of the book – urban production of indigenous vegetables in African cities – and not the importance of livestock rearing for urban livelihoods and environment.

The literature on the subject of urban agriculture has tended to focus on policy and administrative issues of urban planning, on waste recycling and on its contribution to urban livelihoods and nutrition. The literature on urban agronomy is dominated by more or less descriptive studies and is very often poorly quantified and categorized, of a more anecdotal or observing nature. Perhaps this is so because, from the point of view of a crop scientist, a crop growing in an urban environment is not different from one growing anywhere else. Moreover, there is an obvious lack of conventional agronomic studies on urban production systems. This may be due to the *ad hoc* nature of many urban farming systems, essentially consisting of one big 'border effect' and, in some cases, of the somewhat opportunistic nature; as a result, urban agronomy is difficult to study using the conventional methods from crop and plant sciences. It is fair to conclude that beyond the work on heavy metals from various wastes and manures, and the impact of atmospheric pollution on plant growth, there is a vacuum in analytical studies of urban agronomy.

Extent of urban farming

The fact that urban agricultural activity is widespread in most African cities, and that the importance is commonly underestimated, was becoming something of a truism in the literature. Although city planners may not adequately consider and allow for urban vegetable production, numerous reports on the extent and importance of urban agriculture exist. The data are largely incomparable as few systematized efforts to collect the information have been undertaken. Moreover, as described in later chapters, due to the wide national, regional and global variation and the dynamic and *ad hoc* nature of urban farming, common terminologies and definitions are poorly developed and adhered to. For example, what is considered backyard farming in one report may not appear in another. Reported figures may also focus on subsistence or commercialized production, or present agricultural output as a percentage of local consumption. Furthermore, it is not always clear if farming activities are, in fact, urban, or just carried out by the urban population on peri-urban or rural land. One must therefore be cautious when interpreting the reported figures. Reviewing the literature may at least serve to illustrate that although urban farming may in some cases be a marginal or lesser contribution to urban livelihoods (Ellis and Sumberg, 1998), it is

Box 2.1 Indications of the extent of urban agriculture in sub-Saharan African cities

- 90 per cent of locally sold vegetables in Accra (Ghana) are locally produced.
- 90 per cent of all vegetables and 60 per cent of milk supply in Dar es Salaam (Tanzania) is from urban production.
- 90 per cent of maize and 80 per cent of leafy vegetables in urban supply of Yaoundé (Cameroon) are from urban production.
- 80 per cent of families in Libreville (Congo) are engaged in horticulture.
- 68 per cent of citizens in six Tanzanian cities are engaged in urban farming.
- 67 per cent of urban families in Kenya farm on peri-urban and urban land.
- 65 per cent of marketed vegetables in Brazzaville (Congo) are from urban gardens.
- 50 per cent of households are engaged in some form of subsistence production in Accra (Ghana).
- 45 per cent of households in Lusaka (Zambia) practise horticulture or livestock husbandry.
- 37 per cent of urban households in Maputo (Mozambique) produce food in town.
- 36 per cent of families in Ouagadougou (Burkina Faso) are engaged in horticulture or livestock rearing.
- >33 per cent of households surveyed in Harare (Zimbabwe) keep livestock, mainly chickens but also rabbits, pigeons, ducks and turkeys.
- 33 per cent of Kampala's (Uganda) citizens are involved in agriculture.
- 29 per cent of urban population in Nairobi (Kenya) grow food in the urban area.
- Urban agriculture is the second largest urban employer (20 per cent of employees) in sub-Saharan Africa (SSA).

Source: compiled from Egziabher et al (1994); Smit et al (1996); Moustier (1999); Schiere and van der Hoek (2001); Asomani-Boateng (2002); Gerstl et al (2002); van den Berg et al (2003); Cofie et al (2003); Danso and Moustier (2006); Tixier and de Bon (2006)

globally occurring, and involves a significant proportion of the urban population in Africa's cities.

Box 2.1 includes an overview of various estimates of the magnitude and contribution of urban agriculture in different African cities. Although the data are not directly comparable, they serve to provide an indication of the extent of urban agriculture in SSA cities.

Although urban food production in developing countries is often perceived as an activity that contributes to food security and human nutrition, this is not the only explanation for its widespread occurrence. Estimates of prevalence of urban agricultural (horticultural) production in developed countries are not that far from the estimates quoted above. For example 25 per cent of urban families in the US grow vegetables (Smit et al, 1996); one third of the urban gardens surveyed by Smith et al (2005a) in Sheffield (England) had vegetable patches. Although the type and nature of these farming systems may be largely incomparable with those in developing countries, it does serve as a reminder that farming activities may be undertaken for a number of reasons not strictly related to food security, nutrition or income generation.

Urban crops

The very nature of urban production sites, and the opportunities and constraints faced by urban farmers narrow down the choice of crops. In general terms, the characteristics of urban production systems are limited space and high land value, close proximity to markets and, in most cases, proximity to farmers, although some commute to pockets of farmed land within the city, with uncertain tenure when using public land. Risk of theft or vandalism is also high in some places. However, these general characteristics may not be clearly experienced by all individual farmers, as some may have access to more land than they can handle.

Although staple crops such as maize (*Zea mays*), plantain (*Musa* spp.) and tubers can easily be found growing in the centre of many African cities, the urban production sites tend to be dominated by crops that are well adapted to the above-mentioned constraints and opportunities. In response to the lack of space and in order to optimize outputs, farmers tend to engage in horticulture-like production systems with high-value crops, where the individual plant is managed, rather than the field as a whole. The proximity to the market allows for cultivation of perishable vegetables, with short storage time after harvest, particularly in cities with poor infrastructure, where vegetables grown outside town cannot reach the market on time (Ratta and Nasr, 1996). Although the majority (by bulk) of urban production may not be of leafy greens, the short shelf life of these crops determines that the majority for urban consumption must be produced in close proximity to the city or within its boundaries (Moustier, 1997, cited in Tixier and de Bon, 2006).

Besides the advantage of low transportation costs, independence from middlemen and fast access to the market with perishable goods, urban farmers have another advantage in that they can easily monitor prices on the markets and optimize when to sell their produce (Ratta and Nasr, 1996). While efficient communication lines are rapidly developing all over the world, rural farmers are still very dependent upon middlemen to inform them on market prices, and studies have illustrated that the middlemen often take advantage of this position to increase their profit (Rengasamy et al, 2003). For seasonal products, prices are much higher for the first batches to reach the market, and are strongly influenced by supply and demand. Urban farmers have an advantage in being able to closely monitor market fluctuations and to sell their crops at a premium, as seen, for instance, with strawberries in Bamako (Mali).

In the case of the commercial producers, short-cycle crops are often preferred in order to secure continuous production throughout the season or year. Short-cycle crops may also be preferred when land tenure is insecure or the risk of theft and vandalism is high. Input-intensive urban farming systems, for commercial purposes, may also differ from rural farming systems in that they have continuous access to water or other inputs and therefore can deliver a stable production throughout the dry season. In areas with excessive rainfall, infrastructure may worsen during the wet seasons, thus hindering the import of vegetables from outside cities, in this way increasing the value and importance of local production during the rainy season (Tixier and de Bon, 2006).

A study conducted in Nigeria showed that staple crops (cassava, banana, maize, yams and plantain) produced in the cities did not appear on the urban markets due to competition from rural producers; hence, the staples were only grown for home consumption in the city. Meanwhile, there was a profitable industry of urban vegetable production that was sold in the local markets (Ezedinma and Chukuezi, 1999). It was also reported that urban vegetable farming peaked during the dry season due to three factors – namely, fewer problems with pests and diseases; the location of urban fields in flood-prone areas (making them inaccessible during the wet season); and the high availability on the urban markets of vegetables from rural producers during the wet season (Ezedinma and Chukuezi, 1999). Urban vegetable farmers were, to a large extent, rural migrants who commuted to the cities during the dry season; hence, labour shortage during the wet season probably also played a role in the seasonality of urban vegetable production.

Multiple farming strategies may be practised within a single household. For example, in Bamako (Mali), cash crops of strawberries bring in the majority of the income but occupy only a small percentage of the land area. Most of the cultivated area is devoted to staple crops (e.g. maize) for own consumption, which generates no, or very little, profit for the farmer (Eaton and Hilhorst, 2003).

The choice of crops is also influenced by the land tenure and, particularly, tenure security. A study from Cotonou (Benin) showed that high-value labour-demanding vegetables were predominantly cropped on land with secure tenure, or at least security of occupation, whereas more extensive production of staple crops such as cassava and maize took place on plots with lower tenure security (e.g. land awaiting construction) or in corners of land owned by industries or large companies, well out of sight of the main entrance (Brock and Foeken, 2006).

Urban farmers will grow most crops cultivated in Africa, and a detailed accounting of preferred species has little meaning in a general review, varying widely with cultural and biophysical settings. Chapter 3 includes an overview of important indigenous vegetables and their production in Africa. An overview of important urban crops can be found in Tixier and de Bon (2006). In the following sections, specific crops will be mentioned in connection with the cited literature.

Location, tenure and farming practices

Urban farming exhibits a wide variety of cropping practices, ranging from what can be considered conventional horticultural production, in some cases even in greenhouses, to ingenious space and resource-conserving structures, such as the compost mound. As space is the main limitation in several urban farming systems, it is not surprising that many characteristic urban production systems seek to optimize production per unit area, usually by developing vertical structures for plant production (see Plates 2.1 and 2.2).

Urban farming takes place on most land types in cities. The following subsections distinguish between the different types of land used for farming, as this also has an impact upon the nature of the cropping system (Brock and Foeken, 2006).

Private garden farming

Private garden farming is found in all cities throughout the world. It is the most obvious and most widespread form of urban farming. Even when there is no access to a garden, farming may still be practised on rooftops, balconies or on walls. The private garden has a number of distinguishing features. First, there is no transport time to the site, making management convenient, and allowing for collection of minute quantities of harvest (Cockram and Feldman, 1996). Second, the crop is well protected and the risk of theft or vandalism is minimized. Land tenure is usually secure, allowing for longer-term investments of time and space in, for example, fruit trees. Water availability is usually better than for gardens away from the house, although irrigation with domestic water may not be feasible or desirable from a wider perspective. Finally, the farmer has a high level of control over inputs to the system. In most cases, production will also be quite limited due to the constraints in size. Although private gardening occurs in most communities, it is likely to be of lesser importance for the most destitute and resource-poor citizens as they have less access to own land than middle- and higher-income households. Similarly, larger households are often more active in urban farming than small due to better access to land and more available hands (Atukunda and Maxwell, 1996; Gumbo and Ndiripo, 1996; Mboganie-Mwangi and Foeken, 1996).

The lack of land surface has led to the development of various systems that seek to optimize production per unit area, which usually means creating some kind of vertical structure. Compost mounds in various forms are found in many gardens throughout Africa; a mixture of soil and compost are piled and stabilized in the simplest form by putting the mixture into large grain or fertilizer bags with holes down the sides, or, in more elaborate forms, in tall pyramids stabilized by metal wire nets or sticks. Various herbs and vegetables are then grown in the pile and protrude from the sides and top of the pile. Other systems consist of layered boxes, resembling a high-rise building with many balconies, or, in a simple form, used car tyres piled in a pyramid and filled with soil/compost. The tyres stabilize the soil, and the layered structure allows much denser planting with overlapping canopies than if cropped in one level. Vines and cucurbits crawl up walls or posts, and also fully utilize the third dimension.

Finally, farming on rooftops or balconies enables production even in the absence of land. Besides being a very secure place to grow crops, roofs increase the available area for production significantly. Rooftop gardening may provide other services than the actual production, particularly as insulation. Rooftop gardens may reduce energy consumption for cooling by 15 per cent (Wong et al, 2003). In other studies (reviewed by Wong et al, 2003), cooling energy consumption was reduced by 25 to 50 per cent (e.g. only 13 per cent of solar energy was absorbed in the roof soils, while 60 per cent went to the plant and 27 per cent was reflected).

Complete soil cover is not needed for rooftop gardening or gardening on paved or concrete surfaces. Besides farming in beds or pots, shallow soil farming can be used when good-quality soil is a limited resource or in order to reduce the weight load on the roof. Shallow soil farming can be practised in 5 to 15cm thick layers of soil and compost or organic residue on any surface. Provided that the nutrient supply is adequate, the main challenge may be to keep the shallow soil moist; hence, some sort of irrigation system will be required. The shallow soil systems may also be used where the natural soils are too poor or polluted to allow safe crop production.

Community gardening

Community gardening is found in most cities and may undertake a number of forms, ranging from close-knit communities to more loosely organized co-operatives that mainly share access to the same water resources, have common land tenure, or are supported by external agencies such as non-governmental organizations (NGOs) or government officials (e.g. Ward et al, 2004). Other characteristics of community gardens may include common fences, storage facilities, extension service support, inputs such as seeds or fertilizers and, importantly, security or recognition of the site as a farming area. Community gardens form a kind of institutional unit and are, in many cases, supported by public bodies, NGOs or private foundations based on the notion that supporting a community project may be more sustainable than supporting individuals, as well as being an efficient way of distributing support to communities.

Like other forms of urban farming, community gardens are constrained by their access to land. In many countries, community gardening takes place in schoolyards or similar semi-publicly owned grounds. Besides offering an area with secure (and, in many cases, fenced) land tenure, gardening in schoolyards offers a number of opportunities. In some instances, gardening activities are integrated with the school kitchen and teaching activities. The school may have its own vegetable plots in connection with the community garden, and together they produce food for school lunches. Experiences show that engaging the surrounding community in gardening is a way of securing production compared to instances where it is run by the teachers alone. In this way, other members can take care of the garden during school holidays, when the teachers are away. In addition, they may offer a valuable knowledge resource if none of the teachers have much experience in gardening.

Gardening in schools may not only improve the nutrition of schoolchildren, but also serves an important function in introducing urban farming and educating children in vegetable and fruit production. In more progressive examples, gardening activities are integrated within the curriculum and are used in science classes (e.g. seasonal change and biology), mathematics (calculating area, production, etc.) health, and so on (in less progressive learning environments, however, children are sent to weed the gardens as a punishment, creating animosity towards horticulture). Surplus production may also be sold, generating income for the school.

Access to water resources is very often shared in community gardens. The nature of sharing may vary from the individual wells on each plot in the community gardens, as in Bamako (Mali), to elaborate and highly regulated channel or pipe irrigation systems, which in some cases supply numerous extensive community gardens, as in Arusha (Tanzania).

Community gardening may fulfil a number of functions other than providing nutritious food or income for farmers. In Southern Africa, numerous support groups grow vegetables for the HIV sick or for HIV orphans. This form of charity gardening takes place either on existing community gardens or in projects specifically designated for this purpose. In some cases, community gardens are established on hospital grounds, and a group of users comes daily to take care of the garden, providing a nutritious diet supplement for the patients (Smith et al, 2005b).

Farming common land

Limited or no access to tenure-secure land is by no means a deterrence from urban agriculture, as most African cities clearly demonstrate. Farming takes place on a variety of public and privately owned sites, ranging in size from a few square metres to large plots of several hectares. Typical farming sites include vacant plots, and very often urban agriculture is found along the perimeter or fringes of infrastructure (e.g. below power lines, along roads and highways, between highway crossings, and surrounding airports and industrial sites). Although no formal land tenure exists, farming along these structures may provide as 'secure' a land right as any private land, and possibly more secure than community gardens on school property or elsewhere. It is unlikely that any other structures will be erected around airports or along highways; hence, there are few land uses that compete with urban farming on these grounds. Tenure may, however, be insecure in the sense that farmers cannot legally protect the land that they are using from others wanting to take it over or vandalizing crops. Therefore, the social structures surrounding these farming types are important.

Urban farmers from both formal community gardens as well as informal groupings are known to establish watch teams and to take turns in staying overnight in the fields, guarding crops against theft and vandalism. Compared to farmers with secure land tenure, farmers in Nigeria who did not own their land ran a higher risk of losing their crops to thieves, as well as their urban fields turning into a dumpsite (Ezedinma and Chukuezi, 1999). Another study in Nigeria revealed that crop theft was perceived by urban farmers as the most significant problem (Lynch et al, 2001).

In many instances, farming along roads and airports is also a benefit for the owner of the land (usually the state or municipality) as these activities effectively keep the vegetation low. Further benefits to the landowner are discussed below.

Informal tenure rights or land security may also develop through customary rules and tradition. When enough people start planting crops in an area, this becomes the norm and enforces land security, lowering the risk of the land being used for other purposes or the crops of being vandalized. In the community gardens at the outskirts of Durban (South Africa), urban farmers who cannot join the project due to lack of land will plant along the roads surrounding the school, outside of the school fences. Although this gives no formal or physical protection, the fact that the land is close to other recognized farming activities lends a form of security and justification for the land use (Oelofse et al, 2007). In many other cities, clustering of gardens on vacant land helps to build a recognized identity as farming land more effectively than if it were scattered plots between encroaching bushes and wasteland.

Although some urban farming types are temporary (e.g. squatting on vacant land), it is a misconception to categorize all urban farming activities as temporary by nature and to assume that they will inevitably be replaced by 'proper' city functions. Although the scale of agriculture in a peri-urban area diminishes as the city engulfs it, there are notable exceptions. Private gardens, either in connection to the house or on other locations, have secure land tenure and may continue to grow crops. Community or allotment gardens may have a long history and maintain their status even after being completely absorbed within the city, as numerous examples of allotment gardens in Europe and North America illustrate. Finally, even farming on public land may be of a more permanent nature, given that future land use can be predicted – for example, along the side of highways (as in Dar es Salaam), adjacent to airports (Cotonou and Durban), or on swampy or sloped land that is unsuitable for housing and infrastructure (Kampala). In the case of these 'unfavourable' locations, land quality may influence the choice of crops (e.g. the farming of periodically flooded lands restricts the cultivation to crops that tolerate high water levels, such as the cocoyam). In a study from West Africa, it was found that horticulturalists growing ornamentals preferred farming along the side of the roads as this gave the greatest visibility and provided a free exhibition window for their produce (Brock and Foeken, 2006).

The lack of secure land tenure may influence the farming system and choice of crops, as described in the following examples. Lack of tenure makes farmers more reluctant to invest in permanent structures, who therefore do not line their wells with concrete, as is typically the case in Nigeria (Ezedinma and Chukuezi, 1999). Consequently, the farmers must spend days of work maintaining the wells several times throughout the season, although this is still preferred over the investment in concrete wells. Furthermore, farmers with no ownership rights cannot use their land as collateral for credit and have faced problems when they need to raise money for investments in their farming.

In Cotonou (Benin), Brock and Foeken (2006) found that land security was closely connected to the crops on the land. Maize, being a less attractive and lower-value crop, was predominantly grown on tenure-insecure land. When farmers received tenure rights to their land, they shifted to the more input-demanding and higher-yielding vegetable production. A similar mechanism was reported by Cockram and Feldman (1996), who observed that less valuable crops such as maize, cassava and potatoes were grown on the more distant plots.

A slightly different mechanism was observed by Asomani-Boateng (2002) in Ghana, who found that short-cycle vegetables were mainly grown on vacant land not owned by the farmers; in some cases the farmers did not even know who was the owner of the land. Farmers with their own land tended to cultivate fruits, trees and longer-cycle crops such as plantain, cassava, maize and cocoyam on that land.

The high land prices and rents in urban areas will often limit the possibilities of longer-term squatter farming on private land (but not necessarily on public land, as mentioned above); as a result, it is easy to understand a preference for short rotation crops on this land. Planting perennials and trees may be viewed as a sign of occupation of the land, which may be critical if the landowner feels that his or her rights to the land are threatened by invading farmers. On the other hand, perennials may help farmers to win usufruct rights to open-access land by demonstrating their longer-term commitment to the area.

A variety of reasons exist for allowing or encouraging urban farming activities on vacant land – seen from the landowner or neighbouring community's perspective. First, having urban farmers occupy the land is a cheap and efficient way of keeping the area clear of bush encroachment or alien species. Besides aesthetic reasons, this may also be considered effective in preventing pests such as snakes or rodents from breeding on the land (Moskow, 1999; Brock and Foeken, 2006). Land occupation by farmers may also prevent the area from being transformed into a dumpsite, particular in cities with inadequate waste-handling systems. Vacant plots are often used as public toilets if there is no central sanitation system. This is also prevented by farming (Brock and Foeken, 2006). In addition, the community's sense of security may benefit from vacant plots being transformed into farms as they no longer serve as hideouts for thieves and criminals (Rogerson, 1993; Brock and Foeken, 2006). Farmers are there most of the day, which provides surveillance against criminal activities in the area. The fresh produce is also readily appreciated by neighbouring households. On a more abstract level, the sense of identity and coherence within a community may be strengthened by joint activities such as community gardening projects. Finally, farming can also prevent the land from being occupied by squatters and informal housing, a common issue in many growing African cities.

On the biophysical side, urban greening or urban farming has a number of benefits for the local microclimate (Bryld, 2003). Evaporation from the canopy lowers temperatures compared to temperatures associated with typical stone and concrete surfaces found in cities. The vegetation filters dust from the air, as is immediately visible from the thick red layer of dust on the bananas and maize in most African cities. The vegetation may also slow winds and heavy rains and indirectly benefits the urban environment by shortening supply lines, reducing the transportation of goods to the city (Brock and Foeken, 2006). Increasing atmospheric pollution afflicts more than 1.1 billion people, mostly in cities. Nearly 3 million people every year are killed by air pollution, about 6 per cent of all deaths annually, of which 90 per cent occur in developing countries (Iaquinta and Drescher, 2002).

While city planners in some areas have taken a very restrictive approach to itinerant farming on public land, others have a more *laissez faire* position. In some cities, such as eThekwini Municipality (Durban), farming activities are permitted alongside the airport and highways, as well as on other public lands, as long as the land is not needed for other purposes. In many cases, the city municipality tries to assist, together with the Department of Agriculture, in providing inputs, fences, etc. upon requests from community projects and under the condition that the urban gardens must be abandoned once the space is needed for other uses. As reviewed in Chapter 8, most cities (including Durban) initially took a much harsher view of urban agriculture that was considered unsuitable for developing cities, motivated by health and hygiene concerns, aesthetic considerations, and the belief that a modern city should exclude the rural and 'messy' appearance of crops (Drakakis-Smith et al, 1995; May and Rogerson, 1995). The diversity of perception can be illustrated by the two following quotes:

> *Authorities are hesitant to be more proactive on urban agriculture because it is largely seen as resulting from a failure to address rural development adequately. It is creating havoc in urban land-use planning and management. It is holding up city development and redevelopment.* (Mayor of Lusaka, Zambia, cited in FAO, 1999)

> *The city administration of Kampala sees urban agriculture as an efficient way to use urban land optimally ... Urban agriculture has several advantages in Kampala: it increases urban food security, creates sources of income [and] reduces open space maintenance costs to local government ... Good urban agriculture depends crucially on producers being granted proper agricultural extension and advice, including assistance for selecting crops.* (Mayor of Kampala, Uganda, cited in Cityfarmer, 1997)

Latin America exhibits a number of examples that should serve as an inspiration for African city developers. In Havana (Cuba), special economic conditions have fostered a very positive view on urban farming, which is well supported by government policies. Consequently, the main constraints mentioned in these systems were agronomic problems related to crop production, whereas issues such as lack of space, harassment, or lack of knowledge and support from extension services was not mentioned as problematic in the survey (Moskow, 1999).

Mexico City (Mexico) is an example of a city with a well-developed urban agricultural sector, also including extensive livestock rearing. Since the later is officially banned in the city, it may only be practised if concealed and by bribing officials. This example illustrates that strict legislation will not eradicate urban farming or even large livestock from the cities. Therefore, city authorities might as well attempt to get the best out of it and to ensure reasonable hygiene standards, etc. through cooperation, rather than losing contact with urban farmers through marginalization (Losada et al, 1998).

Food security is a problem among the poor families in Accra (Ghana), with many households having no place to cook and living entirely on street foods. City authorities tended to harass farmers who were using wastewater to produce crops. Research activities on Accra's urban agriculture has influenced the Accra Metropolitan Assembly and the Land Commission, leading to a reconsideration of urban agriculture by local government, as evidenced in a central government White Paper on land (Jacobi et al, 2000).

By taking into account cultural norms and traditions, climatic conditions and natural resource settings, the establishment of allotment gardens could help to reduce poverty and improve food security in cities. This is evidenced by the European experience in establishing such systems during times of crisis. Experience shows that garden schemes enable the establishment of participatory community development programmes and allow for the introduction of democratic rules related to the management of plots. Special participatory training programmes can further contribute to community-building. Nevertheless, this process requires monitoring and moderation. In Southern Africa, female-headed households as a group most affected by poverty might be a special target group for such allotment programmes (Drescher, 2001).

Production systems

Urban farming systems cover the range from haphazard, under-managed, poorly producing gardens to high-input systems with well-organized and reliable irrigation, and considerable fertilizer and pesticide use. Due to the abundance of organic wastes in cities, many urban production systems rely on recycled wastes from households or animal production, and also, in many cases, on wastewater – either due to the absence of a clean water source or because the wastewater is preferred due to its nutrient content.

Due to the often small and adaptive nature of urban farming systems, the variety of agronomic practices is enormous and few generalizations can be made. There is an urgent need for a robust and usable typology, and it makes more sense to explore the range of innovations and practices that have been developed under varying conditions. The comparably high value of the crop and labour availability that is often associated with urban farming has motivated the development of innovative and adaptive solutions. In the previous sections, a number of space-conserving measures, unique to urban agriculture, have been mentioned. As described, space is a major limitation, and variations on the theme of multi-storey planting in tyres, hammocks, gutters, etc. are widely employed.

Water is another major limitation for plant growth in many African cities. Most crops require 0.1 to 1 litres of water per square metre per day. In the growing cities of Africa, provision of clean and adequate water to inhabitants is already a challenge that is not likely to diminish. As a result, competition for this water by an agricultural sector is likely to be problematic (Ellis and Sumberg, 1998; Mireri et al, 2007). Drip irrigation systems typically save 20 per cent and up to 50 per cent of the water requirements, and may also reduce labour costs, depending upon the design. Drip irrigation requires relatively clean water so as not to block the nozzles; however, various low-tech solutions such as sand filters in large iron or plastic containers can filter water with suspended particles. Drip irrigation systems range from high-tech automatic fertigation systems that add nutrients to the solution, to low-tech systems consisting of a barrel and a few tubes for the distribution of water. Even the simplest system requires investments, and many local extension services from Senegal to South Africa have given economic support to farming groups using drip irrigation facilities. A major disadvantage is that drip irrigation systems are very prone to theft or vandalism and therefore are not well suited to unguarded sites with easy access – a characteristic of many urban agricultural sites.

A major benefit of drip irrigation systems is that the water can be provided to the soil and roots without coming in contact with the leaves or fruits. When utilizing a biologically contaminated water source, this may be an effective way of preventing transfer of the pathogen to the edible parts of the plant. Depending upon the clay content of the soil, this may also effectively adsorb and remove heavy metals from the irrigation water, making safer use of chemically contaminated sources. On the other hand, micronutrient availability is much lower if provided as a fertilizer to the soil compared to foliar application to the plant. Even organic liquid fertilizers can be supplied through drip irrigation systems. A study in Egypt showed that the nitrogen fertilizing value of plant extracts applied to lettuce and cucurbits was comparable to mineral nitrogen, presumably due to rapid turnover in the hot and humid environment (El-Naggar et al, 2008).

In West Africa, water hyacinth is used for wastewater treatment from latrines in marshy basins. The wastewater is led through a series of basins, where nutrients or sediment are taken up and a certain degree of sanitation takes place. The plant material (water hyacinth) is used for fodder, and the remaining water used for irrigation on the farms (Brock and Foeken, 2006) (see below for a further description of use of waste products in urban agriculture).

Few environments are better suited to rainwater harvesting than the built-up structure of a city with roofs and gutters collecting the water. Given adequate storage capacity, rainwater harvesting has the potential to at least partially overcome periodic water shortages in the (sub-)humid to semi-arid and arid zones of Africa without putting further strains on the already over-utilized municipal supply.

The risks of environmental pollution from urban farming systems are limited as long as such systems are restricted to small-scale subsistence farming. Although mishandling of agrochemicals may occur, as in any farming operation, urban or rural, the chance of farmers overusing chemicals is small as they regularly under-apply in order to save costs. But with the production of high-value crops comes the risk of overuse of water or chemical inputs. When transforming to commercialized horticultural enterprises, the use of inputs dramatically increases, as does the accompanying risk of environmental pollution. Well-educated male farmers with good linkages to extension services are the first to increase the use of agrochemicals (Nkamleu and Adesina, 2000). Widespread use of agrochemicals in urban agriculture is well documented (see Brock and Foeken, 2006), and there is a real risk that misuse occurs either due to lack of knowledge or in the attempt to optimize production for a lucrative market. Experiences from Asia clearly demonstrate considerable environmental risks associated with intensification of urban agriculture, mainly pollution of groundwater and surface water with nitrate due to excessive application of nitrogen in the form of mineral fertilizer or animal manures (Wolf et al, 2003; Huang et al, 2006; Khai et al, 2007). Although no comparable studies from Africa were found in this review, the intensified production systems within many African cities are bound to create similar problems, although costs of inputs are generally higher in Africa than in Asia.

Asia also experiences a strong trend towards intensified systems where animal production is separated from crop production (as is the case with the intensified livestock systems in the North), and livestock rearing is based on imported fodder. This leads to large nutrient surpluses in peri-urban areas where livestock are kept, with consequent over-fertilization on the little arable land in the area. This trend is particularly pronounced in urban and peri-urban areas where space is limited, and near the rapidly growing urban markets for meat (Hedlund et al, 2003; Gerber et al, 2005). Local governments and extension services in Africa have an important role to play in ensuring that Africa does not repeat the nutrient flooding phenomenon that many Asian urban and rural environments are experiencing these years, inflicting major damage on the environment and jeopardizing the provision of potable water in the region.

As discussed in Chapter 5, supply of seeds and germplasm, especially for indigenous species, is a challenge. Although indigenous vegetables are often easier to replant than exotics since their seeds are more viable in the local environment, the lack of commercialization of these seeds poses a problem. For the majority of urban farmers, small quantity seed packages are needed in order to provide affordable quantities of seed. And although their own seed production is significant, particularly for indigenous vegetables, commercial seeds play a vital role in urban agriculture (Prain, 2006).

Urban farmers may play a key role in reclaiming marginalized land and building soil fertility. In Cotonou (Benin), due to the low quality of soils (white sands), the establishment of new gardening activities occurs on waste dumpsites where some soil fertility building has occurred before the initiation of farming activities. Further building of soil fertility is ensured by continued applications of organic matter in the form of urban wastes and animal manures (Brock and Foeken, 2006). In Jos (Nigeria), urban farmers have for years been building soil fertility for intensified urban crop production through the conscious application of household ashes to the soil. The ashes have a high cation content and supply essential nutrients to the soil, as well as increasing the pH of soils to near neutral levels, thus making phosphorus more available for plant uptake (Pasquini and Harris, 2005). There are also experiences from outside the continent. In Mexico City, official attempts to restore degraded land on the surrounding slopes by planting exotic trees such as *Eucalyptus* compared unfavourably with the results from self-organized unsupported peri-urban farmers who reduced erosion and increased soil fertility through conscious farming practices and the application of animal manure and ashes (Losada et al, 1998).

However, when restricted by inputs in subsistence systems, urban farmers struggle to maintain soil fertility. A study of five community gardens in South Africa revealed that farmers' ability to maintain soil fertility was closely linked to the level of support from extension services, individual resourcefulness and the organizational level of the community group – the latter mainly because this had a significant impact upon the access to water and governmental support (Oelofse et al, 2007).

Who are the farmers and what is the importance of urban agriculture?

Just as it makes little meaning to define the typical urban agricultural system, defining the typical urban farmer is a daunting task. As discussed in Chapter 7, and illustrated below, there is an enormous variation across sub-Saharan Africa in terms of the social status, age groups, background and gender of those involved in urban farming. One should be vary careful when generalizing about farming systems, the motivations of farmers or the sustainability of systems; what looks the same may be different, and what looks different across the continent may, in fact, be quite similar (Lynch et al, 2001).

From all the variation in the literature comes one clear picture: the notion of urban farming as being reserved for the poorest of the poor and those with no livelihood alternatives is not true (Flynn, 2001). Therefore, saying that urban farming is solely motivated by the need for increased food security does not hold. On the contrary, most studies have shown that the typical urban farming household may have quite diverse livelihoods, combining various incomes and resources (Dennery, 1996; Mougeot, 1996). As is apparent from this chapter, urban farmers include people who are in it for the money, for recreation, for food and for idealism (e.g. Moskow, 1999; Smith et al, 2005b). Nevertheless, urban farming also plays an important role for local food security and nutrition, as is illustrated by a Kenyan study showing that 77 per cent of urban farmers were producing for their own consumption (Lee-Smith and Memon, 1994). A study from Kampala (Uganda) demonstrated that 11 per cent of the urban farming population was dependent upon this for survival (Maxwell, 1995); a similar study across six cities in Kenya has this figure at 25 per cent (Tinker, 1994).

Household expenditure for food comprises 20 to 50 per cent of income in South African families, demonstrating the large potential importance of own and local production. The expenditure for food is highest for low-income families (May and Rogerson, 1995), reaching 80 per cent of the income for the poorest urban households, compared to 50 per cent for rural households, on average (Argenti, 2000).

Poorer households will focus production on staples such as maize, sweet potato and cassava, whereas better-off households can supplement their diet or income with higher-value products and livestock (Cockram and Feldman, 1996). However, examples of poor households who were forced to sell some of their produce to generate money, even when they were not able to supply the household, have been reported (Maxwell, 1995; Bryld, 2003).

In a study conducted in South Africa, it was found that farming was most pronounced among middle- and higher-income groups, and less in the marginalized households (May and Rogerson, 1995). Thus, urban agriculture was not restricted to a survival strategy for the poorest with no alternatives. This is understandable on the basis of the following:

- The poorest citizens will have very limited access to space to farm, and farming on private land, which often is the most widespread form of urban cultivation, will mainly be within the larger households.
- The pervasive government social grant system in South Africa is regarded by some as a disincentive for poorer households to engage in urban agriculture.

A Brazilian study from Belém showed that the poorer the household, the less successful it was in the attempt to grow food on a regular basis (Madaleno, 2000). Poor households are forced to be opportunistic and may turn from farming if other options arise (Oelofse et al, 2007). Moreover, they are the least shock resilient and have access to the most marginalized land, while using the lowest inputs. As a result, the risk of crop failure due to untimely planting, biological or environmental hazards or theft is much greater for this group. In Dar es Salaam (Tanzania), less than 20 per cent of farmers can be characterized as jobless or even without a stable job (Sawio, 1994), clearly showing that urban agriculture is an activity that does not exclude the poorest, but also involves the more privileged.

The background of urban farmers varies greatly over the continent. Egziabher et al (1994) state that, contrary to some suggestions that urban farmers are a largely immigrating rural population, urban farmers are rarely new immigrants to the city as these individuals typically have poor access to land – both own and common (Sawio, 1994; Atukunda and Maxwell, 1996; Ellis and Sumberg, 1998). To the degree that farmers do immigrate to cities, their integration within farming communities is hindered by poor links to agricultural service providers, extension services and local governance structures, making knowledge transfer difficult (Huang et al, 2006).

Farmers' backgrounds may also differ between the different sectors of urban agriculture. In a Nigerian study, Ezedinma and Chukuezi (1999) found that the commercial vegetable production sector was dominated by poorly educated migrant farmers during the dry season, when they could not farm in the rural fields without irrigation. In contrast, the large horticultural sector of ornamental plants was dominated by well-educated urban residents. Presumably, at least one additional group of farmers was involved in subsistence- and household-based production; but their background was not explored in this study. Both the commercial vegetable production and the ornamentals were economically attractive enterprises, but faced quite different constraints. The vegetable farmers lacked land, credit and technology, whereas the floriculturists were constrained by lack of skilled labour and access to exotic plants that sold well on the urban markets.

As indicated in the Nigerian study, urban farmers vary greatly in educational background. Sawio (1994) reported that urban farmers in Dar es Salaam (Tanzania) are spread over all education levels. In Mekelle (Ethiopia), farmers generally have low educational status (Ashebir et al, 2007). A more detailed study from Brazil found that 10 per cent of urban farmers have a university degree, 6 per cent are illiterate, with the remaining having an intermediate education. Just under half (44 per cent) of the farmers are rural migrants with farming experience; the rest are urban residents (Madaleno, 2000).

The gender distribution of urban farmers is also highly variable from city to city. However, there appears to be a dominance of studies that find more women involved in urban farming. Chapter 7 in this volume includes gender data on traders and producers from 14 African cities. A literature review by Hovorka and Lee-Smith (2006) found that women dominated all parts of the urban production cycle, including farmers, middlemen and traders at markets. Sawio (1994) came to the same conclusion and stated that urban agriculture in Dar es Salaam was generally dominated by women. Maxwell et al (1998) found that 80 per cent of those involved in urban agriculture in Kampala (Uganda) were women.

In Lusaka (Zambia) in all examined compounds, women were more involved in cropping and gardening than men (Drescher, 1996, 1999). Microfarming contributed to household food security in town directly by providing food, and indirectly by generating income. Gathering was still widely practised and contributed significantly to food security and diversity. Women tended to dominate urban cultivation because they were marginalized in other forms of employment in the formal sector of the urban economy (Drescher, 1999).

A more diverse picture was presented by Asomani-Boateng (2002) in Accra (Ghana), who found that the majority of the farmers were men while the majority of traders were women. The majority of the farmers were migrants from neighbouring countries, and this may partially explain this observation as men tend to be more mobile than women in many African cultures.

Contrary to the above situations, a number of studies have demonstrated that urban farmers are predominantly men. Ashebir et al (2007) found a strong dominance of male farmers (94 per cent) in a study from Mekelle (Ethiopia). Similarly, Ezedinma and Chukuezi (1999) found that urban vegetable and ornamental production in Nigeria was mainly a male activity, with 98 and 73 per cent male farmers, respectively. In Kumasi (Ghana), the gender and age distribution of peri-urban farmers in six surrounding villages showed a varied picture, but with a dominance of elderly (women) farmers and few young (women) farmers, with the elder and young males intermediate (Simon et al, 2004). The IndigenoVeg survey (see Chapter 7) also revealed a diverse picture. With South Africa and Senegal as notable exceptions, the farmers were mainly men, whereas women were responsible for the trading. It is important to bear in mind that the IndigenoVeg study focused on the marketed African indigenous vegetables (AIVs) – hence, the commercialized part of the production. In most African cultures, women are responsible for household chores such as shopping, cooking and subsistence AIV production. In rural areas it is often found that women are responsible for the home gardens, whereas the outfields are the men's domain, although this picture is diversifying. Women typically have poorer access to land than men, and in some countries, such as Botswana and Malawi, they are not even allowed to possess land according to the traditional tenure rules. Women may also not have the same access to inputs and household funds, although this depends upon who is the breadwinner of the family, and this picture is also rapidly diversifying. Depending upon the limitation of the sample, studies may therefore find a dominance of women if subsistence activities are included, or a larger dominance of men if focused on commercialized production. However, it is quite clear that women are more involved in urban production and marketing of vegetables compared to the traditional rural farming systems.

Income generation from selling produce is difficult to quantify due to its *ad hoc* nature, payment in kind (Bryld, 2003) and the reluctance of farmers to reveal their income (Drescher, 1996). In Kampala (Uganda), commercialized farmers only constitute 2.5 per cent of all growers, although they are economically important (Maxwell, 1995). In Cameroon, it was estimated that urban farmers could potentially earn the same as the local minimum wage (Gockowski et al, 2003). In Lusaka (Zambia), urban traders doubled their income compared to farmers who did not sell their own produce (Drescher, 1996).

Participating in urban agriculture has been shown to improve child nutrition, indicated by a higher height for age factor (Maxwell et al, 1998). Realizing that human health and performance, particularly for children, is closely linked to proper diet and adequate nutrition, a South African NGO decided to combine healthcare with the promotion of home-grown vegetables. The NGO, operating a number of clinics in and around Durban, hired a number of agricultural extension workers to be posted in the clinics' waiting rooms. Here, they occupied the waiting patients by teaching them about urban vegetable production and the importance of healthy diets. Concrete examples and recommendations were provided in order to stimulate urban farming among the clinics' visitors.

Urban food production supplies products that rural agriculture cannot provide. It can release good rural agricultural land for other uses, reduce pressure to cultivate new rural land, and contribute to the generation of income in all sectors – rural, urban and peri-urban. In addition, it contributes to social welfare and the sustainable development of cities. Urban food production contributes to waste avoidance and recycling of organic waste in cities. There is a need to ensure food security and quality in cities in order to prepare for crisis prevention and to enhance sustainable development (Iaquinta and Drescher, 2002).

Waste handling and urban agriculture

Cities have large nutrient surpluses due to the import of food (Khai et al, 2007), with the main outputs being via waste disposal (in the worst cases, via natural water bodies) or accumulation within the cities themselves. The recovery rates of nutrients such as nitrogen and phosphorus in urban production systems are low and are usually below 10 per cent (Færge et al, 2001). The huge nutrient surplus in cities is not only a challenge for their waste handling facilities and the surrounding environment, but also an opportunity for urban farmers. The cost of urban wastes is often low, or negligible, due to their abundance. In many cases, the use of organic wastes in urban farming can be viewed as an (essential) service offered to the city in providing an environmentally safe, cheap (income-generating) disposal opportunity. Recycling of wastes also comes with a cost, and caution is warranted in order to prevent the spread of diseases or toxic chemicals – to the farmers and to the consumers of their goods. In many cases, although the waste recycling may not offer an optimal solution for safe disposal of the material, it still offers far more benefits than its alternatives – hence, a pragmatic approach will be to alleviate the problems associated with waste recycling in farming, rather than to ban it altogether, if there are no ready alternatives. The subject is extensively reviewed elsewhere (Drechsel and Kunze, 2001); this section seeks to give an overview of the subject.

Recycling of organic solid waste can be an effective and sustainable way of improving soil fertility and preventing high costs and space for disposal. Decentralized composting facilities in public–private partnerships might be another way of handling and sanitizing the wastes (Drescher, 2001). Some cities have developed formalized waste recycling systems with collection, separation and composting. In some cases, these have been promoted by the local authorities; in others, they have developed without external support (Brock and Foeken, 2006). In all cases, even the most effective systems do not have the capacity to handle more than a fraction of the total solid and liquid waste produced in cities. The most convincing example is from outside Africa. In Mexico City, 50 per cent of its sewage is pumped 75km to the north to irrigate 100,000ha of fields for livestock feed (Smit and Nasr, 1992).

Recycling of urban waste products to agriculture includes organic household waste, human urine and excreta, animal manure, wastewater, ashes and industrial organic wastes. The materials may be applied directly or pre-treated in some way (e.g. by composting). Distribution and application may take place along formalized distribution channels (e.g. the use of wastewater for hydroponics in many Asian and, to lesser degree, African cities) or informally by farmers collecting wastes and applying them to their fields. The health problems related to recycling can mainly be divided into chemical risks (heavy metals or other toxic substances in the waste) and biological risks (various pathogens in the waste). The route of transfer and possibilities of mitigating health impacts are quite different for these two, which calls for an adapted strategy.

In most cases, waste recycling will need support in order to fully develop into an efficient and safe mode. Although it may be desirable for farmers to apply municipal wastes to their land in order to build soil fertility, they will most likely not be able to cover costs in connection with collection and sanitation of the wastes. A large proportion of the solid waste consists of biodegradable organic material. Kaseva and Gupta (1996) found that in Dar es Salaam, 60 per cent of the collected waste was organic, mainly vegetable residues. This is very close to a figure from Cuba which found that 60 to 70 per cent of municipal solid waste consisted of organic matter (Körner et al, 2007). From some collection points (e.g. the vegetable markets in town) and from restaurants, hotels and food-processing industries, the waste is of relatively high purity and can easily be used for high-quality compost (Minja et al, 1998). But the main picture is that there are large differences in the quality of the waste with mixtures of inorganic and organic material making handling difficult or even dangerous for the farmers. To the degree that waste is collected, sorted and distributed on the land unofficially by farmers or waste traders, the incentive to get rid of dangerous or non-degradable waste is low, or just unfeasible, particularly if land tenure is insecure (Eaton and Hilhorst, 2003). But local farmers are willing to pay for the waste, as many examples show (Eaton and Hilhorst, 2003). Currently, waste lorries in Bamako (Mali) are paid to illegally dump their waste near farmers' fields, where a certain degree of sorting takes place, before the waste is distributed on the fields. However, in the long term, non-degradable, worthless waste is accumulating at these sites, creating a new waste problem that cannot be solved without intervention from the municipality.

Rather than the municipality completely taking over informal waste management and sanitation, the development of systems (together with local enterprises and farmers under some degree of quality control) may offer a viable model for developing safe and sustainable waste management practices. This will probably be a much cheaper solution for municipalities and minimize demand for dumpsites (Asomani-Boateng et al, 1996; Asomani-Boateng, 2007). Rufisque (Senegal) has implemented a programme built on community and municipality cooperation where solid and liquid waste is handled and treated safely, using composting, microbial decomposition and the use of water ferns, which are later applied to farmers fields (Gaye and Diallo, 1997). This initiative has dramatically reduced refuse dumping on streets at a fraction of the cost of a conventional system.

However, one cannot ignore the fear of local municipalities that loosening the grip on waste recycling and urban agriculture includes the risk of transforming peri-urban fields into dumpsites. By banning urban agriculture or at least waste recycling, the logic seems to be that this will not happen (Eaton and Hilhorst, 2003). However, if the legislation is too far from the current reality, it is usually impossible to implement and enforce it (Asomani-Boateng and Haight, 1999)

Organic municipal waste may only be a second choice for soil improver/fertilizer compared to composted manures, liquid manures or mineral fertilizers, particularly in intensive horticultural systems. However, the low cost and availability may still make it a valuable input. When soils are severely degraded or low in organic matter (e.g. during land reclamation), organic inputs may be necessary, along with more high-grade mineral fertilizers. Farmers may also prefer municipal waste on staple crops such as cereals, rather than on intensive vegetables (Eaton and Hilhorst, 2003) in order to reduce the risk of contamination.

Pre-treatment of wastes can, in many cases, efficiently reduce the transfer of pathogens. Composting is the most widely used method, which serves a number of purposes. First, the volume of the waste is reduced, facilitating transport. The main losses are carbon dioxide (CO_2), although nitrogen and other nutrient losses can be significant if composting is done incorrectly. Composting also leads to higher concentrations of nutrients in the product, in this way producing a higher-quality organic fertilizer. Besides providing nutrients to soil, compost builds soil structural quality, increasing water storage capacity and reducing the risk of erosion. When applied as soil cover, compost may effectively protect emerging crops and soil from erosion under heavy rainfall events.

If composting is carried out as a controlled process, it may also have a significant sanitary effect on the waste. During composting, the temperature in the heap may rise to over 70°C, efficiently killing most pathogens and weed seeds in the compost. Storage alone may also reduce pathogen levels in waste products (e.g. the level of *Ascaris* and *Trichuris* eggs in urine has been shown to decline markedly after a few weeks of storage), especially when increasing pH with urea or ammonium (Ghiglietti et al, 1995; Vinnerås, 2002).

However, composting also comes with problems and disadvantages. Nutrient losses can be significant during the composting process. Volatile nutrients such as nitrogen and sulphur are lost in large amounts during composting, often up to 50 per cent. Losses can be reduced by covering the compost, reducing aeration and reducing pH. If exposed to rainwater, nutrients can also leach from the compost. Under poor management, loss of soluble nutrients may also be well over 50 per cent during the composting process. When exposed to rainwater and leaching, the temperature in the compost heap is also lowered resulting in a reduced sanitary effect.

While the composting process may reduce the biological hazards present in the material, chemical pollutants are more difficult to remove. Toxic chemical compounds, medicines and other pollutants may be partially degraded during composting, but heavy metals are not; rather, they accumulate in the product as CO_2 is lost during composting.

Heavy metal accumulation in urban farming systems is variable, as is to be expected from the diversity of wastes and farming systems. In Dar es Salaam (Tanzania), critically high heavy metal concentrations were found in crops and soils along roadsides, posing a limited but potential risk for consumers (Brock and Foeken, 2006). Similar results were obtained in Kampala (Uganda), resulting in recommendations that plants be grown more than 30m from roadsides in order to minimize airborne pollution by heavy metals (Nabulo et al, 2006). Accumulation of heavy metals has been demonstrated in urban agricultural soils in China, as well as over-fertilization and associated environmental threats. Development of a well-functioning extension service and communication lines to farmers is a prerequisite for ensuring safe and appropriate use and handling of urban wastes and fertilizers with high nutrient and heavy metal concentration (Huang et al, 2006). In parts of Nigeria, the practice of using ashes from burned urban wastes has been widely used to increase soil fertility. Accumulation of heavy metals above the recommended levels was observed in crops grown on these ashes, although the study did not raise a general concern regarding the contamination level (Pasquini, 2006). In Harare (Zimbabwe), wastewater has been used for irrigating crops for more than ten years. Given the current rate of increase, the levels of heavy metal in soil will exceed permitted levels within 5 to 50 years (depending upon the type of metal), raising concern but no alarm over contamination levels (Mapanda et al, 2005). However, heavy metals adhere strongly to clay particles in soil, and once critical levels have been reached, it will take hundreds, if not thousands, of years for the contamination level to drop if untreated. The use of poorly classified solid municipal waste, wastewater, etc. inevitably leads to problems such as suboptimal production due to low fertilizer supply, high nutrient losses to the environment due to excessive application of nutrients (Huang et al, 2006; Khai et al, 2007) or accumulation of toxic elements, such as heavy metals in soil (Mapanda et al, 2005; Khai et al, 2007) and plants (Huang et al, 2006; Mireri et al, 2007). Through training and extension, and by some regulation of the waste-handling process, municipalities can minimize these risks without having to completely replace the existing, partially informal, waste-handling systems.

Safe plant production with polluted soils and water is a central challenge in the development of waste recycling systems. Heavy metals, pathogens and toxic chemicals pose the main problems. The greatest risk is for farmers who handle the waste when it is fresh and who are also exposed to sharp objects in the waste (Harris et al, 2001); but consumers are also affected. From a food safety point of view, there are three possibilities: to produce clean crops; to produce partially clean crops but to use only clean parts; or to alter consumer behaviour (i.e. to encourage consumers to clean crops by washing or boiling before eating).

Heavy metals pose a special problem as they are not removed by cooking. Generative plant organs and storage organs tend to have much lower concentrations of heavy metals than vegetative parts and are better suited for consumption. Contaminated surfaces can be washed; but if the metals are taken up over the leaf or root surface, detailed information on the mobility and pathway of metals in the plant is needed in order to minimize human exposure. Pathogens are usually not transported within the plant; as a result, physical separation of contaminated and uncontaminated parts, and washing or boiling can clean the produce of human pathogens.

Air pollution affects plant growth and quality through a number of mechanisms. Nitrogen oxide (NO_x), sulphur dioxide (SO_2) and ozone (O_3) all affect plant growth negatively and have been shown to decrease yield significantly at concentrations found in urban environments (Heck et al, 1988; Wahid et al, 1995; Agrawal et al, 2003). Aerosols in smog will also lower plant production. SO_2 emissions cause acid rain and the acidification of soils. This leads to increased mobility of most heavy metals and, in combination with high heavy metal concentrations in urban soils, may cause toxicity within plants as well as increased uptake by plants and higher exposure to humans consuming the produce (Jäger et al, 1993; Ashmore and Marshall, 1999). Direct deposition of heavy metals, such as lead (Pb), cadmium (Cd), manganese (Mn), zinc (Zn), copper (Cu) and chromium (Cr), on soils and plant tissue will result in increased heavy metal uptake in plants, as well as higher concentrations on the tissue surface. Heavy metal concentration in plants is strongly influenced by the source (soil or air), plant species and organ (roots, leaves, seeds, etc.) and the type of heavy metal. Therefore, people's exposure to heavy metals will depend upon the choice of crop, the part of the crop that is consumed, the nature of the pollution and the preparation method (Nabulo et al, 2006). As a result, while NO_x, SO_2 and aerosols will reduce productivity in an already constrained production system, heavy metal deposition will influence the quality of the products negatively, potentially leading to human hazard.

The choice of AIVs, which part of them is consumed (leaves versus tubers or grain) and the mode of preparation differs widely between cultures in African cities. Overlaying this spatial diversity in production and consumption patterns is the spatial diversity in pollution exposure. Consequently, a crop that is safe to eat in one area may be hazardous in another.

This calls for a mechanistic and spatially diversified approach to the problem. The combination of detailed knowledge on pollution impacts upon plants, local consumption patterns and pollution exposure will enable accurate predictions of human health hazards in various communities, the possibility of extrapolating results to comparable areas, and recommendations on safe types of crops and preparation methods.

It is unlikely that urban vegetable production will cease, even if banned by local communities, and this would undoubtedly have significant negative effects upon human health and food security (Asomani-Boateng, 2002). Hence, a constructive approach to minimizing the accompanying negative impacts upon food production in urban communities is the way forward.

Priority areas for research

Many of the described challenges in this chapter are of a practical nature, rather than due to a lack of knowledge, and relate to implementing technologies at reasonable costs, as well as the general problems of infrastructure and the organization of cities in developing countries, including tenure. Unfortunately, there is a dearth of agronomic scientific work directly related to urban agriculture; however, the majority of agronomic problems, particularly those related to pest resistance and the development of varieties of crops, are not specific to urban farmers and fall under general breeding programmes. These issues are discussed in Chapter 5. There may be some scope for applied work on developing urban farming systems, such as various forms of intercropping layered systems; but this is probably mainly driven by innovative practitioners, rather than the scientific community.

We have identified three main topics that require further attention in order to facilitate the development of safe urban agricultural systems. First, there is a need for case-specific studies on the accumulation of heavy metals and toxic compounds in urban crops, depending upon source of pollution (soil, water, air or manures), soil chemical properties (which greatly influence the availability of heavy metals), the crop (including distribution to various plant parts) and consumer behaviour (i.e. which part of the plant is consumed and how it is pretreated – e.g. washed, cooked, etc.). Although the general dynamics are well understood, case-specific recommendations on safe crops and handling practices should be developed.

Second, there is a great need for continued development of safe and efficient waste handling and recycling systems, an area that spans the technical development of various sanitizing processes to the practical implementation of these systems in cities. The field is wide, and considerable effort is being put into this area already. A major challenge is the development of decentralized units that can handle solid and liquid waste in cities with no or insufficient waste-handling infrastructure. The handling must produce a sanitized product, free of pathogens, that can be applied to urban fields. Where centralized solid waste and sewage collection systems exist, there is a need to develop separation procedures, which enable recycling of nutrients from the organic and nutrient-rich fractions of the waste to agricultural production. Centralized systems offer some advantages in that the processes are better controlled and can be applied at a greater scale; however, it does increase transportation costs. Even when complete sanitation of wastewater or solid waste cannot be ensured, there is potential for utilization. Currently, numerous safe farming systems utilizing polluted water are under development, mainly based on a physical separation of the edible parts from the polluted substrates.

At a slightly higher level, comprehensive risk analysis must be performed in those cities that greatly depend upon urban production. In order to ensure adequate and efficient intervention, and in order to control risk of pathogen or heavy metal contamination, there must be quantitative mechanistic understanding of the system. If not, we risk losing the benefits of urban farming due to worries over health hazards that might be exaggerated or easily controlled. Clean and healthy food starts at the plate, not at the production end of the cycle. Before we learn to wash our hands in the kitchen, all efforts to produce safe foods on the farm may be wasted.

References

Agrawal, M., Singh, B., Rajput, M., Marshall, F. and Bell, J. N. B. (2003) 'Effect of air pollution on peri-urban agriculture: A case study', *Environmental Pollution*, vol 126, pp323–329

Argenti, O. (2000) *Food for the Cities: Food Supply and Distribution Policies to Reduce Urban Food Insecurity*, FAO, Rome

Ashebir, D., Pasquini, M. and Bihon, W. (2007) 'Urban agriculture in Mekelle, Tigray state, Ethiopia: Principal characteristics, opportunities and constraints for further research and development', *Cities*, vol 24, pp218–228

Ashmore, M. R. and Marshall, F. M. (1999) 'Ozone impacts on agriculture: An issue of global concern', *Advances in Botanical Research*, vol 29, pp32–49

Asomani-Boateng, R. (2002) 'Urban cultivation in Accra: An examination of the nature, practices, problems, potentials and urban planning implications', *Habitat International*, vol 26, pp591–607

Asomani-Boateng, R. (2007) 'Closing the loop – community-based organic solid waste recycling, urban gardening, and land use planning in Ghana, West Africa', *Journal of Planning Education and Research*, vol 27, pp132–145

Asomani-Boateng, R. and Haight, M. (1999) 'Reusing organic solid waste in urban farming in African cities: A challenge for urban planners', *Third World Planning Review*, vol 21, pp411–428

Asomani-Boateng, R., Haight, M. and Furedy, C. (1996) 'From dump to heap: Community composting in West Africa', *Biocycle*, vol 37, pp70–71

Atukunda, G. and Maxwell, D. (1996) 'Farming in the City of Kampala: Issues for urban management', *African Urban Quarterly*, vol 11, pp264–275

Brock, B. and Foeken, D. (2006) 'Urban horticulture for a better environment: A case study of Cotonou, Benin', *Habitat International*, vol 30, pp558–578

Bryld, E. (2003) 'Potentials, problems, and policy implications for urban agriculture in developing countries', *Agriculture and Human Values*, vol 20, pp79–86

Cityfarmer (1997) *Urban Agriculture for Food Security, Jobs and Waste Recovery: Roundtable of Top Local Government Officials*, Second International Colloquium of Mayors on Governance for Sustainable Growth and Equity (UNDP), 28–30 July 1997, United Nations, New York, NY

Cockram, M. and Feldman, S. (1996) 'The beautiful city: Gardens in third world cities', *African Urban Quarterly*, vol 11, pp202–208

Cofie, O. O., van Veenhuizen, R. and Drechsel, P. (2003) 'Contribution of urban and periurban agriculture to food security in sub-Saharan Africa', Paper presented at the Africa session of the 3rd WWF, Kyoto, 17 March 2003

Danso, G. and Moustier, P. (2006) *Local Economic Development and Marketing of Urban Produced Food: Cities Farming for the Future*, RUAF Foundation, IIRR and ETC Urban Agriculture, Philippines, pp137–207

Dennery, P. R. (1996) 'Urban food producers' decision making: A case study of Kibera, city of Nairobi, Kenya', *African Urban Quarterly*, vol 11, pp181–187

Drakakis-Smith, D., Bowyer-Bower, T. and Tevera, D. (1995) 'Urban poverty and urban agriculture: An overview of the linkages in Harare', *Habitat International*, vol 19, pp183–193

Drechsel, P. and Kunze, D. (2001) *Waste Composting for Urban and Peri-Urban Agriculture: Closing the Rural–Urban Nutrient Cycle in Sub-Saharan Africa*, CABI, Wallingford, UK

Drescher, A. W. (1996) 'Urban microfarming in central Southern Africa: A case study of Lusaka, Zambia', *African Urban Quarterly*, vol 11, pp229–248

Drescher, A. W. (1999) 'Urban agriculture in the seasonal tropics of central and Southern Africa: A case study of Lusaka, Zambia', in Koc, M., MacRae, R., Mougeot, L. J. A. and Welsh, J. (eds) For *Hunger-Proof Cities: Sustainable Urban Food Systems*, IDRC, Ottawa, Canada

Drescher, A. W. (2001) 'The German allotment gardens: A model for poverty alleviation and food security in Southern African cities?', *Proceedings of the Sub-Regional Expert Meeting on Urban Horticulture*, Stellenbosch, South Africa, 15–19 January 2001, FAO/University of Stellenbosch, pp159–167, www.cityfarmer.org/germanAllot.html

Eaton, D. and Hilhorst, T. (2003) 'Opportunities for managing solid waste flows in the periurban interface of Bamako and Ouagadougou', *Environment and Urbanization*, vol 15, pp53–64

Egziabher, A. G., Lee-Smith, D., Maxwell, D. G., Memon, P. A., Mougeot, L. J. A. and Sawio, C. J. (1994) *Cities Feeding People: An Examination of Urban Agriculture in East Africa*, IDRC, Ottawa, Canada

El-Naggar, A., El-Araby, A., de Neergaard, A. and Hogh-Jensen, H. (2008) 'Crop responses to 15 N–labelled organic and inorganic nitrogen sources', *Nutrient Cycling in Agroecosystems*, vol 80, pp49–60

Ellis, F. and Sumberg, J. (1998) 'Food production, urban areas and policy responses', *World Development*, vol 26, pp213–225

Ezedinma, C. and Chukuezi, C. (1999) 'A comparative analysis of urban agricultural enterprises in Lagos and Port Harcourt, Nigeria', *Environment and Urbanization*, vol 11, pp135–144

Færge, J., Magid, J. and Penning de Vries, F. W. T. (2001) 'Urban nutrient balance for Bangkok', *Ecological Modelling*, vol 139, pp63–74

FAO (United Nations Food and Agriculture Organization) (1999) *Urban and Peri-Urban Agriculture*, a report of the Fifteenth session of the FAO Committee on Agriculture, FAO, Rome, 25–26 January, www.fao.org/unfao/bodies/COAG/COAG15/default.htm

Flynn, K. C. (2001) 'Urban agriculture in Mwanza, Tanzania', *Africa: Rivista Trimestrale di Studi e Documentazione dell'Istituto Italo-Africano*, vol 71, pp666–691

Gaye, M. and Diallo, F. (1997) 'Community participation in the management of the urban environment in Rufisque (Senegal)', *Environment and Urbanization*, vol 9, pp9–29

Gerber, P., Chilonda, P., Franceschini, G. and Menzi, H. (2005) 'Geographical determinants and environmental implications of livestock production intensification in Asia', *Bioresource Technology*, vol 96, pp263–276

Gerstl, S., Cissé, G. and Tanner, M. (2002) 'The economic impact of urban agriculture on home gardeners in Ouagadougou', *Urban Agriculture*, vol 7, pp12–15

Ghiglietti, R., Rossi, P., Ramsan, M. and Colombi, A. (1995) 'Viability of *Ascaris suum*, *Ascaris lumbricoides* and *Trichuris muris* eggs to alkaline pH and different temperatures', *Parassitologia*, vol 37, pp229–232

Gockowski, J., Mbazo'o, J., Mbah, G. and Fouda Moulende, T. (2003) 'African traditional leafy vegetables and the urban and peri-urban poor', *Food Policy*, vol 28, pp221–235

Gumbo, J. G. and Ndiripo, T. W. (1996) 'Open space cultivation in Zimbabwe: A case study of Greater Harare, Zimbabwe', *African Urban Quarterly*, vol 11, pp210–216

Harris, P. J. C., Allison, M., Smith, G., Kindness, H. M. and Kelley, J. (2001) 'The potential use of waste-stream products for soil amelioration in peri-urban interface agriculture production systems', in Drechsel, P. and Kunze, D. (eds) *Waste Composting for Urban and Peri-Urban Agriculture: Closing the Rural–Urban Nutrient Cycle in Sub-Saharan Africa*, CABI, Wallingford, UK, pp1–28

Heck, W. W., Taylor, O. C. and Tingey, D. T. (1988) *Assessment of Crop Loss from Air Pollutants*, Elsevier Applied Science, London, UK

Hedlund, A., Witter, E. and Xuan An, B. (2003) 'Assessment of N, P and K management by nutrient balances and flows on peri-urban smallholder farms in southern Vietnam', *European Journal of Agronomy*, vol 20, pp71–87

Hovorka, A. and Lee-Smith, D. (2006) 'Gendering the urban agenda', in van Veenhuizen, R. (ed) *Cities Farming for the Future: Urban Agriculture for Green and Productive Cities*, RUAF Foundation, IIRR and ETC Urban Agriculture, Philippines, pp125–144

Huang, B., Shi, X., Yu, D., Oborn, I., Blomback, K., Pagella, T. F., Wang, H., Sun, W. and Sinclair, F. L. (2006) 'Environmental assessment of small-scale vegetable farming systems in peri-urban areas of the Yangtze River Delta Region, China', *Agriculture, Ecosystems and Environment*, vol 112, pp391–402

Iaquinta, D. and Drescher, A. W. (2002) 'Food security in cities: A new challenge to development', in Brebbia, C. A., Martin-Duque, J. F. and Wadhwa, L. C. (eds) *The Sustainable City II: Urban Regeneration and Sustainability*, WIT Press, Wessex Institute of Technology, Southampton, pp983–994, http://library.witpress.com/pages/contact.asp

Jacobi, P., Drescher, A. W. and Amend, J. (2000) *Urban Agriculture: Justification and Planning Guidelines, Urban Vegetable Promotion Project*, MAC, Tanzania, and GTZ, Germany

Jäger, H. J., Unsworth, H. M., De Temmerman, L. and Mathy, P. (1993) *Effects of Air Pollution on Agricultural Crops in Europe*, Commission of the European Communities, Brussels, Belgium

Kaseva, M. E. and Gupta, S. K. (1996) 'Recycling – an environmentally friendly and income generating activity towards sustainable solid waste management. Case study – Dar es Salaam City, Tanzania', Resources, *Conservation and Recycling*, vol 17, pp299–309

Khai, N. M., Ha, P. Q. and Oborn, I. (2007) 'Nutrient flows in small-scale peri-urban vegetable farming systems in Southeast Asia: A case study in Hanoi', *Agriculture,*

Ecosystems and Environment, vol 122, pp192–202
Körner, I., Saborit-Sánchez, I. and Aguilera-Corrales, Y. (2007) 'Proposal for the integration of decentralised composting of the organic fraction of municipal solid waste into the waste management system of Cuba', *Waste Management*, vol 28, no 1, pp64–72
Lee-Smith, D. and Memon, P. A. (1994) 'Urban agriculture in Kenya', in Egziabher, A. G., Lee-Smith, D., Maxwell, D. G., Memon, P. A., Mougeot, L. J. A. and Sawio, C. J. (eds) *Cities Feeding People: An Examination of Urban Agriculture in East Africa*, IDRC, Ottawa, Canada, pp63–78
Losada, H., Martinez, H., Vieyra, J., Pealing, R., Zavala, R. and Cortes, J. (1998) 'Urban agriculture in the metropolitan zone of Mexico City: Changes over time in urban, suburban and peri-urban areas', *Environment and Urbanization*, vol 10, pp37–54
Lynch, K., Binns, T. and Olofin, E. (2001) 'Urban agriculture under threat: The land security question in Kano, Nigeria', *Cities*, vol 18, pp159–171
Madaleno, I. (2000) 'Urban agriculture in Belem, Brazil', *Cities*, vol 17, pp73–77
Mapanda, F., Mangwayana, E. N., Nyamangara, J. and Giller, K. E. (2005) 'The effect of long-term irrigation using wastewater on heavy metal contents of soils under vegetables in Harare, Zimbabwe', *Agriculture, Ecosystems and Environment*, vol 107, pp151–165
Maxwell, D. G. (1995) 'Alternative food security strategy: A household analysis of urban agriculture in Kampala', *World Development*, vol 23, pp1669–1681
Maxwell, D., Levin, C. and Csete, J. (1998) 'Does urban agriculture help prevent malnutrition? Evidence from Kampala', *Food Policy*, vol 23, pp411–424
May, J. and Rogerson, C. M. (1995) 'Poverty and sustainable cities in South Africa: The role of urban cultivation', *Habitat International*, vol 19, pp165–181
Mboganie-Mwangi, A. and Foeken, D. (1996) 'Urban agriculture, food security and nutrition in low income areas of the city of Nairobi, Kenya', *African Urban Quarterly*, vol 11, pp170–179
Minja, A., Kimbi, G., Turuka, F. and Hatibu, N. (1998) 'Potential for nutrient harvesting from agro-waste in urban areas as urban/peri-urban agriculture', *Journal of Management Development*, vol 10, pp83–98
Mireri, C., Atekyereza, P., Kyessi, A. and Mushi, N. (2007) 'Environmental risks of urban agriculture in the Lake Victoria drainage basin: A case of Kisumu municipality, Kenya', *Habitat International*, vol 31, pp375–386
Moskow, A. (1999) 'Havana's self-provision gardens', *Environment and Urbanization*, vol 11, pp127–134
Mougeot, L. J. A. (1996) 'Introduction: An improving domestic and international environment for African urban agriculture', *African Urban Quarterly*, vol 11, pp137–152
Moustier, P. (1999) *Fillières Maraîchères à Brazzaville: Quantification et Observatoire pour l'Action*, CIRAD and AGRISUD–AGRICONGO, Montepellier, France
Nabulo, G., Oryem-Origa, H. and Diamond, M. (2006) 'Assessment of lead, cadmium, and zinc contamination of roadside soils, surface films, and vegetables in Kampala City, Uganda', *Environmental Research*, vol 101, pp42–52
Nkamleu, G. B. and Adesina, A. A. (2000) 'Determinants of chemical input use in peri-urban lowland systems: Bivariate probit analysis in Cameroon', *Agricultural Systems*, vol 63, pp111–121
Oelofse, M., Auerbach, R. and de Neergaard, A. (2007) 'Targeting socially excluded groups: Community gardening in KwaZulu-Natal, South Africa', *Urban Agriculture*, vol 18, pp14–18

Pasquini, M. W. (2006) 'The use of town refuse ash in urban agriculture around Jos, Nigeria: Health and environmental risks', *Science of the Total Environment*, vol 354, pp43–59

Pasquini, M. W. and Harris, F. (2005) 'Efficient use of resources: Urban waste ash and soil fertility on the Jos Plateau, Nigeria', *Area*, vol 37, pp17–29

Prain, G. (2006) 'Participatory technology development for urban agriculture', in van Veenhuizen, R. (ed) *Cities Farming for the Future: Urban Agriculture for Green and Productive Cities*, RUAF Foundation, IIRR and ETC Urban Agriculture, Philippines, pp273–312

Ratta, A. and Nasr, J. (1996) 'Urban agriculture and the African urban food supply system', *African Urban Quarterly*, vol 11, pp154–161

Rengasamy, S., Devavaram, J., Marirajan, T., Ramavel, N., Rajadurai, K., Karunanidhi, M., Prasad, N. R. and Erskine, A. (2003) 'Farmers' markets in Tamil Nadu: Increasing options for rural producers, improving access for urban consumers', *Environment and Urbanization*, vol 15, pp25–34

Rogerson, C. M. (1993) 'Urban agriculture in South Africa: Scope, issues and potential', *GeoJournal*, vol 30, pp21–28

Sawio, C. J. (1994) 'Who are the farmers of Dar es Salaam?', in Egziabher, A. G., Lee-Smith, D., Maxwell, D. G., Memon, P. A., Mougeot, L. J. A. and Sawio, C. J. (eds) *Cities Feeding People: An Examination of Urban Agriculture in East Africa*, IDRC, Ottawa, Canada, pp23–44

Schiere, J. B. and van der Hoek, R. (2001) *Livestock Keeping in Urban Areas: A Review of Traditional Technologies*, Animal Production and Health Papers 151, FAO, Rome

Simon, D., McGregor, D. and Nsiah-Gyabaah, K. (2004) 'The changing urban–rural interface of African cities: Definitional issues and an application to Kumasi, Ghana', *Environment and Urbanization*, vol 16, pp235–248

Smit, J. and Nasr, J. (1992) 'Urban agriculture for sustainable cities: Using wastes and idle land and water bodies as resources', *Environment and Urbanization*, vol 4, pp141–152

Smit, J., Ratta, A. and Nasr, J. (1996) *Urban Agriculture, Food, Jobs and Sustainable Cities*, UNDP, New York, NY

Smith, R. M., Gaston, K. J., Warren, P. H. and Thompson, K. (2005a) 'Urban domestic gardens (V): Relationships between landcover composition, housing and landscape', *Landscape Ecology*, vol 20, pp235–253

Smith, P. M., Yusuf, M. J., Bob, U. and de Neergaard, A. (2005b) 'Urban farming in the south Durban basin', *Urban Agriculture*, vol 15, pp16–18

Tinker, I. (1994) 'Urban agriculture is already feeding cities', in Egziabher, A. G., Lee-Smith, D., Maxwell, D. G., Memon, P. A., Mougeot, L. J. A. and Sawio, C. J. (eds) *Cities Feeding People*, IDRC, Ottawa, Canada, pp7–14

Tixier, P. and de Bon, H. (2006) 'Urban horticulture', in van Veenhuizen, R. (eds) *Cities Farming for the Future*, RUAF Foundation, IDRC and IIRR, Leusden, NL, pp313–346

van den Berg, L. M., van Wijk, M. S. and van Hoi, P. (2003) 'The transformation of agriculture and rural life downstream of Hanoi', *Environment and Urbanization*, vol 15, pp35–52

Vinnerås, B. (2002) *Possibilities for sustainable nutrient recycling by faecal separation combined with urine diversion*, unpublished PhD thesis, Acta Universitatis Agriculturae Sueciae, Agraria, Swedish University of Agricultural Sciences, Sweden, http://diss-epsilon.slu.se/archive/00000332/

Wahid, A., Maggs, R., Shamsi, S. R. A., Bell, J. N. B. and Ashmore, M. R. (1995) 'Air pollution and its impacts on wheat yields in the Pakistan Punjab', *Environmental Pollution*, vol 88, pp147–154

Ward, C., Ballif-Spanvill, B., Fuhriman, A., Soloman, Y. and Widdison-Jones, K. (2004) 'Weeding out failed practices: A case study of community gardens in rural Mali', *Human Ecology*, vol 32, pp509–521

Wolf, J., van Wijk, M. S., Cheng, X., Hu, Y., van Diepen, C. A., Jongbloed, A. W., van Keulen, H., Lu, C. H. and Roetter, R. (2003) 'Urban and peri-urban agricultural production in Beijing municipality and its impact on water quality', *Environment and Urbanization*, vol 15, pp141–156

Wong, N. H., Cheong, D. K. W., Yan, H., Soh, J., Ong, C. L. and Sia, A. (2003) 'The effects of rooftop garden on energy consumption of a commercial building in Singapore', *Energy and Buildings*, vol 35, pp353–364

3

Biodiversity of African Vegetables

Patrick Maundu, Enoch Achigan-Dako and Yasuyuki Morimoto

Africa's flora and vegetables

Africa has numerous ecosystems and vegetation zones that support an immense diversity of native and naturalized plant species, estimated to be at least 40,000 species (Linder, 2001). Approximately 8 to 10 per cent of these are used as food by local communities. About 1000 species are used as vegetables, the majority of which (80 per cent) are leafy vegetables, with the rest made up of vegetables from fruits, seeds, roots and tubers, stems and flowers. Besides the rich flora, the high number of edible vegetable species is attributed to a number of other factors. Among these is the continent's high cultural diversity, as depicted by the over 2000 spoken languages. Other factors include the high dependence of human populations upon locally available resources as opposed to manufactured or market products. In addition to the foregoing reasons, Africa's high use of vegetables is also attributable to the widespread culture of using vegetables as an accompaniment for the main dish. The latter is often starchy, mainly made from cereals but also roots and tubers (such as cassava, yams, potatoes, etc.), starchy stems (such as ensete) or fruit parts (such as plantain). Frequent hardships have also led to local innovations and discoveries of new vegetable species as communities seek to cope with such hardships and enrich their knowledge of local species. The discovery or selection of such new types has involved both introduced species as well as native African species. The majority of African vegetable species are of localized importance either due to their limited distribution (occurrence) or to the fact that the potential as a vegetable is known to specific geographic communities, and the species as a whole are not part of the local food culture of other communities or regions.

Some other species, on the other hand, are widely used across the continent, and this is particularly so for introduced cultivated vegetable species (Chweya and Eyzaguirre, 1999).

In this chapter we look at the diversity of these vegetables not only at a continental scale, but also on a regional scale. It is difficult to provide a substantial list of species in such a space; but attempts will be made to mention the most useful species and also to provide trends. Much of the information available is from sub-Saharan Africa and is mainly drawn from the authors' experience in the region.

Defining the terms

It is important to define a variety of terms, such as 'vegetable', as they are used in this chapter. Other terms that often cause confusion are 'indigenous' and 'traditional', as well as 'native', 'naturalized', 'introduced' and 'exotic'. These too will be defined as they are used in the chapter.

Perhaps a clarification of the use of the terms 'indigenous' and 'traditional' is important to make at the onset. Indigenous (or native) here refers to species, varieties or any taxa known to be native to or to have originated in a specified geographical location – in our case, Africa. Exotic or introduced species refer to species that were brought into the region of focus (in this case, Africa) from other parts of the world. Traditional here refers to exotic or indigenous species, varieties or taxa that have been in use for a sufficient length of time to be part of the local food habits, knowledge systems and customs of communities (Maundu et al, 1999). Through many years of selection, communities may develop varieties that are unique and adapted to their local ecological and cultural landscape from introduced species. It then follows that exotic species such as cassava, sweet potato and cocoyams (taro) are also traditional. Most species in the indigenous category are also traditional. Naturalized refers to introduced species that are able to reproduce and sustain their populations in their new environment without the help of humans. Many introduced vegetable species have that potential so long as they are in the right environment. A number of weedy vegetable species (such as some species of *Amaranthus* and *Bidens*) are naturalized in parts of the continent, even though they may have truly indigenous conspecifics. For example, *Amaranthus dubius* and *A. cruentus* are introduced species, which are widely used alongside the indigenous *A. graecizans* and *A. sparganiocephalus*; but the newcomers were easily adopted and used in a similar manner as the local types.

The term 'vegetable' is non-scientific and, hence, difficult to define precisely. The term in this chapter is limited to plant parts (underground or shoot parts, including stem, leaves, fruit and flowers) usually eaten fresh or processed in any way (cooked, steamed, dried, marinated, etc.) and that are not used as conventional fruit, nut, root/tuber crop, pulse or staple, but as an accompaniment of the main dish or alone as a snack mainly for their micronutrient benefits. This can be simplified to any plant part that is consumed usually fresh or after minimal processing to facilitate consumption of the main dish, as well as for its micronutrients. It is a difficult term on which to put rigid restrictions as there will be crops that may behave more as vegetables in certain situations (e.g. pulses) and as non-vegetable in others; but, generally, vegetables are consumed for their freshness and for their being good sources of micronutrients (especially vitamins A and C), some minerals and roughage. Linking this with the characteristics of indigenous or naturalized species results in the concept of African indigenous vegetables (AIVs) as defined in Chapter 1.

Diversity of African indigenous vegetables

What determines diversity of species used?

The diversity of AIVs in a location or community is the number of species and varieties or cultivars within them that are used as a vegetable, and is determined by a number of factors:

- *The number of plant species in the local environment.* This is a significant contribution to the diversity of AIVs consumed within a geographical area or in a community because the more species that are available, the greater the number people have to select from, as well as a wider diversity of desirable traits. An impoverished species range means that local communities have less to choose from. How the proportion used as AIVs varies across different environments and relative to the total number of plant species available is unknown. White (1983) recognized eight African phytochoria – namely, the Congolian, Guinean, Sudanian, Somalian, Zambezian, Sahelian, Saharan and Kalahari. These phytochoria are characterized by uniqueness in a large number of local taxa (endemicity). Species richness as well as composition vary across this range; but a number of species may occur across several phytochoria. Certain regions are known to have extraordinarily high species numbers: the so-called mega-diverse regions. The East African coastal region and the southern Cape region are good examples.
- *A culture and traditional knowledge of AIVs.* Even though many different plant species may be encountered in a region, they only become AIVs once they are used by local communities. For this to happen requires innovation and willingness to experiment, and reasonable local knowledge on how to use the species. Such knowledge is handed down from generation to generation, but is also shaped through daily experiences by members of local communities. Occasionally, the community is faced with challenges such as hardships resulting from lack of food; in such cases, the community's innovations to, for example, try new plants or methods of cooking become critical for their well-being or survival. It may well be that if a culture of AIV use is developed, communities will be more willing to experiment and use additional species. Several studies have shown that some communities know and use dozens of species. For example, the Mijikenda of coastal

Kenya and the Ani community of Benin use over 100 leafy vegetables (Maundu, 1997; Fondo et al, 2005). Keding et al (2007), working in the coastal district of Muheza in Tanzania, recorded 74 vegetables. In south-east Burkina Faso, Mertz et al (2001) recorded 77 species across two separate villages. Shackleton et al (1998) recorded the use of 45 different species in eastern Mpumalanga Province, and Ogle and Grivetti (1985) reported over 200 species in the small kingdom of Swaziland.

- *Introductions of new species, varieties and knowledge.* These introductions add to the existing knowledge and species complement, offering new products to the local community that were not previously known or available. Historically, such introductions were facilitated by travel, trade and exchanges between communities. In the modern day, they are more frequently facilitated by researchers, extensionists and public media.
- *Domestication and crop improvement by farmers.* Innovative farmers take old and new species and may select the most desirable traits, such as high yield, sweet taste, early flowering, etc., which ultimately provides a new variety or crop for local consumption. Improvement might also refer to the farming practices and their own skills. Examples include fruits of several species, some of which are also used as AIVs, such as *Irvingia gabonensis* and *Sclerocarya birrea* (Leakey et al, 2004, 2005).
- *The type of farming systems and cropping systems.* The methods by which farmers tend their fields and gardens influence the availability of species. This includes whether or not they intercrop, whether they maintain or remove weedy species, crop rotation and the like. Mixed cropping as opposed to mono-cropping, for example, tends to provide a higher diversity of vegetables as well as improved yield (see Chapter 5).

Introduced crops as African indigenous vegetables

Local cultures may take time before accepting new foods; but many are also known for their versatility in trying new foods when faced with hardships such as food shortage. A number of introduced crops came to Africa for uses other than vegetable. Over the years, however, a vegetable use for these initially non-vegetable food crops has been developed through innovative processing, often combined with variety selection. For example, sweet potato (*Ipomoea batatas*) was introduced mainly for the root crop; but the leaves are now an important food in several parts of Africa, particularly Tanzania, Malawi (*kholowa*), Mozambique (*domutomu*) and Mali (*sakassaka*). In these countries, special varieties have been developed by farmers for their leaves alone. Sweet potato leaves are also used in several other countries in Southern Africa and parts of West Africa, but are unimportant in Uganda, Kenya and Ethiopia, where the crop is grown for its tubers only.

Cassava plant (*Manihot esculenta*) was also introduced from South America. In Benin, the leaves of cassava represent an important vegetable for communities throughout the country. The leaves are the most important vegetable in the Congo Basin, where they are known by the name *pondu*. They are prepared in a rather complex way involving pounding and use of traditional salt obtained from the ash of the palm tree inflorescences, plantain peelings or rice inflorescences. In several other areas, including Tanzania (*kisamvu*), Malawi (*ntapasya*), Mozambique (*mutchcobwe*), Cameroon, Nigeria and Senegal (*nambi*), the leaves are important, but to a lesser degree. The leaves of the related ornamental species, the tree cassava (*M. glaziovii*), called *mpira* in East and Southern Africa, are also used for its leaves in that region and also in parts of West Africa, including south and central Benin (*kpatindehouin*).

Taro or cocoyam (*Colocasia esculenta*) and tania or elephant ear (*Xanthosoma sagittifolium*) are two related species with more or less similar uses and habits. Both species were introduced to Africa for their corms; but like cassava and sweet potato, their leaves are now also used as an AIV, although not widely. The two are important in Southern Africa where they are known as *madumbe*. Both are very important in the Congo Basin and in humid West Africa (Nigeria to Guinea).

Among the most widely used introduced AIVs are the pumpkins *Cucurbita pepo* (see Plate 3.1), *C. maxima* (see Plate 3.2) and *C. moschata* (see Plate 3.3), originally from South America, which are important for their fruit and leaves. Pumpkin leaves are sold and used in their fresh form throughout Africa; but in Southern Africa they are also extensively sold in the dried form, often with their flowers. The three species of pumpkins are widely grown; but *C. moschata* and *C. maxima* are more commonly used as a leaf vegetable in Southern Africa. In the Bougouni region of Mali, we have encountered home gardens of *Cucurbita pepo* exclusively for leaf production. In northern Benin, the leaves of *C. pepo* (*gboro*) are widely used by the Bariba communities.

Several other introduced species are used as AIVs. The cherry tomato (*Solanum* spp.), although originally from South America, was widely distributed in much of humid, sub-humid and semi-arid Africa long before the modern cultivars were developed and is considered a traditional food in most communities. Although the tomatoes are the main food, the leaves of this variety as well as the modern tomatoes are used as a vegetable by a few communities in the continent (e.g. northern Uganda, central Mozambique, southern Malawi, northern Benin and coastal Kenya). Other tropical American members of the tomato family are the chilli peppers (mainly *Capsicum annuum* and *C. frutescens*) whose fruit are used as an AIV and for flavouring foods. The leaves are also picked by various communities in Africa (e.g. Uganda, Kenya, Tanzania, Mozambique and Malawi) and used as a leafy vegetable, mainly mixed with others. Young bean leaves are widely used, mainly in their dried form, in Malawi, Zimbabwe, Mozambique, Zambia and the Democratic Republic of Congo, but only rarely in other areas of the continent. The leaves of the pawpaw plant (*Carica papaya*) are used in isolated areas such as coastal parts of Kenya, in southern Malawi and in Benin, but are quite minor in importance. The young fruits, too, are often used as a vegetable in parts of the continent such as northern Uganda and parts of Kenya.

Occasionally an introduced species may easily turn to an AIV due to close resemblance with an already existing indigenous species. The native amaranths are *Amaranthus thunbergii*, *A. graecizans*, *A. sparganiocephalus* and perhaps *A. lividus*. These were originally used by local communities as vegetables; so when the newcomer amaranths arrived at different times, they too became vegetables, such as *A. cruentus*, *A. hybridus*, *A. dubius*, and spiny amaranth (*A. spinosus*). *A. cruentus* is today the main leafy vegetable in Benin, Togo and Sierra Leone and very important in many other lowland areas in Africa.

Cultural linkages: Variations across community groups and regions

As mentioned earlier, besides the size of the flora, AIV diversity is influenced by cultural diversity and historical influence. A recent survey conducted in Nairobi showed a strong correlation of the species used by different ethnic communities living in the city with the traditional food species in their rural areas (Kimiywe et al, 2006; Imbumi and Maundu, 2008). Some species, therefore, may have widespread geographical distribution and could be important food to some communities, while deemed inedible by others. In many regions, communities are attached to their vegetables, which may otherwise be unimportant or not used at all in other regions. *Launaea cornuta* (*mutsunga*), for example, is amongst the most important leafy AIVs in coastal Kenya; but it is only sparingly used elsewhere despite its wide distribution. Keding et al (2007) ranked the most important AIVs used in four districts of north-west Tanzania. None of the districts had similar preferences as the others. African nightshade, despite being important in one district (Arumeru), did not appear among the six most important in the other districts. In two villages only 30km apart in south-east Burkina Faso, almost 50 per cent of all the species recorded where used within only one of the two villages (Mertz et al, 2001). Vegetable consumption is more of a culture in some communities than in others.

In other cases, a particular AIV species has a wide distribution and its use as an AIV is also widespread. This category includes what is considered the most important vegetables in the continent, such as cowpea (*Vigna unguiculata*), okra (*Abelmoschus esculentus*), African eggplants (*Solanum* spp.), some species of leaf amaranth (Amaranthus spp.), spider plant (*Cleome* spp.), some species of nightshades (*Solanum* spp.) and pumpkins (*Cucurbita* spp.) (see Chapter 7). These will be mentioned further in this chapter.

The diversity of AIVs consumed varies significantly from one community group to another, and the distinction may be significant even between groups living in similar conditions due to cultural reasons. In other words, there are communities with a strong culture of consuming a diversity of AIVs, while others do not. Many communities use AIVs as an accompaniment to starchy foods, particularly the stiff porridge commonly prepared all over Africa and known by such names as *ugali*, *sima* (Eastern Africa), *pap* (South Africa), *sadza* (Zimbabwe), *atap* (north Uganda) and *foufou* (*fufu*) (West Africa).

Communities with a preponderance of a variety of AIVs often combine several types in mixtures in one preparation. Among the Giriama of coastal Kenya, as many as seven leafy vegetables and three fruit vegetables have been recorded in one dish. As one would expect, vegetable consumption is generally more established among arable than pastoral groups. Consumption of vegetables among pastoral groups such as the Maasai and the Fulani is limited; in many cases, it is only recently taking root. The conventional vegetable consumption among Somali pastoralists is also almost absent or low. AIV species used by other communities are often seen growing among pastoralist groups but are not used or known. Good examples include African nightshade and leaf amaranth among the pastoral Samburu and Somali communities of northern Kenya. Much of the plant material among pastoral groups is taken as medicine or tonic boiled in soups or milk. Hardships, such as the loss of animals, may also force communities to depend a little more upon AIVs – hence, the discovery of more species or more use of the community's pool of knowledge to solve the current crisis. This use and knowledge may persist long after the end of the crisis. The Ngikebootok of Kenya who are affiliated with the pastoral Turkana, for example, lost animals and turned to farming and collecting vegetables, and have a high diversity of local vegetables (Maundu, 1996).

Diversity of plant parts used

According to the definition of an AIV or even a conventional vegetable, consumption is not restricted to just the leaves of the plant concerned. AIVs include tubers, flowers and seeds, alongside the more common leafy species.

Leaf vegetables

This group constitutes the bulk of the AIVs used throughout the continent. Leafy vegetables are dominated by such plant families as those of the amaranth (Amaranthaceae), tomato (Solanaceae), sunflower (Asteraceae), pumpkin (Cucurbitaceae), cabbage (Brassicaceae), bean (Fabaceae) and jute (Tiliaceae). Some are harvested from tree species (e.g. baobab, moringa and vitex), while others are from herbaceous annuals and perennials. For many species, only the young new leaves are harvested as the older ones are bitter or hairy; but there are no hard-and-fast rules. Different communities favour different tastes and practise varying harvesting and cooking techniques to either reduce negative characteristics or to enhance the more desirable traits.

Fruit vegetables

The fruit vegetable category is dominated by the tomato family (Solanaceae) and the pumpkin family (Cucurbitaceae). The fruit of many species are sliced and cooked in sauce, while a few can be eaten fresh. Species such as *Lagenaria siceraria*, *Cucurbita* spp., *Momordica charantia*, *Momordica balsamina* and *Abelmoschus esculentus* are used as AIVs.

Flowers

As expected, flowers are less used and less important as an AIV compared to leaves and fruits. Often they are mixed in with the leaf portions of the plant and form a relatively insignificant proportion. Flowers are a sign of maturity of the plant and therefore are, in most cases, avoided while picking leaves. In some species, such as *Cleome gynandra*, *Lippia multiflora* and *Solanum* spp., flowers are often part of the harvest as they appear when the leaves are still palatable. Pumpkin flowers, however, are often picked intentionally as part of the vegetable harvest. They are particularly popular in Southern African states where the tradition of drying leaves is most rooted. In this area, dried pumpkin leaves are mixed with their dry flowers both for food and for brightening up the rather dull dried leaves. Other plants with edible flowers include *Carissa* spp., *Bombax* spp. and *Adansonia digitata* (baobab). In West Africa, flowers of *Bombax costatum* are an excellent dish for the Boko and Bariba communities of Benin. In Mali, both the leaves and the flowers are used as a vegetable. The young calixes of roselle (*Hibiscus sabdariffa*) are commonly used from Senegal to Niger as a cooked vegetable or are finely cut and used in sauce.

Seeds

In West Africa, seeds (*egusi*) (see Plate 3.4) of a number of species in the melon family (Cucurbitaceae), including *Cucumeropsis mannii*, watermelon (*Citrullus lanatus*), *Citrullus colocynthis*, bottle gourd (*Lagenaria siceraria*) and melon (*Cucumis melo*) (Zoro Bi et al, 2003; Achigan-Dako et al, 2006; Djè et al, 2006), are consumed in a number of ways. They are prepared for consumption by parching and pounding to free the seed coat from the kernel, which can be eaten either raw or cooked, or more usually when ground to a powder and added to soups and stews (Burkill, 1985). *Telfairia occidentalis* and *Parkia biglobosa* seeds serve a similar function. Seeds of the fluted pumpkin (*Telfairia occidentalis*) are cooked as a special accompaniment by Yoruba communities. The preparation of the seed of *Parkia biglobosa* yields products such as '*afintin*' in Benin and '*soumbala*' in Mali. The seed of *Irvingia gabonensis* is widely used in Ghana, Togo, Benin and Nigeria, as is the kernel of the marula (*Sclerocarya birrea* subsp. *caffra*) in Southern Africa.

Stems

The stems (and flowers) of various species of *Aponogeton* are consumed in different parts of the continent.

Roots and tubers

As mentioned earlier, there are many tuber crop species, of which the leaves are used as a vegetable (e.g. sweet potato, cocoyam and ensete); but the tubers themselves are generally regarded as a carbohydrate staple rather than as a vegetable.

Multiple-use species

There are many particularly useful species for which nearly all parts are edible, some of which may be used as vegetables. Typical ones include some bean species, where both the leaves and seeds are eaten, some tree fruit species, of which the leaves may also be consumed, and cucurbits, where both fruits and leaves (and in some regions also flowers) are consumed.

Diversity within species and naming

Most communities maintain specific varieties of AIVs species for various reasons, including taste, marketability, growth form, maturation period and so on. Cultivated species tend to be morphologically more diverse. Cowpeas, for example, are maintained in diverse forms. A community group in Kitui District in Kenya was found to maintain up to 23 different farmer varieties. Cultivar or variety names often correspond to a distinctive characteristic of the AIV type such as colour or origin. The creeping type, according to Kyanika Women's Group in Kitui, is named '*ndamba*', while a similar creeping type in north-east Tanzania is called '*tambaa*' (Keding et al, 2007), with both names meaning creeping or spreading.

Regional diversity of vegetables

In this section we examine the diversity of AIVs in five selected regions of Africa: North, humid and Sahelian West, Central, East and Southern Africa. For the purposes of this chapter, the regions are not related to regional political and socio-economic groupings, but rather to geographical and, to some extent, eco-geographical assemblages. North Africa includes the Sahara and areas north of it. West Africa includes Nigeria to Senegal. Central Africa includes the area in tropical Africa bounded by the Atlantic Ocean to the west and south-west Sudan and the Great Lakes to the east, and between the Congo Basin and southern Chad. East Africa refers to the region between Eritrea and Tanzania and includes the Horn of Africa and southern Sudan. Southern Africa is the region of Angola to Mozambique, south to South Africa as well as the neighbouring Indian Ocean islands.

North Africa

A large proportion of North Africa is taken up by the Sahara Desert; but the northernmost region has a Mediterranean climate, while the western part has the high Atlas Mountains, both home to a high floral diversity. A number of local greens are collected. One important local AIV in the Mediterranean region is the spiny chicory, or *stamnagathi* (*Cichorium spinosum* – Asteraceae), a somewhat bitter vegetable, usually collected in the wild in winter or spring. A popular AIV of the hot humid lowlands is jute mallow (*Corchorus olitorius*); it is used all over Africa, but is particularly important in Egypt and the Nile Valley, in general. Here it is known by the name *molokhiyya* (also spelled in several different ways). It is mainly a summer AIV and is used to make the *molokhiyya* soup. *Molokhiyya* is also commonly used in Sudan and most of North Africa. Okra, which is almost as mucilaginous, is also popular in North African dishes. North Africa has great influence from European, Middle East and Asian vegetables and the use of spicy vegetables (e.g. onion, garlic, ginger, pepper, chilli, cinnamon and paprika) in cooking is more established here than in the rest of Africa.

West Africa

West Africa has quite varied ecosystems, mainly due to extremes in precipitation and humidity. The southern areas near the Atlantic coast are humid, with some experiencing an equatorial climate; but humidity decreases rapidly northwards towards the Sahara Desert. This variable climate also supports a changing plant community and agricultural crops. Although in many cases the change is gradual, five distinct zones are easily discerned: the Guineo-Congolian zone, the Guinean zone, the transitional Sudano-Guinean zone, the Sudanian zone and the Sahelian zone. In addition, a major bio-geographical gap (the Dahomeen gap) exists in West Africa between the Congolian and the Guinean forest blocks. It is a broad savannah corridor that reaches the coast of southern Ghana, Togo and Benin, and which interrupts the zonal West African rainforest. This diversity of ecosystems offers to the region an important AIV biodiversity of both wild and cultivated species, estimated at between 100 to 200 species in Togo and Benin alone (Batawila et al, 2007; Dansi et al, 2008) and about 40 species in Senegal (Diouf et al, 1999). Some of the AIV species are limited to one or a few zones, while the hardier species may have a widespread distribution that cuts across humid and drier zones. *Corchorus olitorius*, or its wild relative *Corchorus tridens*, is a common dish in humid as well as in dry West Africa.

Humid West Africa

This is a long belt stretching from the Gambia and south Senegal, and running along the Atlantic to the east of Nigeria where the zone expands into the Central African humid zone. Differences in humidity exist along this zone, particularly from Ghana to Benin, which is relatively drier; but this does not affect the species used in this zone significantly. The area enjoys a tropical climate that is conducive not only to indigenous species, but also species from two other key centres of crop origin – tropical Asia (South Asia/South-East Asia) and tropical America (Brazil to Mexico). It is no surprise that crops such as cassava and cocoyams (*Colocasia esculenta* and *Xanthosoma* spp.) have taken root in this zone and are now amongst the most important local AIVs. Tania is relatively newer in the region than taro; hence, it is sometimes referred to as the new cocoyam. Its importance is on the increase. It is represented by two species, *Xanthosoma sagittifolium* (tania or yautia) and *X. poeppigii* (*X. mafaffa*). Some of the indigenous and widely used AIVs in this zone include bitter leaf (*Vernonia amygdalina*), Lagos spinach (*Celosia argentea*) and okra (*Abelmoschus esculentus*). A relative of okra (*Abelmoschus caillei*) is also present in the region. The use of palms as vegetables is also quite unique to this region, which has about four palms used as a vegetable. The region has a strong culture of using seeds of cucurbits (the so-called *egusi* crops) for sauces; hence there are a number of *egusi* species in the region, including *Cucumeropsis mannii*, which is unique to this region. Others *egusi* species include *Citrulus lanatus* subsp. *mucosospermus* and *Lagenaria siceraria*, known as bottle gourd but encompassing non-hard coat varieties known as *Accra-kakoun* or *aklampa* in Benin, *seregbe* in Côte d'Ivoire or *apentra kessa* in Gondja communities in Ghana. Others species of the Cucurbitaceae family used include *Momordica charantia*, *M. cissoids*, *M. foetida* (by the Ani communities) and *Telfairia occidentalis* (see Plate 3.5). Many tree species, including *Irvingia gabonensis* and *Blighia sapida*, are widely used. The use of the seed kernel of *Irvingia gabonensis* is particularly widespread as it is employed as a substitute for okra. The three most frequently consumed species in Burkina Faso include chilli (*Capsicum frutescens*), African locust bean (*Parkia biglobosa*) and okra (Mertz et al, 2001).

Dry West Africa

The dry zone borders the humid zone to the south and the Sahara Desert to the north. It is a low-lying region stretching from Senegal along the Atlantic Coast to the Red Sea and Somalia's Indian Ocean coast. For West Africa, we will limit the eastern extent to Chad. The main vegetation types are savannahs. The savannah region of West Africa is particularly known for the use of tree and shrub species as vegetables. Widely used species include *Vitex doniana*, *Bombax costatum*, *Rodognaphalon brevicuspe*, *Grewia mollis* and *Senna* species. Woody plant species, for example, are employed a great deal by the Gourounsi people in south-central Burkina Faso (Kristensen and Balslev, 2003). These plants remain important options for low-income rural communities. The use of the baobab (*Adansonia digitata*) as a fruit and vegetable is widespread. Other tree species commonly used include moringa (*Moringa oleifera*), fig trees (*Ficus* spp.), desert date (*Balanites aegyptiacea*) and *Strychnos spinosa*. In Benin, *Vitex doniana* has a great market value and is found pre-boiled in all main markets of the countries. The species is also widely used in some communities in Mali. Because young leaves of *Vitex doniana* may become rare in certain periods of the year, people in southern Benin use young leaves of *Daniellia oliveri* (zaman) as a substitute. It is a common practice to dry leaves and to keep them for use during the dry periods, especially from Niger to Senegal where *Moringa oleifera* or sliced okra are dried and kept for use in lean periods. Some important non-tree AIVs include the cowpea leaves (*Vigna unguiculata*), roselle (*Hibiscus subdariffa*) and eggplant (*Solanum aethiopicum* – Kumba group), and *Corchorus olitorius*. Fruits of pumpkins (*Cucurbita maxima*, *C. pepo* and *C. moschata*) are important market products in savannah regions and may be encountered in large quantities in Niamey along the Niger River. The leaves of *Cucurbita pepo* ('*gboro*') are particularly appreciated by Bariba communities in northern Benin. *Luffa acutangula* and *Trichosanthes cucumerina* (see Plate 3.6), said to originate from Asia, are naturalized in West Africa. In Burkina Faso, the Moré people used to name them '*napokumana*' (the queen's okra). In the arid region from northern Benin to Senegal, the two species are cultivated in home gardens or allowed to trail on fences. The young fruits in soup or sliced in okra soup are a delicacy.

Central Africa

Central Africa is characterized by high humidity for its greater part. Much of it lies within the tropical rainforest zone. The extreme south of the region, which extends towards the border of the Democratic Republic of Congo with Angola and Zambia, as well as north towards Chad and north-east towards Sudan are a bit drier and have savannah vegetation. AIVs in this region are therefore dominated by warm humid climate species. As is the case with humid West Africa, tropical American and Asian crops are at home here, including taro (*Colocasia esculenta*), tania (*Xanthosoma sagittifolium*), cassava, sweet potatoes and *Amaranthus cruentus*. Cassava leaves (*pondu*) are the dominant vegetables in the Democratic Republic of Congo. AIVs occupy an important place, especially in Cameroon and Congo. The forest climbers, locally known as *eru* in Cameroon (*Gnetum africanum* and *G. buchholzianum*) are harvested in tonnes from the forest and distributed far away from their source, as well as being exported to the Cameroonian and Nigerian Diaspora in Europe and America. Bitter leaf (*Vernonia amygdalina* and its relative *V. hymenolepis*) is used extensively in Cameroonian cuisine. These two have coarse leaves and, hence, are cooked with the small succulent pan-tropical species *Talinum fruticosum* to soften them. The giant African nightshade (*Solanum scabrum*), locally known as *jamajama*, has its home here and it is extensively cultivated in western Cameroon. Other important indigenous vegetables include *Corchorus olitorius*, okra (*Abelmoschus esculentus*) and *Hibiscus sabdariffa* (*sawa-sawa*).

Eastern Africa

Eastern Africa here includes Sudan east to Djibouti and south to Tanzania. The region has many different climatic zones – from humid forests and alpine zones to deserts. The topography of the region is characterized by highlands with high mountain peaks (e.g. Ethiopian Highlands) and hot dry lowlands. These varied habitats have resulted in high species diversity.

Dry northern East Africa

Northern East Africa is predominantly savannah (Sudan); but humidity generally decreases northwards towards the expansive Sahara Desert. The relatively cooler and more humid Ethiopian Highlands break this low expansive dry zone, which is continuous with the West African savannah and Sahel. Many plant species are therefore shared between the dry north East Africa and the dry West Africa.

People in the Sahelian region and adjoining dry savannah are known for their use of tree and shrub vegetables in the form of leaf, fruit, seed or flower. In the dry northern parts of Kenya (among the Turkana and Pokot) and Uganda (among the Karamoja and Iteso), south-east Sudan (among Toposa, Bahr el Ghazal and Jonglei) and southern Ethiopia (among the Merille), it is traditional to boil tree fruits for a meal or part of a meal. Common plants used include *Dobera glabra*, *Balanites* spp. (*B. aegyptiaca, B. pedicellaris* and *B. rotundifolia*) and *Maerua* spp. (e.g. *M. subcordata*). In the same areas, the tender leaves of *B. aegyptiaca* are a major leafy AIV. In Djibouti and northern Somalia, *Aponogeton nudiflorus* (Aponogetonaceae) (*tibilkih* in Afar language) is an important plant collected in the wild. The shoot of the plant is eaten as a vegetable while underground parts are a source of flour used in baking local bread. It grows in flood areas and, hence, can provide continuous supply for a considerable period. During famines, the Somali people in Dikhil and most of the communities in the dry horn of Africa boil the seeds of *Balanites aegyptiaca* and related *Balanites* species and eat them as food while the seeds of *Dobera glabra* are eaten fresh or boiled (Demissie, 1991; Kiambi, 1999).

Humid and sub-humid East Africa

Ethiopia and the Horn of Africa comprise one of the eight Vavilov centres of crop origin. A number of AIVs have their home here. The region has a number of local moringa species, the best known being *Moringa stenopetala*, which is used frequently in southern Ethiopia, particularly among the Konso. The Ethiopian kale (*Brassica carinata*), *Brassica nigra*, shallots (*Allium oschaninii*) and other *Allium* spp., as well as enset (*Ensete ventricosum*) whose roots are often eaten as a vegetable, are used a great deal.

The East African coastal region has an exceptionally rich biodiversity. For its size, the region, comprising the coastal area of Kenya and Tanzania, perhaps uses more AIV species than any other region in Africa, comparable only to Southern Africa. Here, the bitter lettuce (*Launaea cornuta*), although collected from the wild, is the most important AIV. It is also used for medicinal purposes. The Kenyan region has some unique species, such as *Crotalaria* spp. (*C. brevidens* and *C. ochroleuca*) and a type of stinging nettle (*Urtica massaica*) commonly used as an AIV. The fig-leaved or malabar gourd (*Cucurbita ficifolia*) from Central America is an important AIV in the central part of the country, but the fruit is not used (Maundu and Imbumi, 2003).

The Ugandan area represents a transition zone to the Central Africa rainforest climate and therefore has many AIVs that are allied to those of Central Africa. Some of the widely used, but unique, AIVs are the fruit and leaf vegetables of *Solanum* species. *Solanum aethiopicum* Gilo group, locally known as *ntula* or *enjagi*, *S. aethiopicum* Shum group, locally known as *nakati*, and *S. anguivi*, locally known as *kantunkuma*, are all typical of Uganda, and this becomes obvious as one gets into the local markets. A local amaranth (*Amaranthus lividus*) which has green to reddish purple shiny leaves is also typical of Uganda. The northern part of the country is known for its use of the jute mallow and its related *Corchorus* species, as well the use of a number of *Hibiscus* species for AIVs. Tanzania and Kenya have a high diversity of *Solanum* species of the so-called *Solanum nigrum* complex. Leading among these is the orange-berried solanum (*Solanum villosum*), the green-berried solanum (*Solanum eldorettii*) and a few others, all common in the highland parts of the region. Tanzania represents a transition to Southern Africa and therefore some of the vegetables found here are also commonly used in Southern Africa. East Africa is, however, one region that has been most affected by the decline in the use of local vegetables; but thanks to the efforts of both local and international organizations, the situation is changing since the turn of this century and local vegetables are increasingly becoming part of the local dishes again.

Southern Africa

Southern Africa is characterized by vast savannah and miombo woodland areas, the Kalahari Desert and a unique Mediterranean climate in the region's southern-most tip. The island of Madagascar has tropical rainforest on the eastern side, which gives way to savannah and deciduous forest towards the west. Due to these varied climatic conditions, the Southern Africa region has more plant species diversity than any other. Dependence upon wild vegetables is high in this region.

The diversity of useful cucurbits is high – some introduced and others indigenous. The pumpkins (*Cucurbita maxima*, *C. moschata* and *C. pepo*) are all widely used in the region, and except for *C. pepo*, the leaves of the others are perhaps more important than the fruit in the region. The region has a variety of cucumbers (*Cucumis sativus* has particularly large fruit), which are an important source of water and food in Malawi and Mozambique. The young fruits and leaves of sponge gourd (*Luffa cylindrica*) are used here more than in other regions. In West Africa where the species grows wild, reports on its use as a vegetable are scanty. Some of the local indigenous cucurbits include the horned gourd (*Cucumis metuliferus*), the Balsam pear (*Momordica balsamina*) with a pan-tropical distribution, edible forms of the common gourd (*Lagenaria siceraria*) and local varieties of the watermelon (*Citrullus lanutus*). In South Africa and Botswana, a local bitter variety of the watermelon, referred to as the bitter melon (*Citrullus rehmnii*), is popular. This is a wild relative of the watermelon and an important source of water and food to some of the Kalahari communities.

Also important in the region are the Brassicas. Although all are introduced to the area, they are part of the traditional AIV eating culture. Among the most important in the region are *Brassica rapa* (Perkinensis group) and *Brassica rapa* (Chinensis group) (Toxopeus and Baas, 2004). The use of wild AIVs is also unusually high in Southern Africa (see Chapter 7). An important wild AIV that is not popular in the other regions is the blackjack (*Bidens pilosa*). The use of wild AIVs, especially in their raw form, is very common among communities living in the Kalahari and neighbouring areas. The local Koi and San people have a deep knowledge about such plants in their environment.

The culture of drying AIVs is well established in Southern Africa, probably due to the long dry spell experienced in a mono-modal rain regime. Most markets are full of these dried AIVs. These are dried in the sun either soon after harvesting or after brief boiling, usually in salt (blanching) (see Chapter 5).

A selection of important African indigenous vegetables

Here we present an alphabetical listing and description of important AIV species. This has been developed using country AIV priority lists, where available, marketed species, ranking results of ethnobotanical surveys in specific communities, and our own experience with specific AIVs and regions. It will already have been noticed that certain families have contributed more AIVs than others. The Asteraceae, Solanaceae, Cucurbitaceae and Malvaceae families are most commonly used for leafy vegetables. The Cucurbitaceae and Solanaceae families dominate with respect to fruit AIVs.

Abelmoschus esculentus (L.). Family: Malvaceae. Synonyms: *Hibiscus esculentus* (L.). Arabic: *bamia*; English: okro, okra, lady's finger; French: *gombo, bamie-okra*; Portuguese: *quiabo, bamia* (Amharic). *Abelmoschus esculentus* is an important African fruit vegetable used from Senegal to Ethiopia and south to South Africa. It is particularly popular in West Africa and in Southern Africa. The fruit is used for making slimy soups and sauces. In Central Africa, it is a favourite food from Cameroon to the Democratic Republic of Congo, where the plant (*gombo* in French; *dongodongo* in Lingala) is widely grown in the country, particularly in lower parts of the country (Bas-Congo). It is a favourite of the Bakongo people and the Baluba of Kasai region, who use it with *fufu*. In East Africa, the plant is more commonly used in the western and coastal parts of Kenya and most of Uganda and Tanzania (Bamia). It is grown on a small scale by farmers in Ethiopia in home gardens. Sliced dried okra fruit (Nhaguguzire in central Mozambique) is common in Southern African markets.

The leaves are much less important as a vegetable than the fruit, although their use is widespread in most regions. The Mijikenda of coastal Kenya, communities in southern Malawi and central Mozambique, and some communities in West African countries, including Benin, Ghana and Senegal, use the leaves, which occasionally are sold in local markets. A great determinant of local regional variation in its use is cultural preferences rather than climatic factors. Some communities like the mucilage in the okra while others do not. In South Africa, while many people in the north like the mucilaginous texture of the vegetable, people in the south dislike it (Vorster et al, 2002; Jansen van Rensburg et al, 2007).

Abelmoschus is a genus of the Old World and includes up to 15 species mainly in Africa and Asia; but okra is the most widely used, being grown in the tropics and subtropics of the world, an indication of its high level of adaptability. In Africa, *Abelmoschus caillei*, the West African okra, is also used. It has a much limited range of cultivation. It is commonly seen in cultivation in West Africa, but also in East and Southern Africa.

Adansonia digitata (L.). Family: Bombacaceae. Synonyms: baobab, *calebassier du Sénégal*. The African baobab grows in low, hot sub-humid to dry savannahs of Africa, from Senegal to Eritrea and Ethiopia's Tigray region, and from Somalia to southern Mozambique and the Limpopo Province of South Africa. It is plentiful in central Tanzania, in the south-west dry parts of Malawi, Mozambique near Tete, south-west Zambia, and Angola north to the mouth of the Congo. The species is not recorded in the humid forests of the Congo Basin and West Africa. The fruit, which is high in vitamin C, is the most widely used resource; but the leaves are used as a vegetable in most places where the plant grows. The pro-vitamin A-rich leaves are particularly important in the Sahelian region of West Africa, where they are picked in the short period when the plant is in leaf, sun-dried and pounded into powder, which is used as part of the daily sauce in the long dry season. In Benin, the leaves are used to produce a slimy soup eaten with starchy food (sorghum, maize or yam-made *foufou*) as a substitute to okra. It is a good thickener. Increasingly, there is more interest in growing baobab trees, particularly in Senegal and Mali. The flowers as well as the seed are also used in the Sahel, particularly in parts of Burkina Faso where the plant is called *koun-nou*. Throughout its range, the species is extremely variable. Morphological traits of the capsules and their yield have been recently characterized in Benin (Assogbadjo et al, 2005). The African baobab is one of eight recognized in the genus, six of which are in Madagascar alone and one in North-West Australia. The Madagascan baobabs are relatively less important, perhaps with the exception of the Malagasy baobab, *Adansonia grandidieri*, a huge tree with a cylindrical bole whose seeds are a source of oil for the locals.

***Amaranthus* spp.** Family: Amaranthaceae. Amaranths are among the most important leafy vegetables in the continent, a fact attributed to their hardiness and, hence, wide occurrence, low susceptibility to disease, low pest incidence, low labour input (making them easy to cultivate), ease in cooking and high nutritional value. Close to 20 species are known on the continent, and most are used as a vegetable, although to varying degrees. Many of the more important species are introduced, but are part of the local vegetable-eating tradition. The more important species include *Amaranthus cruentus* (synonym *A. hybridus* subsp. *cruentus* (L.)), which is widely distributed and used throughout the continent; *A. cruentus*, which is cultivated widely in humid West Africa, mostly in urban and peri-urban areas, and sold fresh in all city markets; *A. dubius*, which is important in East and Southern Africa, occasionally with large leaf forms; *A. lividus* (*A. blitum*), commonly used from northern Tanzania, western Kenya and Uganda to Cameroon; and *A. hypochondriacus*, which, together with *A. cruentus*, constituted most of the local seed amaranths. The above species are wild as well as cultivated.

Other less important species that are not normally cultivated include the spiny amaranth (*Amaranthus spinosus*), which as the name suggests usually has spines on its stem. It is commonly used in parts of East and Southern

Africa. *A. caudatus*, with a pendulous inflorescence, is used from the wild in humid West Africa. It is well known for its use as an ornamental and has forms that are used as a vegetable. The species has been reported as being used in Ethiopia. *Amaranthus viridis* is a very cosmopolitan and relatively small amaranth, usually common in urban areas. It is frequently used in both humid and Sahelian West Africa. With the exception of possibly *A. lividus*, all of these species are not native to Africa.

Many other less important species are consumed on the continent. *Amaranthus graecizans*, one of the indigenous vegetables, is a small amaranth, erect or prostrate, hardly 30cm high, very variable and widely distributed throughout the continent. It is important in the Sahel and other dry parts, from Senegal to the northern part of Kenya and south to Southern Africa. It quickly shoots up after rains, grows fast, produces seeds and dies off – hence, it is handy when other vegetables are in low supply (Maundu and Grubben, 2004). Appearing similar in habit is *Amaranthus thunbergii*, which is mainly used in Southern Africa. A rare indigenous amaranth is *A. sparganiocephalus*, associated with pastoralist animal kraals, from Tanzania north to Ethiopia and Somali.

Amaranthus is a relatively difficult group to master due to its cosmopolitan nature and similarity between species. Some scholars prefer to put what is now known locally as *A. cruentus* and *A. hybridus* (and even some forms of *A. hypochondriacus*) under one species, *A. hybridus*, with ssp. cruentus and ssp. hybridus.

Balanites aegyptiaca (L.). Del. Family: Balanitaceae. The desert date is a widely distributed thorny tree, occurring in most of Africa's sub-humid and dry savannahs – from Senegal east to Somalia, and south to Zimbabwe and Angola. It does not grow in desert and compared to other *Balanites* species, it is not as hardy as the English name suggests. The fruit pulp, the cooked seed and leaves are used for food and are particularly important in the Sahelian region through to Sudan, northern Uganda, northern Kenya, Djibouti, Somalia and dry parts of Ethiopia. The species particularly comes in handy at times of food shortage. In West Africa, the leaves are used a great deal by communities in the northern part of Ghana and across in Burkina Faso. The ripe fruit pulp is edible and the seeds are cooked and eaten. *Balanites aegyptiaca* is the most commonly used member of the genus. Also found in the dryer zones are *B. pedicellaris* and *B. rotundifoli*, both common in the Sahel. Both species have edible seeds that have to be cooked for long periods with constant removal of water.

Basella alba (L.). Family: Basellaceae. Synonyms: *B. lucida* (L.), *B. rubra* (L.), *B. cordifolia* (L.), ceylon spinach, Indian spinach, *baselle*. *Basella alba* is an evergreen creeping plant with heart-shaped leaves that is widely used as a vegetable. It is sometimes planted on hedges or staked near homesteads where the leaves are accessible; but it is mainly picked from the wild. It is commonly used from West Kenya (where it is called *nderema*), through Uganda to the Democratic Republic of Congo (*baselle* or *epinard*) and West Africa, where it is occasionally cultivated. The species is found in two forms, a green form (*alba*), which is the more widely distributed, and a red type (*rubra*), which is more commonly used as an ornamental. In the Democratic Republic of Congo, both types are grown. The crop does better during the dry weather and is especially suited to the eastern part of the country (Okulungu, 2003).

Bidens pilosa (L.) Family: Asteraceae. *Bidens pilosa* is a cosmopolitan herb, widely distributed in Africa as a weed of cultivation and disturbed areas. It is eaten throughout its range in Africa, rather sparingly, but is one of the most important vegetables in Southern Africa, particularly in Malawi (where it is known as *chisoso*) and Mozambique (*nhungunila*). In these countries it is a vegetable of commerce, sold in both dry and fresh forms; but it is not cultivated. In West Africa; the use of the species as a vegetable is not widespread. However, many records exist about the medicinal utilization of *B. pilosa* in Benin, Côte d'Ivoire and elsewhere in West Africa. The related species *B. biternata* is also used as a vegetable. It is distinguished by its many leaflets (five to seven) compared to one to three of *B. pilosa* (usually three).

Blighia sapida König. Family: Sapindaceae. Synonyms: *pommier finsam*, *blighia savoureuse*, *ackee*. *Blighia sapida* is a tree species native to West Africa where the tree is found wild as well as domesticated. *Blighia sapida* is also present in Central Africa and has been introduced elsewhere in Asia and tropical America. Other species of the genus *Blighia*, both from tropical Africa, are *B. unijugata*, which has edible leaves, and *B. welwitschii*, which has medicinal uses. The most used part of the tree is the aril. The edible aril of *B. sapida* is eaten cooked, but must be mature, or fresh, and harvested when the fruit opens naturally. Immature arils, the outer rind of the fruit, the pink membrane under the seeds and the seeds contain hypoglycins, which are toxic and can cause death (Burkill, 2000). When harvested in good time and prepared correctly, the arils are delicious and safe to eat. In Benin and Nigeria, Yoruba communities prepare the aril of *B. sapida* together with a leafy vegetable such as *Solanum macrocarpon*.

Bombax costatum Pellegr. and Vuillet. Family: Bombacaceae. Synonyms: *Bombax andrieui*, *kapokier à fleurs rouges*. *Bombax costatum* is a native of Africa and is distributed from Senegal across Ghana and Nigeria to southern Chad and Central Africa. Its tuberous roots act as water and/or sugar storage facilities during long drought periods. It is a tree species that flowers after leaf fall in November to February. The calyxes of the flowers are used to prepare sauces and the young fruit is cut, dried and used for the preparation of meals. Leaves are highly digestible and are eaten by livestock, but are a delicacy in Mali. The bark is used for the treatment of skin diseases, yellow fever and headache, and leaves and immature fruit are used as an emollient. Various parts are used for fever or to promote lactation and as a tonic for fatigue.

Brassica carinata A. Braun. Family: Brassicaceae. Synonyms: Ethiopian kale/mustard. *Brassica carinata* is a kale, often distinguished from others by its purple tinged stem and thinner softer leaves. Its origin is believed to be the Ethiopian Highlands where the plant *Yabesha gomen* (Amharic) is grown in home gardens but is also found wild in the fields. The leaves are widely used in Ethiopia especially during fast days when meat is prohibited. The plant is also traditionally grown in western Kenya where it is known as *kanzira* or *kandhira*, as well as occasionally being found wild. Also typical of this region is *Brassica nigra* (senafitch in Amharic), whose leaves, too, are used as a vegetable. Like the kale, this plant is grown in home gardens and also occurs as a weed in north-west Ethiopia (*shewa*, *harar*, *gonder*). It is also widely used as a spice. The assumed origin is the Horn of Africa (Demissie, 1991). *Brassica carinata* now has widespread use, particularly in East and Southern Africa where it has been a subject of promotion. Large-leaved forms are now available.

Brassica rapa (L.) ssp. *chinensis*. Family: Brassicaceae. Chinese cabbage is one of the traditional vegetables of Southern Africa, particularly in Malawi, Zimbabwe, Mozambique and South Africa, where it is known by the names *Chinisi, repu*, (Malawi), *Sjinese kool* in Afrikaans and *mutshaina* in Tshivenda and other local African languages (Jansen van Rensburg et al, 2007). The related type, *Brassica rapa* (Perkinensis group), commonly known as *pok choy* or *bok choy*, is more commercialized. Another related species *B. napus*, or rape kale, is very popular in the Limpopo and Mpumalanga provinces of South Africa and Zimbabwe, where it is known as *murhodisia* (Schippers, 2000; Toxopeus and Mvere, 2004; Jansen van Rensburg et al, 2007). It is also used in neighbouring Malawi and Mozambique.

Cajanus cajan (L.) Millsp. Family: Fabaceae. Synonyms: pigeon pea, *pois cajan*, *pois d'Angole* and *Yewof ater* (Amharic). *Cajanus cajan* is widely grown in most of Africa from humid to semi-arid zones. The seeds are an important pulse, made into stews when green or dry and also into a host of other preparations. Pigeon peas are particularly important in East Africa where they are commonly mixed and cooked with polished maize (without testa) to make a preparation known as *muthokoi*.

Celosia argentea (L.). Family: Amaranthaceae. Synonyms: *célosie*, *cockscomb*, *soko* (Yoruba). Lagos spinach is a beautiful herb, usually about 1m high with greenish red mottled leaves and a flower head like a cock's comb, hence its name. Outside Africa, it is used mainly as an ornamental. It is commonly cultivated in humid Central and West Africa, particularly from Nigeria to Ghana. The species is generally not well known in East Africa, although it is used in northern Tanzania and recently introduced to the coastal part of Kenya. This species is in the same family as the amaranth and is used basically like leafy amaranth in soups and stews, along with other ingredients. Like amaranth, the species is hardy, with little problems of disease, and so it is easy to cultivate but requires a relatively high humidity. Just like most amaranths, it has the potential to become a weed. The related species and probably the better known is *Celosia trigyna*, a small plant, mainly of semi-arid and sub-humid areas and commonly used as a vegetable throughout Africa. The species accumulates raphides (needle-shaped crystals of calcium carbonate or calcium oxalate found in specialized plant cells called idioblasts) on its leaves (Abbiw, 1990).

Ceratotheca sesamoides. Family: Pedaliaceae. *Ceratotheca sesamoides* is a herbaceous plant in the sesame family, widely distributed in the continent mainly in semi-humid to semi-arid land. It is used throughout the Sahel and is particularly important in Senegal (*yoroxlan*), northern Ghana, Benin and Burkina Faso, Chad and west Sudan. It is also used in East Africa south to Malawi. In Tanzania and Malawi, the plant is often picked, dried and preserved for use in the dry season. The related genus, *Sesamum*, where sesame (*Sesamum orientale* or its synonym *S. indicum*) belongs has important vegetable species as well. *Sesamum angustifolium* (Oliver) Engl., very much like *Ceratotheca*, is used in East Africa. Its use is particularly common in western (*onyulo*) and coastal Kenya (*mlenda*), Tanzania and Malawi (*chewe*). Leaves have a strong odour and therefore are often mixed with others when cooking. Sesame leaves are also used as a vegetable in parts of Africa, including coastal Kenya and Tanzania. *Sesamum alatum* Thonn. is also widely used as a vegetable particularly in the Sahel (i.e. Senegal to Chad and Sudan). It also occurs in East and Southern Africa.

Citrullus lanatus (Thunb.) Matsum. & Nakai. Family: Cucurbitaceae. *Citrullus lanatus* is well known for its fruit: the watermelon. The species is of African origin and there still exist many landraces and other primitive types in various parts of the continent. It is important in both humid and Sahelian West Africa, both as a fruit and source of seed (*egusi*) used in sauces. Known subspecies are subsp. *lanatus*, subsp. *vulgaris* and subsp. *mucosospermus*. There are cultivars of which the only edible portion is the seeds, as the pulp of the fruit in these cultivars is usually too bitter for human consumption. In West Africa these cultivars are generally known as *egusi* (Burkill, 1985; Schippers, 2004; Achigan-Dako et al, 2006), a term which in Yoruba language refers to a wider group of cucurbit species that produces protein and oil-rich seeds (Burkill, 1985). The group consists mainly of species such as C. *lanatus* subsp. *mucosospermus* Fursa, known as *egusi* watermelon, and *Cucumeropsis edulis* (Hook. f.) Cogn., considered as the true *egusi* (Burkill, 1985).

In the literature, *C. lanatus* has often been confused with its wild relative *C. colocynthis* (L.) Schrad., which occurs distributed in the Sahelian region and the western Kalahari Desert. *Egusi* forms of *Citrullus lanatus* are widely cultivated from Nigeria to Côte d'Ivoire, where the crop has a very substantial market prospect. *Citrullus lanatus* subsp. *vulgaris* is mainly used as fruit, although the seed can also be used as a snack. Bitter, more primitive, forms are widely distributed in Southern Africa, particularly in and around the Kalahari

Desert – hence their common name, Kalahari watermelon or bittermelons. Primitive forms are also found in the rest of the continent, including northern Kenya, Sudan and the Sahel. In the Kalahari, the wild watermelon fruit (*tsamma* in Khoisan) is an important source of water. The *tsamma* are a bit smaller than cultivated melons. The seeds are a delicacy. Larger-fruited cultivated forms are cultivated by farmers and are known by the Tswana name *makataan*. The leaves and young fruits are used as a vegetable while the outer soft shell of the mature fruit is also cooked (van Wyk and Gericke, 2000).

Cleome gynandra (L.). Family: Capparaceae. Synonym: *Gynandropsis gynandra* (L.) Briq. *Cleome gynandra*, also commonly known by the queer names of spider plant, spider flower and cat's whiskers, is an Old World species. It is a herbaceous vegetable, rarely exceeding 1m high. *C. gynandra* is an important leafy vegetable throughout sub-Saharan Africa, being used in all the four regions. Increasingly, it is now particularly cultivated in East Africa; but the bulk of it is still picked from the wild. The spider plant is particularly important to Nilotic-speaking communities who live in and around the Nile Basin in Kenya, Uganda and Sudan. It is grown by those communities around homesteads in rural areas and urban centres. It is one of oldest leafy vegetables in many countries of West Africa (e.g. Benin and Mali), and among the Iteso community of Kenya and Uganda (*ecadoi*) it was the vegetable served to chiefs and important people in the olden days. It may be planted but usually grows spontaneously, although managed and harvested on a regular basis. Both leaves and flowers are used as a vegetable. It is commonly dried in the sun and preserved. In Ghana, the plant is also used as a medicine. It is one of the most popular vegetables in Malawi and Mozambique where it is known as *luni*. Here it is usually harvested, sun dried and used in the long dry seasons. Spider flower is known as *oorpeultjie* in Afrikaans; *lude* in Ndebele; *amazonde* in Zulu; *lerotho* in Sepedi and Sotho; *murudi* in Tshivenda; and *rirhudzu*, in Xitsonga (Jansen van Rensburg et al, 2007). The plant is quite variable throughout its range. The most noticeable difference is the colour of the stem, flower and fruit, which may vary from green to purple. It is a slightly bitter to very bitter vegetable; but its bitterness is appreciated by many. It is normally cooked with other less bitter vegetables such as amaranths. In the Central African country of the Democratic Republic of Congo, spider plant is known as *bilolo* in Lingala from its bitter taste, and it is widely grown in the region.

Other relatively less important *Cleome* species include *C. monophylla*, *C. rutidosperma* and *C. hirta*, which are also widely distributed but used only sparingly as a vegetable. The leaves are much smaller and time consuming to harvest (Schippers et al, 2002).

Coccinia grandis (L.) Voigt. Family: Cucurbitaceae. Synonym: ivy gourd. This is a widely distributed species found in sub-humid to semi-arid areas from Senegal to Somalia and most of Eastern Africa. The ripe bright red fruits are edible. This is the species grown widely in Asia, particularly India, for its fruits which are used as a vegetable. It is also cultivated in parts of East and Southern Africa for the Asians living in large cities in these regions.

Colocasia esculenta (L.) Schott. Family: Araceae. Taro, also known as cocoyam, is widely grown in sub-Saharan Africa for its starchy corms (underground stems) from humid to semi-arid zones, but restricted to wet habitats such as along watercourses and marshlands. It is frequently used in humid West Africa and in Ghana, Benin and Côte d'Ivoire, where it has names such as *eddoes*. The Akans appreciate it. The leaves are used as a vegetable, but rather sparingly. Raphides occur in the leaves. The species is very variable as seen from the colour of the corms and the leaf stems. *Colocasia* is related to *Xanthosoma* and the two even frequently share the local names taro and cocoyam. Here, these two terms have been reserved for *Colocasia*, which has been known for a longer period in Africa, and tania, yautia or malanga for *Xanthosoma*. *Colocasia*, which originates from the South-East Asian and Pacific region, reached Africa well before *Xanthosoma*. In West Africa, however, *Xanthosoma* has been gaining importance as it can more easily substitute for yam in the preparation of *foufou* and can be grown away from water. In most other preparations, the tubers are prepared in a similar manner. In Eastern Africa, taro is still the most important crop. *Xanthosoma*, which was introduced through West Africa from the tropical Americas, is unknown to many people in the region. *Xanthosoma* has, however, taken root in the highlands of northern Tanzania and parts of Southern Africa. Both tania and cocoyams are popular breakfast foods. *Colocasia* and *Xanthosoma* are easily distinguished by the way in which their leaf stalk attaches onto the leaf blade. For *Colocasia*, the point of attachment is inside the blade. They can also be distinguished by their corms.

Corchorus olitorius (L.). Family: Tiliaceae. Jute, or Jew's mallow, is used all over Africa. It is an erect annual that is usually not more than 1m high, although certain varieties can get up to 2m or more. In many cases, it is spontaneous; but increasingly it is being planted. It does best in low, warm to hot humid areas with well-drained soils. As with most African vegetables, it is popular with some communities and not with others. Leaves, including young shoot tips, are an important vegetable, which is mucilaginous, a quality appreciated by many communities but not by others. The use of *Corchorus* is most rooted in the cultures of communities living in the Nile Basin, from Lake Victoria Basin north to Egypt, where it is a prized vegetable and known as *molokiya*. The species is popular in Benin and Togo where the Mina and Ewe communities call it *ademe*. In Mali the crop is a delicacy and is eaten with rice. The Iteso community of north-eastern Uganda is particularly fond of these vegetables. *Atigo* is the general term they give to slimy vegetables. In northern Uganda and across in southern Sudan, *C. olitorius* occurs in two forms: the normal type with long regularly toothed leaves and another with lobed leaves that are deeply and irregularly toothed at the edges, forming numerous whisker-like outgrowths. Among the Teso, the former is called *atigo apaulina* (or *atigo apio*); the other is called *atigo ajaye* (*ajaye* means leaves are like those of cannabis). Tall forms of the same species (jute), over 2.5m high, are a source of fibre mainly used for gunny bags and rope in Asia. The genus *Corchorus* in Africa is represented by over a dozen species; but most important is *C. olitorius*. Other important ones include the wild relatives, *C. tridens*, which is tolerated in crop fields, *C. trilocularis* and *C. aestuans*. The fruits or capsules of *C. olitorius* are stout, with clear ridges, usually splitting into five parts. The other species are usually split into three parts (rarely four). *C. tridens* is very widespread in Africa, from humid to semi-arid areas. The plant has a more spreading form, usually not more than 0.5m high, with relatively thin leaves. Capsules, split into three parts, are thin, faintly ridged and end in three short slightly spreading protrusions. *C. trilocularis* is more common in East and Southern Africa and can attain a height of close to 1m. It has relatively long and thin capsules with a pointed tip and is split into three to four parts. Leaves are long, to 10cm or more, but are thinner than those of *C. olitorius*. *Corchorus aestuans* is also widely distributed in most of Africa in humid and sub-humid to semi-arid zones. It is distinguished by its spreading nature and plump, deeply ridged capsules with three to four wide-spreading protrusions at the tip. It is particularly useful in humid West Africa. These species are mainly found in grasslands and cultivated areas as weeds. Leaves are cooked fresh or dried and are appreciated a great deal. Drying of vegetables is common in Southern Africa due to the long seasons without fresh vegetables, but also performed in West Africa, especially in Cameroon. Dried vegetables are common in the market throughout the year in Southern Africa.

Cucumeropsis mannii Naud. Family: Cucurbitaceae. Synonym: *Cucumeropsis edulis* (Hook. f.) Cogn. *Cucumeropsis mannii* is said to be the true *egusi* (Burkill, 1985). The species is native to West Africa, where it has similar uses as *Citrullus lanatus* subsp. *mucosospermus*. In addition to the use of the seeds, young leaves of *Cucumeropsis mannii* are used as leafy vegetable. This has been reported in the south-west in the Adja community and in the Bariba community of Benin.

The production of *Cucumeropsis mannii* is strongly declining. *Cucumeropsis mannii* used to be an important vegetable in West and Central Africa at a time when there was plenty of forest to practise shifting cultivation. In Benin today, *Cucumeropsis mannii* is rarely cultivated in monocultures for several reasons. First, *Cucumeropsis mannii* is a climbing species and requires stakes for growing. Farmers grow it in association with yam, which also needs staking. Both crops can then benefit from the support of the same stakes and farmers save on resources and time for stake searching. However, farmers also sow *Cucumeropsis* near dead trees at the edge of a garden. Second, the cropping cycle of this species covers seven to eight months, and in some regions, mainly in northern Benin, farmers use complementary watering if necessary. In addition, *Cucumeropsis mannii*, known as *egusi tchigan* (the prestigious *egusi*), is used as a gift in recognition of the social status of the giver. This behaviour is related to friendship and is also an indicator of good harvest (Achigan-Dako et al, 2008).

Cucumis metuliferus E. Mey. Ex Naudin. Family: Cucurbitaceae. Synonyms: horned cucumber, horned gourd. This is a creeping or climbing plant naturally found from West to Southern Africa. It bears uniquely magnificent fruits, resembling cucumbers but with stout protrusions scattered on its surface. It is light green, ripening to yellow or orange. It is common in markets in Southern Africa, especially Zambia, Zimbabwe, Malawi, Mozambique and South Africa. It is also sold in parts of Tanzania and Uganda, but is generally unimportant in East Africa. The fruit has long-keeping qualities. In West Africa it is mostly used for medicinal purposes.

***Cucurbita* spp.** Family: Cucurbitaceae. Among the most widely used introduced vegetable species are the pumpkins or squashes (*C. pepo* (L.), *C. maxima* Duschesne and *C. moschata* Duschesne), used for fruit and leaves. Pumpkin leaves are sold in their fresh form throughout Africa, but more so in Southern Africa where they are also sold in the dried form, often with their flowers. The three species of pumpkins are widely grown; but the former two are most commonly used for vegetable. Pumpkins are among the most important vegetables in sub-Saharan Africa as they provide fruit and leaves, both of which are heavily utilized. A fourth species, *Cucurbita ficifolia* Bouché (fig-leaved gourd or Malabar gourd), also known as *kahurura* in Kenya, is not a pumpkin and its fruits are not used in East Africa; but it is grown for its leaves in Central Kenya. All four species are of American origin. *C. moschata* is represented by the butternut squash, which is popular in Southern Africa as well as a few other forms of pumpkins. Butternuts are relatively recent in East Africa, but are apparently well known in Southern Africa. *C. pepo* is distinguished by its short trails and is represented by zucchini, gem squash and some pumpkins. It is important in most of West Africa, especially Niger, Burkina Faso, Ghana and Mali. *Cucurbita maxima* leaves (*feuilles de courges* in French) are grown all over the Central African region, although the fruit is preferred. The use as a leafy vegetable is popular among the Baluba of central Democratic Republic of Congo. With an exception of perhaps *C. pepo*, all of the others may, in some cases, be grown for the leaf vegetable. In Southern Africa, pumpkin leaves are extremely important. The leaves are eaten fresh or dried. When picking the leaves, farmers also pick the flowers and sometimes the young fruits as well. It is common to see pumpkin leaves being sold mixed with the flowers. The flowers are both a food and a decoration.

Ensete ventricosum – (Welw.) Cheesman. Family Musaceae. *Enset*, in Amharic, is the false banana or wild banana. It is an important food for inhabitants of the southern central highlands of Ethiopia. The banana-like plant is widely distributed in sub-Saharan Africa, from Nigeria and Cameroon, east to

Ethiopia and south to South Africa. It is found naturally along wet valleys and ravines; but in cultivation as a crop, it is only found in Ethiopia, where it is grown in moist highlands to 3000m. The stem is a mass of overlapping leaves – hence the use of the term pseudo-stem. While the main starchy food comes from the trunk and the underground parts (corm), young soft leaf stalks and the soft pith are boiled like cabbage. The cultivated *enset* is extremely variable and farmers can recognize a number of farmer varieties.

Gnetum africanum Welw. Family: Gnetaceae. Synonyms: *fumbwa* in Congo and in Lingala; *eru* in Cameroon; *koko* in Nigeria. The family Gnetaceae in Africa is represented by two members: *Gnetum africanum* and *G. buchholzianum*. These climbers are found in humid forests, where locals harvest them both for home use as a vegetable and for trade, including export (Lowe, 1984; Laurie and Sunderland, 2004). Leaf products are taken out of Cameroon by air and sea in thousands of tonnes a year. Most *Gnetum* from Cameroon, Gabon and the Central African Republic is transported by road to Atlantic ports mainly to the port of Idenau, near the Nigerian border, and is then exported. A recent survey recording *eru* consignments passing through the South-West Province of Cameroon revealed that approximately 3600 tonnes are shipped annually to Nigeria and exported to European countries, as well as to the US (Asaha et al, 2000; Nkefor, 2000; Ndam et al, 2001) for use by the Central and West African diaspora.

The two species are found from Nigeria through Cameroon to the Congo Basin forest. *Gnetum* is known as *eru* in Cameroon and parts of Nigeria, and *koko* in other parts of Nigeria and the Central African Republic. *Eru* leaves are a typical non-timber forest product. The Bakongo people of west Democratic Republic of Congo like this vegetable and have influenced other communities living in the capital, Kinshasa, to like it too.

The family Gnetaceae is a gymnosperm (ancient group of vascular plants bearing seeds not by flowers but spores – for example, pines) and therefore needs special conditions (humid forests) for reproduction; as a result, cultivation by farmers has been difficult. *Gnetum* species are known to take long periods to germinate. Most trials to germinate seeds under nursery conditions have not been successful, and seed germination has not been pursued in propagation studies (Okafor, 1997; Shiembo, 1999; Ndam et al, 2002). There are reports of farmers transplanting wild *Gnetum* seedlings from the forest and forest clearings, farmers leaving *Gnetum* in farms that have been newly cleared from the forest, and even the collection of seeds for large plantations (Fondoun and Tiki-Manga, 2000). Cuttings from the leafy vines have proved to be the most successful means of vegetative propagation and a simple methodology has now been developed. The species are good examples of overexploited species as domestication is yet to pick up.

Hibiscus sabdariffa (L.). Family: Malvaceae. *Bissap* (Senegal), or roselle, is an important vegetable in the Sahel. In Senegal, this is the leafy vegetable most commonly and widely eaten by many communities. Some sour types are used for flavouring food such as rice. It is cultivated for both vegetable and calyxes. The calyxes are dried and used to prepare a type of drink popular in most of Sahel and North Africa. The species is important in northern communities of Uganda (where it is known as *malakwang*) and in south Sudan. Variability in this species is seen in leaf shapes (shallowly or deeply lobed, number of lobes, etc.) and leaf colour, which may range from green to reddish-purple. In the Democratic Republic of Congo, *Hibuscus sabdariffa* is known as *ngaingai* (in Lingala) and is a common vegetable all over the region, both in cultivation and also on wasteland. It is a favourite vegetable of the Baluba people of central Democratic Republic of Congo. It is a sour vegetable and, hence, is used with fish and also as a local medicine for various ailments. Several other species of *Hibiscus* are also important. *Hibiscus cannabinus* (L.) (kenaf) is cultivated and grows wild in humid West Africa and also in Uganda (*egwanyira* in Iteso). *Hibiscus asper* (Hook. f.), another wild relative, is used for its sour leaves.

Ipomoea batatas (L.) Lam. Family: Convolvulaceae. Sweet potatoes tubers are an important food in most parts of Africa; but the vegetables are important only among specific community groups and regions. Leaves are widely consumed in coastal Tanzania south to Malawi (*kholowa*), Mozambique (*domutomu*) and South Africa. Here, special varieties with thin leaflets and occasionally without tubers, depending upon the variety, have been developed by farmers. Sweet potato leaves are also used in several other countries in Southern Africa and parts of West Africa where they are popular among the Joola of Senegal (*pataas*) and some communities in Ghana. Sweet potato leaves are, however, not important in Uganda or Kenya where the crop is grown for its tubers only. In the Democratic Republic of Congo (where it is known by names such as *patate douce* in French, *mbala* in Lingala, *viazi* in Kiswahili and *tamba* in Kitetela), the crop is widely grown mainly for its tuber; but its leaves are also important and it is a staple species for some communities. It is mainly grown in cleared forest areas and in the savannah. A related, but indigenous, species is *Ipomoea aquatica*, a creeper with beautiful purple flowers found in marshlands. It is used as a vegetable throughout sub-Saharan Africa. In the same family of sweet potato is another creeper, *Jacquemontia tamnifolia*, with heart-shaped leaves and small blue to purple flowers resembling those of the morning glory. It is a cosmopolitan weed, just like morning glory (*Ipomoea violacea*). It is common in cultivated and fallow land. It is used as a vegetable in various parts of the continent, particularly coastal Kenya and also in Senegal (*mefer*).

Irvingia gabonensis (Aubry-Lecomte ex O'Rorke) Baill. Family: Irvingiaceae. Known as wild mango, *mango sauvage* and *pomme sauvage*, this species could grow up to 40m high. It is distributed throughout tropical Africa from Senegal to Angola. In West Africa, the plant is considered one of the most important trees of the wild. The fruit of *Irvingia gabonensis* has an edible pulp, slimy or thick depending upon the variety. The main use of *I. gabonensis* as a vegetable is the seed kernel. The edible kernels are used for culinary purposes and are traded widely enough to be quoted on the weekly commodity lists in West Africa. In Nigeria, it is known as *ogbono* and is used as a soup thickener. The fruit is also used to make dika cake in Central Africa, particularly in Gabon. Another species in the genus is *I. smithii* (Hook. f.), which has similar uses as *I. gabonensis*.

Lagenaria siceraria (Molina) Standl. Family: Cucurbitaceae. The calabash, or bottle gourd, is indigenous to Africa. Both edible and inedible bitter fruit forms occur. Edible types are associated with some community groups and not with others. They may have a smooth or rough surface, round or elongated and so on. These species are important in most of West Africa and among some communities in East and Southern Africa. Young and tender fruit of *Lagenaria siceraria* are sliced and prepared in soup, fried or used in stew. They may also be mixed with some other leafy vegetables. In West Africa (Nigeria, Benin, Togo, Ghana and Côte d'Ivoire), there is a particular variety known as *aklamkpa* in Benin, *sêrêgbe* in Côte d'Ivoire, or *apentra kesa* in Ghana, which has a soft seed coat and which is used as *egusi*. The seed of this variety is used to cook meatballs in soup or to thicken tomato soup, or may be prepared together with leafy vegetables.

Launaea taraxacifolia (Willd.) Amin ex. C. Jeffrey. Family: Asteraceae. African lettuce, or wild lettuce, occurs from Senegal to the Ethiopian Highlands. It has been domesticated in Nigeria and cultivated in home gardens in Benin and Senegal. The leaves are eaten fresh or cooked in soup in Benin and Nigeria where Yoruba communities call it *efo yanrin*. Many therapeutic uses of *Launaea taraxacifolia* have been reported in Benin and Ghana. *L. cornuta* is a related species found mainly from Central to Eastern Africa. It is probably the most important vegetable in coastal Kenya and Tanzania. Communities here cook and eat it mainly with other vegetables such as amaranth, pumpkin or cowpea leaves. *L. cornuta* is also used for medicinal purposes.

Leptadenia hastata (Pers.) Decne. Family: Asclepiadaceae. This is a climbing plant often found on bushes and hedges in dry areas. It is found in the Sahel and savannah from Senegal east to Sudan and northern Kenya. The leaves are used as a vegetable within its distribution range. It is known as *cahatt* in Wolof and *ekamong'o* among the Turkana of Kenya. It is an important leafy vegetable among the Gourounsi of Burkina Faso.

Luffa cylindrica (L.) M. J. Roem. (*Luffa aegyptiaca* Mill.). Family: Cucurbitaceae. *Luffa*, or sponge gourd, is the extensive climber often seen on hedges and it is the source of the fibrous sponge used for scrubbing utensils and bathing. Immature fruits and young leaves are used as a vegetable in West, East and Southern Africa. It is used especially in Ghana, Ethiopia, coastal Kenya, Malawi and Mozambique. In West Africa, it grows mostly wild and its utilization as a vegetable is limited. People prefer *Luffa acutangula* (L.) Roxb., which is used in dry areas such as northern Benin, Ghana, Burkina Faso, Mali and Senegal. In Burkina Faso, the Moré people call it *napokumaana* (the queen's okra). The plant is common in home gardens in the drier parts of West Africa.

Manihot esculenta Crantz. Family: Euphorbiaceae. The cassava plant, originally from Brazil in South America, is very important in most of sub-Saharan Africa. In French-speaking countries, it bears the names *manioc*; elsewhere it is known as *nzete ya pondu* (the plant in Lingala), *pondu* (the leaves in Lingala), *matamba* (the leaves in Tshiluba) and *dzese* (the leaves in Kitetela). The tuber is named *songo* (Lingala), *muwoko* (Swahili) and *mange* (Kitetela). The leaves are the most important vegetable in the Congo Basin, where they are known by the name of *pondu* (Lingala) (Okulungu, 2003). They are prepared in a rather complex way involving pounding and use of traditional salt obtained from ash of the palm tree inflorescences, plantain peelings or rice inflorescences. Cassava root leads all other starchy food in importance in the Democratic Republic of Congo. Cassava roots and leaves form the staple food for many communities in the country. Dry cassava leaves are made by pounding leaves and letting them dry in an open area for about six days. The leaves are used, but to a lesser degree, in several other areas, including Tanzania (*kisamvu*), Malawi (*ntapasya*) and Mozambique (*mutchcobwe*), and parts of West Africa including Senegal (*nimbi* in Wolof). In Benin the leaves are used in all communities from south to north. The leaves of the related ornamental species, the tree cassava (*M. glaziovii*), called *mpira* in East and Southern Africa, are also used in Benin. Cassava is important along the coastal areas of East Africa, northern Uganda among the Iteso and among the Luo of Lake Victoria Basin and their relatives in Uganda, the Acholi, Langi and Alur. Dry cassava can be roasted or boiled or eaten fresh.

Momordica balsamina (L.). Family: Cucurbitaceae. Synonym: balsam apple. This is a creeping or climbing plant with narrow stems and bearing small green fruits that turn orange upon ripening. The fruits taper to one end and are covered with lines of short prickles. The plant is widely distributed throughout the tropics of the Old World. In Africa it is found from Senegal to Southern Africa; but its use as a food is highest in Southern Africa where both the leaves and the fruit (*nkaka* or *cacana* in Mozambique; *laloentje* in South Africa) are used as a vegetable and are often sold in markets. The plant is also used in the Sahel – for example, Senegal (*mberboof*) – where the leaves are used as a vegetable. Other related species include *Momordica foetida* Schumach, whose leaves are used as a vegetable in Benin and in East and Southern Africa, particularly in Malawi, Tanzania and Ethiopia. *Momordica charantia* (L.) (Balsam pear, bitter melon and bitter gourd) is also a vine with warty and usually pointed fruits that are eaten only when young and immature before they ripen to an orange colour. It thrives wild in most of Africa and is domesticated in some countries for its fruit, which is both a vegetable and medicine. It is popular with the Asian community and is found in markets throughout sub-Saharan region. Another species of the genus used as a vegetable is *M. cissoids* Planch, which is well appreciated among the Kotafon and Adja communities in south-west Benin.

Moringa oleifera Lam. Family: Moringaceae. *Moringa* is a genus with approximately 12 species, mainly in North-East Africa and West Asia. The most important is *Moringa oleifera* (horse radish or drumstick tree). *Moringa* is a small tree or bush. The leaves and the young pods (drumsticks) are used as a vegetable. *Moringa* leaves are particularly important in Senegal (called *saap saap*), where it a vegetable for most communities. The drumsticks are more commonly used by Asian communities in cities. In northern Kenya and southern Ethiopia in dry climates is another important species, *M. stenopetala* (*aleku* in Amharic). The leaves are used in southern Ethiopia, particularly by the Konso people who have been maintaining it in their fields and the edges of their farms for its leaves, which they use as a vegetable.

Phaseolus vulgaris (L.) (Fabaceae). Bean leaves (*feuilles d'haricot* in French) is a common vegetable among the Baluba and Basonge people of central Congo and also among communities in the Great Lakes region. Young bean leaves are also used extensively in Southern Africa, especially in Malawi (*kwhanya*), Zimbabwe, Mozambique and South Africa, where they are also dried in the sun and stored or sold in local markets.

Plectranthus esculentus N. E. Br. (*Coleus esculentus* (N. E. Br.) G. Taylor). Family: Lamiaceae. The Livingstone potato is a perennial herb with soft, ridged stems commonly found in rocky savannah bushland or woodland. It bears long, lightly coloured finger-like edible tubers from underground stems. These are boiled and eaten; but some communities, especially pastoral groups, may eat them raw. Tubers are used in East and South Africa and are occasionally seen in the market. In Tanzania, the tubers are normally sold when boiled.

Portulaca oleracea (L.). Family: Portulacaceae. Purslane is a fleshy succulent herb, mainly seen creeping on the ground at roadsides and also in cultivated land. It is found in both dry and humid lands in most of Africa. The fleshy shoots, including leaves, are used in northern Kenya, south Sudan, south-west Ethiopia, parts of the Sahel and humid West Africa; but sparingly. The leaves can be eaten cooked or in salads and are occasionally sold in markets. The related species *Portulaca quadrifida* is also used as a vegetable but is a bit difficult to deal with due to its relatively smaller leaves.

Senna tora (L.) Roxb. Synonyms: *Cassia tora* (L.), foetid senna, sicklepod. *Senna tora* is a small sub-shrub or herb widely distributed in the tropics and subtropics as a noxious weed. It is used as a vegetable throughout the Sahel and is popular in Senegal and Mali. Related species are *S. obtussifolia* (L.) H. S.

Irvin and *S. occidentalis* (L.) Link; both are used as a vegetable as well. Within the genus many species are used for medicinal purposes.

Solanum aethiopicum (L.). Family: Solanaceae. Eggplants are a bushy perennial or annual, usually slightly more than 1m high, often with large leaves. The species is a complex of four inter-fertile taxa (groups), now believed to have arisen from a single progenitor, *Solanum anguivi* Lam, but originally treated as many species.

Solanum aethiopicum 'Gilo group' Synonyms: *Solanum gilo* Raddi. (Gilo); white brinjals (Kenya); *nyanya msumaa* (Swahili); *ntula* (Luganda). Many eggplants sold in markets in humid West Africa, Central Africa, Eastern Africa and Southern Africa belong to this group. It is cultivated a great deal in Uganda, Tanzania and the Democratic Republic of Congo. In Uganda, this is known as *entula* (*ntula*) or *njagi* (Rubaihayo, 1997; Rubaihayo et al, 2004). Most important in this group are its large fruits, which may be white or cream with green stripes to red and are eaten raw or cooked. The leaves are hairy and are not used as a vegetable.

Solanum aethiopicum 'Shum group'. This has relatively smaller, nearly round, greenish fruits that turn red upon ripening. The leaves are not hairy and therefore are used as a vegetable, especially in Uganda, where they are known as *nakati*. This group is found from Uganda west to humid West Africa.

Solanum aethiopicum 'Kumba group' (*jaxatu* in Senegal) has large leaves and usually cream to yellow fruits that are more broad than long, with marked longitudinal ridges like those of a pumpkin. It is typical of the Sahelian region and common in markets in Burkina Faso. The fruits as well as the leaves are used as a vegetable. The last group, 'Aculeatum', has inedible but ornamental fruits and prickly leaves (Lester, 1986; Lester and Niakan, 1986).

Solanum leucopersicum. The tomato, although originally from South America, is widely used all over the continent and is an important ingredient in salads and stews. The small-fruited cherry tomato was widely distributed in much of humid, sub-humid and semi-arid Africa long before the modern cultivars were developed. This is attributed to birds that eat the fruit and disperse them as they move. It is therefore considered a traditional tomato in most communities. Although the fruit are the main food from the tomato plant, the leaves are also used as a vegetable by a few communities on the continent (e.g. northern Uganda, central Mozambique and southern Malawi). Another tropical American member of the tomato family is the chilli pepper (mainly *Capsicum annuum* and *C. frutescens*), whose fruit are used as a vegetable and for flavouring foods, depending upon the cultivar. The leaves are also picked by various communities in Africa and are used as a vegetable, mainly mixed with others (e.g. Senegal, Ghana, Uganda, Kenya, Tanzania, Mozambique and Malawi).

Solanum macrocarpon (L.). Family: Solanaceae. Synonyms: eggplants; *gboma* in Benin. This is a bushy perennial, about 1m to 1.5m high, with large non-hairy, shallowly lobed leaves. Fruits are small, the size of eggplants, green to yellow-green, often with a dull purple tinge and ripening to a red colour. The plant is grown for both its leaves and fruits. The young leaves and fruits are harvested and cooked together in East Africa. The species is widely cultivated in East Africa's coastal region and in Central Africa, as well as in humid West Africa, where the leaves and fruits of *gboma* are a common market fresh vegetable. The fruit types are mainly restricted to coastal areas of West Africa.

Solanum melongena (L.) is the common eggplant also known as *brinjal* and aubergine. It is widely cultivated throughout the region from humid to semi-arid areas for its fruit. In its range it is extremely variable, and this can easily be noticed from the fruit's shape. Eggplants are commonly used in stews and are important in humid West Africa.

***Solanum nigrum* complex**. This is a complex of leaf species with at least eight species commonly used as leafy greens in much of sub-Saharan Africa (Edmonds and Chwewa, 1997). Notable members of the complex include the following:

- *Solanum americanum* Mill has relatively thin branches and is easily distinguishable from the rest in its small, usually glossy leaves and small shiny green fruits turning purple-black when ripe. The species is mainly found in warm humid areas, particularly coastal environments. It is found in all zones, but mainly in humid areas.
- *Solanum 'eldorettii'* – this species, whose taxonomic nomenclature is yet to be finalized, is restricted to the Kenyan Highlands and highlands in northern Tanzania. Fruits are small. Mature ones are green unlike in other *Solanums* of this group, which turn purple-black. Mature fruits fall off easily. It is frequently grown by farmers in Kisii District and in the highland parts of the Rift Valley Province of Kenya.
- *Solanum retroflexum* Dunal (*muxe* in Tshivenda) – this species is found across the Sahel to the Horn of Africa and East Africa (Tanzania) to Southern Africa. Its use as a vegetable is more widespread in South Africa. It is grown by farmers in the Limpopo Province of South Africa in their home gardens as well as in irrigation schemes (Jansen van Rensburg et al, 2007).
- *Solanum scabrum* Mill – within the complex, this is the largest species, growing to 1.5m, with large fruit and leaves. It has large purple-black fruits but is a very variable species – some forms have small leaves while others have large leaves; some types have green leaves and stems, in others these are tinged dark purplish in colour; some types are very leafy, while others bear large quantities of fruit but relatively less dense foliage. This species was introduced into East and Southern Africa perhaps from humid West

Africa. It is the most frequently cultivated nightshade in West Africa. In East Africa it has overtaken all others in areas under cultivation. The species is cultivated intensively in Cameroon. Varieties with leaves as large as those of common kales are now available.

- *Solanum villosum* Mill – this is the orange-berried nightshade. It is distinguished from the rest by its orange to yellow berries, which are edible. It is more common in dry environments. The species is best known in East Africa, where it is both picked from the wild but also cultivated.

Solenostemon rotundifolius (Poir.) J. K. Morton. Family: Lamiaceae. Synonyms: Zulu potato, *fra-fra*, Hausa potato. *Solenostemon rotundifolius* is a small soft perennial herb with ridged stems and aromatic leaves. It bears small round to elongated tubers that are boiled or roasted and eaten. Potatoes are also sold in the market, prepared or in their raw form. The plant is widespread in Africa in both humid and sub-humid areas. It is commonly used in Sahelian West Africa from Mali west to the Sudan.

Talinum fruticosum (L.) Juss. (*T. triangulare* Willd.). Family: Portulacaceae. Synonym: water leaf. *Talinum* consists of succulent, usually perennial, herbs found both in humid and dry environments. *Talinum fruticosum* is the most widely used. It is cultivated but also grows spontaneously in humid West Africa and Central African region. The species is found throughout the humid tropics of the world. In recent years, farmers have begun to grow it in the Democratic Republic of Congo and Cameroon, and it is usually mixed with the more coarse vegetables such as *Gnetum* spp. It is widely consumed by communities found in the central part of Democratic Republic of Congo, especially the Baluba. The crop is also good pig fodder.

Tamarindus indica (L.). Family: Fabaceae. Synonyms: tamarind, *sonhon-non* (Burkina), *mkwaju* (East Africa). *Tamarindus indica* is a large tree of hot areas in low and middle altitudes. It is common in the Sahel and most of dryland Africa. The sour fruits are used for flavouring food but the sour young leaves (reddish) can also be chewed fresh. The leaves and fruits are used in Senegal (*daqaa*).

Telfairia occidentalis (Hook. f.). Family: Cucurbitaceae. Synonym: fluted pumpkin. This is an extensive climber, often seen climbing to tree tops. It is found in humid areas from West Africa through Central Africa to Uganda and Tanzania. The huge fruit has large seeds that are a source of oil. The leaves are cooked as a vegetable in humid West Africa.

Tylosema esculenta Burch. Family: Fabaceae. This is the marama bean. It is an extensive climber with heart-shaped leaves, yellow flowers and a huge underground tuber up to 200kg or more in mature plants. The plant is found in and around Kalahari where it is an important food (van Wyk and Gericke, 2000).

Young tubers are edible; the seeds, which are large, can be roasted as a nut and eaten or pounded and made into a drink, and young pods are food. The plant is extremely drought resistant. The related species *Tylosema fassoglense* (Kotschy ex Schweinf.) Torre and Hillc. (*T. fasoglensis*) has a similar habit and uses, but is more widely distributed from South Africa to Eastern Africa and the Shaba region of the Democratic Republic of Congo, from semi-arid to sub-humid land.

Vernonia amygdalina Delile. Family: Asteraceae. Bitter leaf is a shrub that grows to 4m high. It is widely distributed in humid parts of Africa and is used extensively in Central and West Africa, where it is also cultivated for its leaves. The related species *V. hymenolepis* is used in Central Africa, especially Cameroon, and also in West Africa. Bitter leaf is extremely important in Cameroon.

Vigna unguiculata (L.) Walp. Family: Fabaceae. Synonyms: cowpea, *niébé* (in Senegal) *adenguare* (in Amharic), *kunde* (in Swahili). The cowpea is among the most important African crops as both a pulse and a source of leaf vegetable as well as fodder. Young pods can also be eaten as a vegetable. It is grown from Senegal to Ethiopia and south to South Africa, mainly in warm sub-humid to semi-arid areas. Within each region, the importance attached to either the seed or leaves varies. Some communities may grow it for the seed and not the leaves, while others may grow it for its leaf only. In Senegal (*nebbe* in Wolof), leaves are popular with the Pulaar, the Soninke and the Souse. In Kenya, the Luhya community grow it more for the leaf than the seed. This is an old crop with many forms across the globe. Most conspicuous morphological characters include the colour and size and shape of seeds and pods and the habit of the plant (prostrate or climbing or erect). Creeping varieties in Africa are normally more lasting on the farm, are drought resistant and therefore are good as a source of leaves. Tender green cowpea pods, mainly of a long-pod variety from Asia (subsp. *sesquipedalis* (L.) Verdc.) are grown for commercial purposes only as they are popular with the Asian communities living in East and Southern Africa. The wild relatives of cowpea are common in Africa and are also harvested.

Vitex doniana Sweet. Family: Verbenaceae. Synonyms: *Prunier noir*, (*fonma* in Benin). *Vitex doniana* is a widespread deciduous forest tree largely found in coastal woodlands and savannah, but also in wetter areas at lower altitudes. It is widespread in tropical West Africa, extending eastward to Uganda, Kenya and Tanzania in savannah and also high rainfall areas. Fruits are edible. The ripe black fruit pulp is eaten raw and has a sweet taste. Fruits are occasionally sold when in season. In certain areas the fruits are cooked before consumption. The leaves of *Vitex doniana* are a delicacy in Benin where the young leaves are pre-cooked and sold in most city markets.

Xanthosoma sagittifolium (L.) Schott. Family: Araceae. Tania or elephant ear (*mawole* in Lingala and *atabala* in Kitetela) is related to taro (*Colocasia esculenta*) and both have more or less similar uses. Both species were introduced to Africa for their corms, but like cassava and sweet potato, their leaves are also now used as a vegetable, though still at low levels. The two are important in Southern Africa where they are known as *madumbe*. Both are very important in the Congo Basin and in humid West Africa, as well as Mozambique and Zambia, but only rarely in other areas of the continent. Tania is grown all over the Central African region for its tubers and leaves. The area is ideal for its growth as it prefers light, well-loosened soils and good rains. It is grown in the forest as well as in the less humid areas, but not in swampy or very humid soil. Only the young leaves at the growing tip are picked. The leaves, however, must be well cooked because they contain silicate crystals, which irritate the throat and other internal body linings if half cooked. In East Africa, *Xanthosoma* is popular in Tanzania (called *marumi* in the Meru region of northern Tanzania), especially among the Chagga, Ameru and Arusha peoples. It is also commonly used in Uganda, but it is largely unknown in Kenya. A related species, *Xanthosoma poeppigii* Schott (*Xanthosoma mafaffa*) – also known as African yautia, *mafafa* and *ewe koko* (in Yoruba) – is found in West Africa. Like tania, it was introduced from tropical America. The young leaves of *Xanthosoma mafaffa* are popular in Ghana and are marketed extensively throughout the country (Abbiw, 1990).

Research and development gaps

Country inventories of vegetables

As mentioned earlier in this chapter, the diversity of AIVs is enormous. Unfortunately, however, few countries have well-documented inventories of their AIV diversity. Inventories are an initial step in planning vegetable programmes and also in understanding the full potential of AIVs. Inventories should cover cultivated, weedy and wild species both at the species and infra-specific levels. Developing such inventories calls for partnership with local herbaria and their personnel; therefore, countries need to have access to good herbaria. Fortunately, most of the regions in the continent have at least one well-maintained herbarium with regional collections; several good ones are located in Nairobi, Addis Ababa, Arusha, Limbe, Abomey-Calavi, Accra, Dakar, Harare, Zomba and Pretoria.

Conservation concerns

Poor conservation of certain AIV species or varieties is a major concern. While in the cultivated species the major loss is at varietal level, for wild species the main concern is loss of their habitats and overexploitation (e.g. *Gnetum africanum*), leading to diminished populations that are not viable and even local extinctions. Research on easy and cheap technologies of cultivating the plants should be a priority. Other species that are increasingly becoming rare in the wild include *Chrotalaria brevidens* and *C. ochroleuca*, with current populations being generally maintained by farmers. Rare farmers' varieties (landraces) are encountered in many cultivated species, including the bottle gourd and watermelon. *Ex situ* conservation, mainly in gene banks, coupled with specific strategies of *in situ* conservation and the promotion of use are needed for many of these rare farmer varieties with restricted distribution.

Conservation through use

In spite of the huge diversity, only a few species have widespread use in the continent or regions. Species or varieties that are not useful to a community would be more likely to be neglected and forgotten in favour of useful ones. Many AIV species and varieties can thus be saved if communities continue to attach some value to them. Declining use of less common AIVs is likely to lead to loss of knowledge and, finally, loss of the species or variety (Maundu and Morimoto, 2008). Awareness of the value could contribute to conservation.

The wealth of AIV species on the continent, combined with large cultural diversity, provides a rare opportunity for the continent to further diversify the AIVs used. The high cultural diversity provides an opportunity for tapping into localized knowledge on useful species. Increasing vegetable diversity would have positive effects on nutrition and the health status of people.

Many of the minor underutilized species, particularly those of African origin, however, have to be subjected to research, improvement and promotion for more widespread consumption. Simple selection work can often result in a huge improvement of desired characteristics such as leaf size and, hence, yields (see Chapter 5). Previous work has shown that selection alone could considerably improve the leaf size and other characteristics of African vegetables within a relatively short period of time.

Probably the biggest current constraint in AIV cultivation concerns access to good seed stock (see Chapter 5). Many farmers cite this as a bottleneck to growing AIVs. Developing seed systems is particularly important in cases where AIVs are picked from the wild. Improvement of existing local seed systems should, however, be done in a manner that takes biodiversity conservation issues into account as improvement may often lead to the loss of original biodiversity. In addition to seed, farmers need support with agronomic skills. For many underutilized AIVs, value chains are undeveloped and existing marketing is not normally conducted as serious business. Marketing can be enhanced by diversifying products and adding value to the AIVs (see Chapter 7). The culture of drying AIVs is well rooted only in a few areas, particularly Southern Africa and parts of East and West Africa. Expanding the culture of drying can benefit the rest of the continent. Drying can enhance the diversity of AIVs consumed during times of seasonal shortage.

On the consumer side, utilization can be enhanced significantly by increasing awareness of the benefits of AIVs – particularly their health benefits. Nutritional profiles are thus crucial in promotion work (see Chapter 4). For African vegetables, such profiles are scanty and often limited to a few nutrients

– hence the need for more complete nutritional profiles. It is also vital for consumers to access recipes that they can test and improve upon to fit their own cultural context. This calls for the documentation of local AIV recipes and wide dissemination of information. Research and promotion of local AIVs, however, calls for concerted efforts by local and international research and development organizations, as well as by relevant government agencies. Additionally, this should be supported by research and the right policies. Government departments and ministries responsible for food, health and agriculture need to spearhead the development and implementation of relevant policies that recognize the potential role that AIVs can play in enhancing nutrition, health and general livelihoods (see Chapter 9). Policies should also recognize the need to conserve the diversity of AIVs by promoting different diets and methods of preparation in order to avoid loss of varieties as a result of a few species or varieties dominating the markets and dishes.

Problematic taxonomy

The taxonomy of a few taxa is still a major hindrance to information flow. Many leafy *Solanum* species continue to be referred to as *Solanum nigrum*; yet it is a complex of many species. The delineation of some, furthermore, is still a problem. While this may be partly a problem of information flow, the taxonomy of the genus with respect to African nightshade, for example, still needs to be worked on further. The species occurs as diploids, tetraploids and hexaploids. Thanks to the current work on *Solanums* going on at Radboud University Nijmegen in The Netherlands, we may understand more about the taxonomy of the group better in the near future. *Amaranths*, on the other hand, present another problem of identification – particularly the *Amaranth hybridus* group, including *A. cruentus* and *A. hypochondriacus*. This calls for institutions and programmes committing funds and personnel to the study of such genera.

Diversity studies

Diversity studies among wild as well as cultivated vegetables can help us to understand better the potential within a species. Use of morphological characteristics such as colour and even molecular markers in more advanced studies can help to link with useful traits such as taste, nutritional value, cooking and keeping properties, drought or pest resistance, etc. Farmers' selection criteria, such as cultural reasons, income potential, nutritional value, etc., are worth understanding as these are directly related to the conservation of specific varieties. Diversity studies can be further linked to breeding and promotional campaign programmes.

Finally, these AIVs are associated with certain environments and land-use practices, as well as farming and cropping systems, which are important for the maintenance of specific diversity. This includes intercropping, mixed farming and leaving fallow land as well as wild lands around homesteads. Research, therefore, should also focus on such environments.

Conclusions

Africa is a vast continent with diverse ecosystems that support a large diversity of AIVs, both native and introduced. While some AIV species or varieties have a limited distribution, a large number are distributed across several regions. All of the different regions – North, West, Central, East and Southern Africa – have some unique food cultures and local AIV diversity. The diverse cultures are a vital resource for local knowledge on the use of various species. Utilization of a great majority of these AIVs is on the decline. Neglect of some species or varieties leads to loss of both the species and the associated knowledge. We advocate for conservation through promoting use as this can safeguard both the knowledge and the species. There is a need for concerted efforts by government as well as non-governmental organizations to handle the myriad issues relating to the use and conservation of AIVs. Major issues concern research, promotion and integrated conservation techniques. Conservation through use requires nutritional, agronomic and seed research, as well as the promotion and development of value chains. This will not only achieve the conservation of species and varieties, but will also improve the well-being of local populations.

References

Abbiw, D. K. (1990) *Useful Plants of Ghana*, Intermediate Technology Publications and Royal Botanic Gardens, Kew, London, UK

Achigan-Dako, G. E., Fanou, N., Kouke, A., Avohou, H., Vodouhe, S. R. and Ahanchede, A. (2006) 'Evaluation agronomique de trois espèces de Egusi (Cucurbitaceae) utilisées dans l'alimentation au Bénin et élaboration d'un modèle de prédiction du rendement', *Biotechnologie Agronomie Société Environnement*, vol 10, pp121–129

Achigan-Dako, G. E., Fagbemissi, R., Avohou, T. H., Vodouhe, S. R., Coulibaly, O. and Ahanchede, A. (2008) 'Importance and practices of egusi crops (*Citrullus lanatus*, *Cucumeropsis mannii* and *Lagenaria siceraria* cv. 'Aklamkpa') in socio-linguistic areas in Benin', *Biotechnologie Agronomie Société et Environnement*, vol 12, pp393–403

Asaha, S., Tonye, M. M., Ndam, N. and Blackmore, P. (2000) 'State of knowledge. Studies on Gnetum africanum and Gnetum buchholzianum', report for the Central African Regional Programme for the Environment, Limbe Botanic Garden, Limbe, Cameroon

Assogbadjo, A. E., Sinsin, B. and van Damme, P. (2005) 'Caractères morphologiques et production des capsules de baobab (*Adansonia digitata* L.) au Bénin', *Fruits*, vol 60, no 5, pp327–340

Batawila, K., Akpavi, S., Wala, K., Kanda, M., Vodouhe, R., Akpagana, K. (2007) 'Diversité et gestion des légumes de cueillette au Togo', *African Journal of Food, Agriculture, Nutrition and Development*, vol 7, no 3, www.ajfand.net/Vol7No3.html

Burkill, H. M. (1985) *The Useful Plants of West Tropical Africa*, 2nd edition, vol 1, Families A–D, Royal Botanic Gardens, Kew, London, UK

Burkill, H. M. (2000) *The Useful Plants of West Tropical Africa*, 2nd edition, Vol 1: Families A–D, Royal Botanical Gardens, Kew, London, UK

Chweya, J. A. and Eyzaguirre, P. B. (1999) *The Biodiversity of Traditional Leafy Vegetables*, IPGRI, Rome, Italy

Dansi, A., Adjatin, A., Adoukonou-Sagbadja, H., Faladé, V., Yedomonhan, H., Odou, D. and Dossou, B. (2008) 'Traditional leafy vegetables and their use in the Benin Republic', *Genetic Resources Crop Evolution*, vol 55, pp1239–1256

Demissie, D. (1991) 'Potentially valuable crop plants in a Vavilovian centre of diversity: Ethiopia', in Attere, F., Zedan, H., Ng, N. Q. and Perrino, P. (eds) *Crop Genetic Resources of Africa, vol 1*, IPGRI, Rome, Italy, pp89–98

Diouf, M., Diop, M., Lo, C., Drame, K. A., Sene, E, Ba, C. H., Gueye, M., Faye, B. (1999) 'Senegal', in Chweya, J. A and Eyzaguirre, P. B. (eds) *The Biodiversity of Traditional Leafy Vegetables*, IPGRI, Rome, Italy

Djé, Y., Kouonon, L. C., Zoro Bi, A. I., Gnamien, G. Y., Baudoin, J. P. (2006) 'Etude des caractéristique botaniques, agronomiques et de la biologie florale du melon africain (*Cucumis melo* L. var. agrestis Naudin, Cucurbitaceae)', *Biotechnology, Agronomy, Society and the Environment*, vol 10, no 2, pp109–119

Edmonds, J. M. and Chweya, J. A. (1997) *Black Nightshades:* Solanum nigrum *L. and Related Species – Promoting the Conservation and Use of Underutilised and Neglected Crops*, IPGRI, Rome, Italy

Fondo, K., Kahindi, B., Morimoto, Y. and Maundu, P. (2005) 'Documenting the diversity of leafy vegetables used by the Mijikenda communities of coastal Kenya: A community-led initiative', Presented at the Nutrition Safari 2005 Meeting, 18th Congress on Nutrition, 19–23 September 2005, Durban, South Africa

Fondoun, J. M. and Tiki-Manga, T. (2000) 'Farmers' indigenous practices for conserving *Garcinia kola* and *Gnetum africanum* in southern Cameroon', *Agroforestry Systems*, vol 48, pp289–302

Harlan, J. R. (1969) 'Ethiopia: A centre of diversity', *Economic Botany*, vol 23, pp309–314

Imbumi, M. and Maundu, P. (2008) 'Promotion of African leafy vegetables: Is knowledge of use and potential health benefits the key determining factors to success?', Presented to the XI ICE Conference held June 2008, Cusco, Peru

Jansen van Rensburg, W. S., van Averbeke, W., Slabbert, R., Faber, M., van Jaarsveld, P., van Heerden, I., Wenhold, F. and Oelofse, A. (2007) 'African leafy vegetables in South Africa', *Water*, vol 33, pp317–326

Keding, G., Weinberger, K., Swai, I. and Mndiga, H. (2007) *Diversity, Traits and Use of Traditional Vegetables in Tanzania*, Technical Bulletin no 40, AVRDC – World Vegetable Center, Shanhua, Taiwan

Kiambi, D. (1999) 'Assessment of the status of agrobiodiversity in Djibouti: A contribution to the National Biodiversity Strategy and Action Plan', Unpublished report prepared as part of the Djibouti National Biodiversity Strategy and Action Plan for the Bureau Nationale de la Diversité Biologique, Direction de l'Environment, Ministere de l'Environment, du Tourisme et de l'Artisanat

Kimiywe, J., Waudo, J., Mbithe, D. and Maundu, P. (2006) 'Nutraceutical benefits of traditional foods: A food-based strategy for alleviation of hunger and malnutrition', in *The Role of African Universities in the Attainment of the Millennium Development Goals*, Kenyatta University, Nairobi, Kenya, Proceedings of MDG Conference held in Nairobi, Kenya, 14–18 November 2005, pp2–11

Kristensen, M. and Balslev, H. (2003) 'Perceptions, use and availability of woody plants among the Gourounsi in Burkina Faso', *Biodiversity and Conservation*, vol 12, pp1715–1739

Laurie, E. C. and Sunderland, T. C. H. (eds) (2004) *The Key Non-Timber Forest Products of Central Africa, State of the Knowledge*, African Rattan Research Programme Technical Paper no 122, US Forest Service, USAID, Washington, DC

Leakey, R. R. B., Tchoundjeu, Z., Smith, R. I., Munro, R. C., Fondoun, J.-M., Kengue,

J., Anegbeh, P. O. and Ukafor, V. (2004) 'Evidence that subsistence farmers have domesticated indigenous fruits (*Dacryodes edulis* and *Irvingia gabonensis*) in Cameroon and Nigeria', *Agroforestry Systems*, vol 60, pp101–111

Leakey, R. R. B., Shackleton, S. E. and du Plessis, P. (2005) 'Domestication potential of marula (*Sclerocarya birrea* subsp. *caffra*) in South Africa and Namibia: I. Phenotypic variation in fruit traits', *Agroforestry Systems*, vol 64, pp25–35

Lester, R. N. (1986) 'Taxonomy of scarlet eggplants, *Solanum aethiopicum* L.', *Acta Horticulturae*, vol 182, pp125–132

Lester, R. N. and Niakan, L. (1986) 'Origin and domestication of the scarlet eggplant, *Solanum aethiopicum* L., from *S. anguivi Lam*', in D'Arcy, W. G. (ed) *Solanaceae: Biology and Systematics*, Columbia University Press, New York, pp433–456

Linder, H. P. (2001) 'Plant diversity and endemism in sub-Saharan tropical Africa', *Journal of Biogeography*, vol 28, pp169–182

Lowe, J. (1984) 'Gnetum in West Africa', *Nigerian Field*, vol 49, no 1–4, pp99–104

Maundu, P. (1996) 'Utilization and conservation status of wild food plants in Kenya', in Maesen, L. J. G., van der Burgt, X. M. and van Medenbach de Rooy, J. M. (eds) *The Biodiversity of African Plants*, Proceedings of the XIVth AETFAT Congress, 22–27 August 1994, Wageningen, The Netherlands, pp678–683

Maundu, P. M. (1997) 'The status of traditional vegetable utilization in Kenya', in Guarino, L. (ed) *Genetic Resources of Traditional Vegetables in Africa: Conservation and Use*, IPGRI, Rome, Italy, pp66–75

Maundu, P. M. and Grubben, G. J. H. (2004) '*Amaranthus graecizans* L.', in Grubben, G. J. H. and Denton, O. A. (eds) *PROTA 2: Vegetables/Légumes*, PROTA, Wageningen, The Netherlands, pp76–78

Maundu, P. and Imbumi, M. (2003) 'The food and food culture of the peoples of East Africa (Kenya, Uganda, Tanzania, Rwanda, Burundi, Sudan, Ethiopia, Eritrea, and Somalia)', in Katz, S. H. (ed) *Encyclopaedia of Foods and Culture*, vol 1, Charles and Scribner's Sons, New York, NY, pp27–34

Maundu, P. and Morimoto, Y. (2008) 'Reconciling genetic resources and local knowledge conservation and livelihoods enhancement in research and development: Experiences of Biodiversity International in sub-Saharan Africa', *Journal of Life on Earth*, vol 9, no 1–2, pp56–60

Maundu, P., Kabuye, C. H. S. and Ngugi, G. W. (1999) *Traditional Food Plants of Kenya*, KENRIK, National Museum of Kenya, Nairobi, Kenya

Mertz, O., Lykke, A. M. and Reenberg, A. (2001) 'Importance and seasonality of vegetable consumption and marketing in Burkina Faso', *Economic Botany*, vol 55, pp276–289

Ndam, N., Nkefor, J. P. and Blackmore, P. (2001) 'Domestication of *Gnetum africanum* and *G. buchholzianum (*Gnetaceae*)*, over-exploited wild forest vegetables of the Central African region', *Systematics and Geography of Plants*, vol 71, no 2, pp739–745

Ndam, N., Nkefor, J. P. and Blackmore, P. (2002) 'Domestication of *Gnetum africanum* and *G. buchholzianum* (Gnetaceae), over-exploited wild forest vegetables of the Central African Region', in Robbrecht, E., Degreef, J. and Friis, I. (eds) *Plant Systematics and Phytogeography for the Understanding of African Biodiversity: Proceedings of the XVIth AETFAT Congress*, National Botanic Garden of Belgium, Meise, Belgium, pp739–745

Nkefor, J. P. (2000) 'From the forest to the Kitchen: Following the market chain of Eru, an important non-timber forest product in Cameroon', mimeo

Ogle, B. M. and Grivetti, L. E. (1985) 'Legacy of the chameleon: Edible wild plants in the Kingdom of Swaziland, Southern Africa – a cultural, ecological and nutritional study. Part 2: Demographics, species availability, and dietary use', *Ecology of Food and Nutrition*, vol 17, pp1–30

Okafor, J. C. (1997) 'Conservation and use of traditional vegetables from woody forest species in southeastern Nigeria', in Guarino, L. (ed) *Traditional African Vegetables. Proceedings of the IPGRI International Workshop on Genetic Resources of Traditional Vegetables in Africa: Conservation and Use*, 29–31 August 1995, ICRAF, Nairobi, Kenya

Okulungu, C. (2003) *Selected Traditional Leafy Vegetable Recipes of the Democratic Republic of Congo (DRC)*, Unpublished manuscript

Rubaihayo, E. B. (1997) 'Conservation and use of traditional vegetables in Uganda', in Guarin, L. (ed) *Genetic Resources of Traditional Vegetables in Africa: Conservation and Use*, IPGRI, Rome, Italy, pp104–116

Rubaihayo, E. B., Hart, T., Kakonge, E., Kaaya, A., Kawongolo, J., Kabeere, F., Mugisha, J., Tumwiine, J. and Rubaihayo, P. (2004) *Development of Mechanisms for Sustainable Production and Utilisation of Indigenous Vegetables and Management of Their Genetic Diversity in Uganda*, Unpublished report

Schippers, R. R. (2000) *African Indigenous Vegetables. An Overview of the Cultivated Species*, Natural Resources Institute/ACP-EU Technical Centre for Agricultural and Rural Cooperation, Chatham

Schippers, R. R. (2004) 'Cucurbitaceae', in *Légumes Africains Indigènes: Présentation des Espèces Cultivées*, Margraf Publishers, CTA, Wuerzburg, pp113–182

Schippers, R., Maundu, P., Imbumi, M. and Obiero, H. (2002) *How To Grow and Use Spider Plant*, Horticultural Development Services, Baarn, The Netherlands

Shackleton, S. E., Shackleton, C. M., Dzerefos, C. M. and Mathabela, F. R. (1998) 'Use and trading of wild edible herbs in the central lowveld savanna region, South Africa', *Economic Botany*, vol 52, pp251–259

Shiembo, P. N. (1999) 'The sustainability of eru (*Gnetum africanum* and *Gnetum buchholzianum*): over-exploited non-wood forest product from the forests of Central Africa', in Sunderland, T. C. H., Clark, L. E. and Vantomme, P. (eds) *Current Research Issues and Prospects for Conservation and Development. Based on the Outcome of the International Expert Meeting on Non-Wood Forest Products in Central Africa*, held at the Limbe Botanic Garden, Limbe, Cameroon, 10–15 May 1998, FAO, Rome, Italy, pp61–66

Toxopeus, H. and Baas, J. (2004) '*Brassica rapa* L.', in Grubben, G. J. H. and Denton, O. A. (eds) *PROTA 2: Vegetables/Légumes*, PROTA, Wageningen, The Netherlands, CD-ROM, pp146–151

Toxopeus, H. and Mvere, B. (2004) '*Brassica napus* L.', in Grubben, G. J. H. and Denton, O. A. (eds) *PROTA (Plant Resources of Tropical Africa / Ressources végétales de l'Afrique tropicale)*, Wageningen, The Netherlands

van Wyk, B. and Gericke, N. (2000) *People's Plants: A Guide to Useful Plants of Southern Africa*, Briza Publications, Pretoria, South Africa

Vorster, H. J., Jansen van Rensburg, W. S., van Zijl, J. J. B. and van den Heever, E. (2002) 'Germplasm management of African leafy vegetables for the nutritional and food security needs of vulnerable groups in South Africa', Progress Report, ARC-VOPI, Pretoria

White, F. (1983) *The Vegetation of Africa: A Descriptive Memoire to Accompany the UNESCO/AETFAT/UNSO Vegetation Map of Africa*, UNESCO, Paris, France

Zoro Bi, I., Koffi, K. K. and Dje, Y. (2003) 'Caractérisation botanique et agronomique de trois espèces de cucurbits consommées en sauce en Afrique de l'Ouest: *Citrullus* sp., *Cucumeropsis mannii, Lagenaria siceraria*', *Biotechnologie, Agronomie, Société et Environnement*, vol 7, no 3–4, pp187–199

4

Nutritional Contributions of Important African Indigenous Vegetables

Ray-Yu Yang and Gudrun B. Keding

Introduction

Insufficient vegetable and fruit consumption causes 2.7 million deaths annually worldwide and belongs to the top ten risk factors contributing to mortality (Ezzati et al, 2002). Malnutrition is rampant in the tropics where per capita vegetable supplies in most countries fall far short of the minimum recommended 73kg per person per year. In sub-Saharan Africa (SSA), per capita vegetable supplies are only 43 per cent of what are needed, leading to widespread malnutrition. These conditions are intolerable, yet they are projected to worsen. The International Food Policy Research Institute (IFPRI) predicts an 18 per cent rise in the number of malnourished children in SSA from 2001 to 2020 (IFPRI, 2001). While micronutrient deficiencies mainly causing vitamin A, iron and iodine disorders remain widespread, the chronic diseases, including cardiovascular disease, cancer, chronic respiratory diseases and diabetes, that caused 60 per cent of all deaths in 2005 are increasing globally (WHO, 2007). About 80 per cent of the deaths from chronic diseases occur in low- and middle-income countries. Increased fruit and vegetable consumption has been widely promoted because of the health benefits of micronutrients, as well as the many non-nutrient phytochemicals associated with health maintenance and prevention of chronic diseases (Steinmetz and Potter, 1996). Greater fruit and vegetable consumption can help to address the double burden of micronutrient deficiencies and chronic diseases.

Furthermore, diets rich in micronutrients and antioxidants are strongly recommended to supplement medicinal therapy in fighting HIV/AIDS (Friis et al, 2002). Vegetables rich in micronutrient and antioxidant phytochemicals could improve nutrition and help to alleviate HIV/AIDS.

Globally, there are hundreds of plant species consumed as vegetables; but only about 20 crops are produced in intensive cropping systems (Siemonsma and Kasem, 1994). Indigenous vegetables (IVs) are native to a particular region or introduced to the region from another geographical area, but have been used over a long period of time (Engle et al, 2003). They are grown locally on a small scale, are often resistant to diseases and tolerant of environmental stresses, are nutritious and contain a vast range of phytochemicals; however, most are neglected or underutilized. Indigenous vegetables have the potential for introduction or greater use as cash crops in peri-urban systems, as vegetables for daily sustenance in home gardens, and as a means of diversifying production systems and diets.

Vegetable species are diverse in Africa. The Plant Resources of Tropical Africa (PROTA) initiative has documented at least 275 plant species that are primarily used as vegetables and 528 species that are used for food, medicinal or ornamental purposes (Grubben and Denton, 2004; PROTA, 2005). The primary-use vegetables encompass 53 different botanical families of which more than 60 per cent are pan-African. About 75 per cent of the primary-use vegetables are indigenous; 16 per cent were introduced long ago and are widely adapted. Only 8 per cent were recently introduced and are regarded as exotic (PROTA, 2005).

Factors affecting the nutritional contribution of vegetables to human nutrition and health include:

- per capita consumption;
- nutrient and phytochemical content; and
- nutrient bioavailability.

This chapter discusses how African indigenous vegetables contribute and could better contribute to human nutrition from the above three aspects, as well as the overall nutritional outcome from current African diets.

Consumption of African indigenous vegetables

While many vegetable studies focus on production and commercial importance, less comprehensive data is found on vegetable consumption; therefore, examples are given from only a few African countries.

Traditional leafy and fruity vegetables were established long ago as important components in African diets (Chweya and Eyzaguirre, 1999). They form a significant part of the traditional diets of agricultural communities in Africa, while their consumption is usually less significant among pastoral communities. Obviously, the significance of African indigenous vegetable (AIV) consumption depends upon the community in question (Maundu, 1997); therefore, traditional vegetable consumption in urban and peri-urban communities will be characterized by different influencing factors.

In general, vegetable consumption in traditional African societies has undergone substantial changes since pre-colonial days, mainly through interaction with other cultures (Maundu, 1997). With urbanization, these interactions and changes become more frequent and rapid and great differences between different urban consumer groups emerge. Thus, it is not possible to refer to an 'average' urban consumer, but to consider different cultural backgrounds, socio-economic status, lifestyles and, consequently, different consumer needs and behaviours (Delisle, 1990).

Recipes: Preparation and preservation

At first glance, AIVs seemed to serve as food only, as a component in meals besides starch and protein, as important suppliers of vitamins and minerals, and as taste improvers (Oomen and Grubben, 1978). However, traditional vegetables have manifold uses in medicine and even cosmetics as well. In Kenya, for example, cooking of leafy vegetables is especially important among agricultural communities and hunter-gatherers, while among traditional pastoral systems, an important use of plant parts is in soups and milk for flavour and good health (Maundu, 1997). In South Africa, different culture, ethnicity and gender affect crop choice and their use (Vorster et al, 2007).

Preparation

There are different ways in which AIVs are consumed – as side dishes, processed into soups, sauces or pastes, included with the main course (i.e. mixing leaves with the starch component before cooking), or as edible packaging material for meat dishes that are packed and steamed in vine, nightshade or cabbage leaves (Oomen and Grubben, 1978).

In Benin, vegetables are usually seen as an accompanying dish (e.g. a cooked vegetable or spicy tomato sauce to accompany a starchy staple, which is most commonly a maize paste) (Sodjinou et al, 2009). Similarly, in Kenya and South Africa, boiled or fried vegetables are usually eaten along with stiff porridge made from cereals (see Plate 4.1). Mashing the vegetable with maize or with a starchy tuber is widespread in Kenya, while mixing of leaves into the stiff maize porridge is also done by the Xhosa in South Africa. Cooking the vegetable with flour made from cereals or dried tubers is also possible (Maundu, 1997; Vorster et al, 2007). Traditional leafy vegetables are usually boiled or fried for up to 10 or 15 minutes. There are some exceptions, such as bitter lettuce (*Launea cornuta*), which was reported to be cooked in plenty of water for 40 minutes in Tanzania (Keller, 2004). A recipe from Senegal recommends cooking cowpea leaves (*Vigna unguiculata*) together with fish and other ingredients for 90 minutes (Seck et al, 1999). In this case, if vegetables are prepared together with starchy staples, meat or fish, the cooking time will be prolonged and valuable nutrients might be lost.

Common preparation methods for AIVs include boiling, steaming or frying for all leafy or root vegetables, and also roasting for fruity vegetables – for example, African eggplant (*Solanum* spp.). The ingredients used are very much adapted to what is locally available. In Tanzania, for example, coconut milk is included in dishes primarily in coastal areas, while groundnuts are a common ingredient in semi-arid central Tanzania. In peri-urban villages close to Arusha town, recipes are characterized by exotic ingredients such as carrots, sweet pepper, tomatoes and onions, as well as meat, fish or fresh milk (Keller, 2004). In Senegal, cooking ingredients are different in urban and rural areas – in urban areas, exotic vegetables are used more frequently (Seck et al, 1999).

Food preparatory practices have changed over time. From the 1930s to 1960s in East Africa, boiling and occasional roasting were the principal cooking techniques that are superior to current practices favouring frying and deep-frying of foods (Raschke et al, 2007). Besides the shift towards frying instead of boiling vegetables and other foods in urban areas, there seem to be no distinctive urban recipes. People tend to stick to food habits; thus, preparation techniques acquired in early childhood have a long-lasting influence even if people move from villages to towns (Sodjinou et al, 2009).

Preservation

Drying of traditional vegetables is common in areas where they cannot be grown year-round due to dry spells and lack of water, and where a seasonal surplus is produced and cannot be used. In South Africa most of the leaves are dried in the direct sun, sometimes blanched before drying (Vorster et al, 2007). Also in Zimbabwe the common preservation method is sun-drying with some leaves boiled before drying, with others cut into small pieces (Chigumira and Mvere, 1999). However, open drying methods do not ensure against re-wetting of the material by rain, contamination by dust and dirt, the possibility of microbial spoilage or mould, and predation by birds, rodents and insects. Furthermore, the amount of beta-carotene (β-carotene) in the final product is appreciably higher when leaves are dried photo-protected in a solar drier than if conventionally dried. However, the potential of preservation by solar-dehydration of food does not seem to be sufficiently recognized (Mgoba et al, 1993).

In Burkina Faso, the common drying method is also to spread surplus vegetables on the roof of a hut or on straw mats on the ground in the direct sun. However, a local non-governmental organization (NGO) implemented a solar-drying programme in the country with the aim of utilizing seasonal overproduction of vegetable so that farmers do not have to sell below cost or even waste produce. The promoted solar drier is reliable and ensures a sufficiently good quality, so that products can even be sold on the market (Legay, 2004).

In urban areas, preserved vegetables will be of less concern as better storage facilities are available as well as a more consistent year-round vegetable supply. However, the question of whether the urban poor also have access to these facilities should be considered.

Table 4.1 *Estimated regional intake of fruit and vegetables*

Region	*Gender*	*Estimated regional mean intake of fruit and vegetables (grams per person per day) Age group (years)*							
		0–4	*5–14*	*15–29*	*30–44*	*45–59*	*60–69*	*70–79*	*≥80*
Africa (high child and adult mortality)*	Male	144	296	288	413	419	439	446	476
	Female	140	279	302	345	305	355	349	382
Africa (high child and very high adult mortality)**	Male	94	193	192	278	294	325	333	380
	Female	91	181	201	236	214	257	244	245

Notes: Countries in region:
* Algeria, Angola, Benin, Burkina Faso, Cameroon, Cape Verde, Chad, Comoros, Equatorial Guinea, Gabon, Gambia, Ghana, Guinea, Guinea-Bissau, Liberia, Madagascar, Mali, Mauritania, Mauritius, Niger, Nigeria, Sao Tome and Principe, Senegal, Seychelles, Sierra Leone and Togo.
** Botswana, Burundi, Central African Republic, Congo, Côte d'Ivoire, Democratic Republic of the Congo, Eritrea, Ethiopia, Kenya, Lesotho, Malawi, Mozambique, Namibia, Rwanda, South Africa, Swaziland, Uganda, United Republic of Tanzania, Zambia and Zimbabwe.
Source: Lock et al (2005)

Patterns: amounts, frequencies and seasonality of AIV consumption

Amounts

In general, little is known about the actual quantities of AIVs consumed (Smith and Eyzaguirre, 2007) and especially about the structure of urban demand (Gockowski et al, 2003). Since vegetables are usually considered a relish or even a condiment or spice, and always as a side dish accompanying other (usually starchy) food, the amount consumed can be rather small. Overall figures exist for daily fruit and vegetable consumption (see Table 4.1) and daily vegetable consumption alone (see Figure 4.1) in different African countries.

From Table 4.1 it can be calculated that in Tanzania, women between 15 and 44 years of age have a mean intake of fruit and vegetables of about 220g per person per day (Lock et al, 2005). While this is the amount of combined fruit and vegetable consumption, it was estimated by Ruel et al (2005) that in Tanzania about 100g of vegetables alone per person per day are consumed on average (see Figure 4.1). In another study in Tanzania, it was measured by a semi-quantitative 24-hour recall approach that, on average, about 270g of vegetables per person per day were consumed in two rural and one peri-urban districts (Keding et al, 2007a), with no differences between the districts. However, the average amount of exotic vegetables consumed was higher in the peri-urban district (see Table 4.2), which suggests an increasing availability and demand for exotic vegetables with increasing urbanization. Nevertheless, generating vegetable consumption data is complex and, as can be seen for the data from Tanzania (generated by different studies), amounts vary substantially. Variation can be due to season, survey method, interviewer, gender and age of interviewee, and also if rural or urban dwellers are considered.

In general, vegetable (and fruit) consumption in different African countries is higher in urban areas compared to rural regions (see Figure 4.2). However,

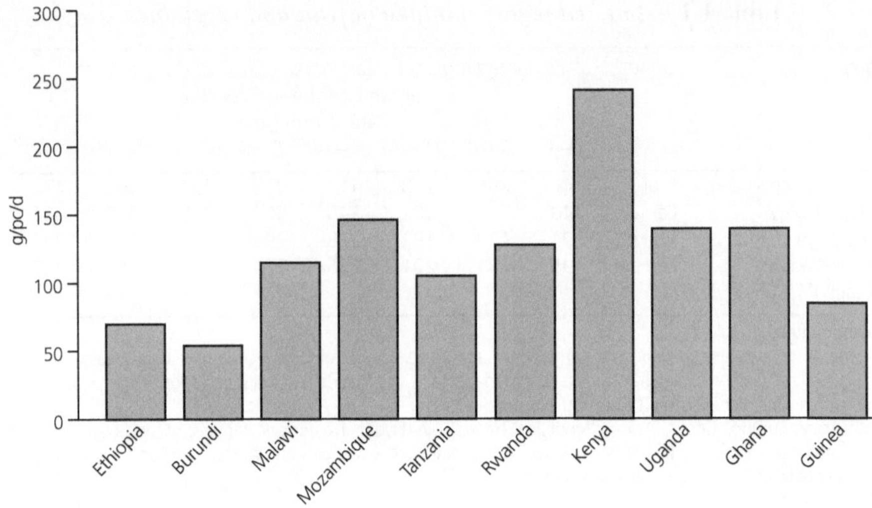

Source: adapted from Ruel et al (2005)

Figure 4.1 *Vegetable consumption per capita and day in sub-Saharan Africa*

differences are relatively small in all researched countries except in Burundi and Kenya, where vegetable consumption in urban areas seems particularly high. Factors influencing the higher consumption of vegetables and, in general, the more diversified diets of urban residents compared to rural dwellers are manifold. They include change in lifestyle and cultural patterns, the availability of storage facilities, and the accessibility of a wider variety of foods in urban markets (Ruel et al, 2005). However, not all urban residents automatically have access to the mentioned facilities. Therefore, vegetable consumption should also be compared between people of different wealth groups.

Socio-economic differences in total vegetable consumption do exist, with a steady increase in consumption as income increases. In Mozambique, for example, the magnitude of the differences between the lowest and highest income quintiles was twofold while it was even eightfold in Burundi (Ruel et al, 2005). In a study in Tanzania, participants from the lowest expenditure quintile consumed only half the amount of vegetables per day (154g) compared to those in the highest expenditure quintile (317g) (Weinberger and Swai, 2006). Consequently, it was shown that poor groups with low food expendi-

Table 4.2 *Average amount of indigenous and exotic vegetables consumed (in grams) by women in one day during the dry season in Tanzania*

District in Tanzania	*African indigenous vegetables (AIVs) only*	*Exotics only*	*AIVs + exotics*	*Range*
Kongwa (rural, semi-arid)	330	120	320	0–1000
Muheza (rural, humid)	210	180	210	0–430
Singida (peri-urban, semi-arid)	300	240	290	0–850

Source: Keding (2006)

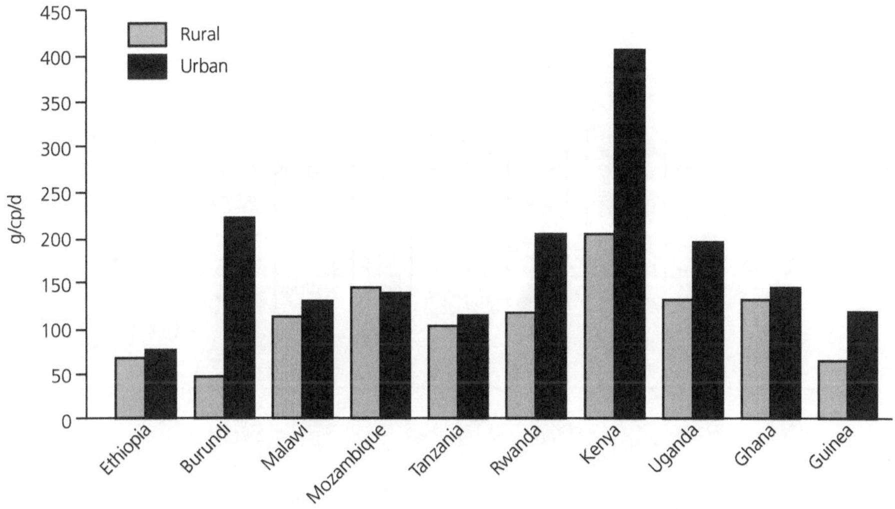

Source: adapted from Ruel et al (2005)

Figure 4.2 *Vegetable consumption by rural–urban area*

ture cannot meet the recommended daily intake of 400g of fruit and vegetables per capita for a healthy diet (WHO, 2003a), of which usually two-thirds should be vegetables. While this trend between poor and wealthy groups was not clear for traditional vegetables in terms of the amount consumed, the percentage of traditional vegetables as a share of the value of all food consumed was about 8 per cent for poor households and only 2.7 per cent for wealthy households (Weinberger and Swai, 2006). Cocks (2006) found that poor urban households in South Africa consumed significantly more AIVs than their wealthier counterparts (41kg and 29kg per household per year, respectively). However, in rural areas, Shackleton and Shackleton (2006) and Dovie et al (2007) found no differences in the amounts consumed between wealthy and poor households.

Frequency

The frequency of vegetable consumption depends upon the frequency of meals. If this frequency is rather low, with only one or two daily meals, vegetables are only consumed once or twice a day and portion sizes have to be adjusted to meet the recommended daily intake. While incorporating vegetables in every meal and, thus, eating vegetables daily is recommended (Burgess and Glasauer 2004), for some people it is not easy or possible to follow this suggestion.

For example, a study in Tanzania showed that 13 per cent of survey participants from two semi-arid districts had not consumed any vegetables on the previous day; in a humid district, even 33 per cent of all participants did not eat any vegetables on the previous day (Keding, 2006). The possibility that certain dietary patterns exist which restrict vegetable consumption in some communities needs to be considered. Ruel et al (2005) calculated the percent-

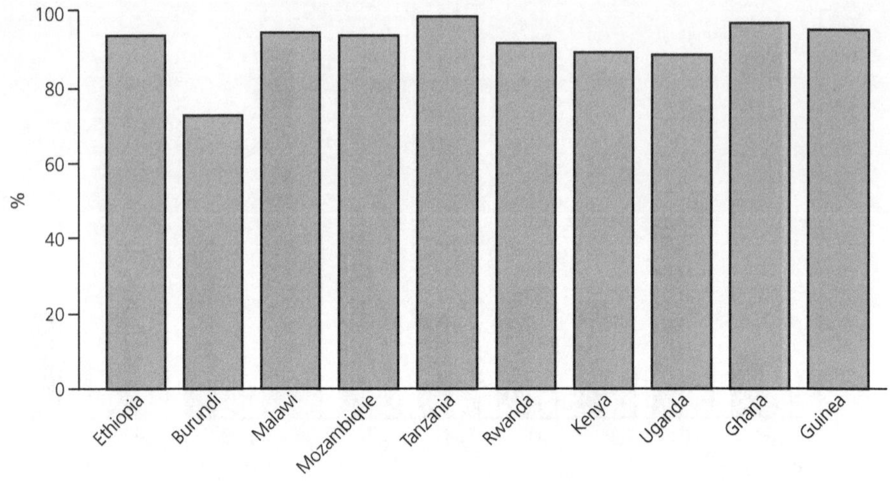

Source: adapted from Ruel et al (2005)

Figure 4.3 *Percentage of households consuming vegetables*

age of households that generally consumed vegetables in different sub-Saharan African countries (see Figure 4.3). While in seven out of ten countries more than 90 per cent of all surveyed households did consume vegetables, these figures need to be interpreted with caution as the reference period of the survey varied between 7 and 30 days (Ruel et al, 2005). In South Africa, consumption of AIVs was variable, with some households consuming them daily, but others only every few days (Shackleton et al, 1998; Dovie et al, 2007).

Seasonality

It is generally acknowledged that seasonal malnutrition in rural areas in Africa is widespread (SCN, 1988) and is caused to a great extent by seasonal fluctuations in food availability and the ability to acquire food (UNU, 1985). In urban areas, however, one would expect a more constant supply of vegetables year round due to food supply from different agro-ecological zones and the greater availability of irrigation facilities for urban agriculture. Nonetheless, in Yaoundé (Cameroon), a decline in vegetable supply was noticed during the dry season, which is a food security concern especially for the very poor (Gockowski et al, 2003). In Zimbabwe, wild and semi-wild vegetables were only consumed during December and January when they were available (Benhura and Chitsaku, 1992). Year-round production of AIVs was observed in Tanzania in areas with water availability throughout the year. Yet, in other regions with prolonged dry spells, AIVs were mainly grown in home gardens during the dry season but also played an important role during rainy seasons, when they are readily available from the wild (Weinberger and Msuya, 2004). Similarly, in South Africa, AIVs were observed to be important during the rainy season, but even more during the dry season, although often in the dried form (Shackleton et al, 1998; Vorster et al, 2007).

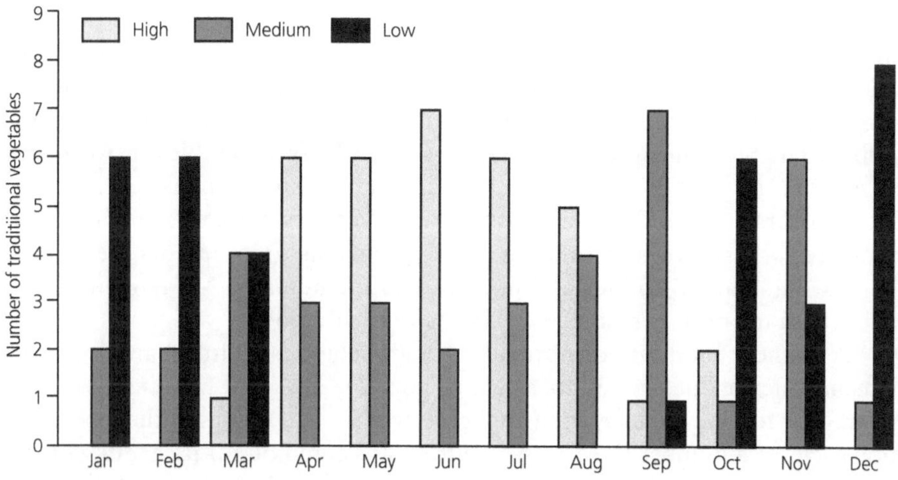

Source: adapted from Nekesa and Meso (1997)

Figure 4.4 *Availability calendar for traditional vegetables in Kenya*

As observed in Kenya, most AIVs displayed distinct seasonal patterns, while availability for most vegetables is only medium or low for the rest (see Figure 4.4). While low AIV availability refers to the fresh product, preserved vegetables can still be available. This was also found in South Africa, where food is very expensive during the dry season and where dried traditional vegetables form the basis of nutrition in most rural households (Vorster et al, 2007). This situation will be different for urban or peri-urban households. It was noted in South Africa that in areas where transport to the markets or towns with shopping facilities was infrequent or expensive, households relied much more on traditional vegetables (Vorster et al, 2007).

Behaviour: Nutrition transition of consumers

For sub-Saharan Africa (SSA) there are only minimal data on dietary trends and no systematic national surveys accessible that obtain dietary and activity data. Neverthless, data on nutrition-related non-communicable diseases is available which shows that even among the poorest, large proportions of the population are shifting towards a diet high in total fat, cholesterol and sugar, and low in polyunsaturated fatty acids and fibre, often accompanied by an increasingly sedentary life (Popkin, 2002). At the same time, knowledge of traditional cooking practices stays mainly with old people and, if not recorded, may become lost over time. In South Africa, for example, the lack of traditional knowledge by children was attributed to compulsory schooling; small boys have to go to school, do not herd livestock anymore, and do not rely on their knowledge of wild plants for survival (Vorster et al, 2007).

Traditional knowledge of AIVs is not only about preparation and preservation, but is also characterized by knowledge of food taboos and medicinal values. In the Democratic Republic of Congo it was found that women from an agricultural society have more food taboos than men and these are particularly high during pregnancy and lactation (Bentley et al, 1999). Similarly, most taboos concerning AIV consumption in Tanzania applied to pregnant or lactating women, but also to people with certain diseases or men, in general (Keller, 2004). The fact that men would not eat green leafy vegetables in Arumeru District (Tanzania) could be traced back to their history as pastoralists. For most pastoralists, besides meat, fresh blood mixed with milk was an important food item (Maundu and Imbumi, 2003), whereas green plants were only regarded as animal fodder and not suitable for humans. Many men of the amaXhosa in South Africa regard AIVs as women's food.

Permanent food taboos normally include religious, cultural and hygienic reasons, such as the prohibitions against pork for Jews and Muslims, whereas temporary food avoidance often applies to women and relates to the reproduction cycle. In a number of African countries, it is common for pregnant women to avoid green vegetables (den Hartog, 2003). It could be assumed that due to food prohibitions, people and, especially, vulnerable groups of a community such as women and children could become malnourished. However, it was reported from the Democratic Republic of Congo that Lese women had several strategies to minimize the ecological and cultural constraints on their nutrition and were, thus, avoiding most food taboos (Bentley et al, 1999).

While traditional culture, beliefs and practices influence dietary health, changing dietary patterns, especially when occurring at a rapid pace, have an even more intense impact upon the nutritional health of people. The so-called nutrition transition is occurring in all but the poorest countries of the world. Within its course, traditional plant-based diets that are rich in fruit and vegetables are replaced with diets rich in animal fats and sugar and low in complex carbohydrates (Popkin, 2002). While some argue that obesity, as a diet-related disease, remains rare in SSA (Martorell, 2002), others claim that in most developing nations, obesity has emerged as a more serious health threat than hunger (Popkin, 2007). In South Africa, for example, more than half the adults are either overweight or obese; in North Africa, at least one out of four adults is overweight; and even in highly poor countries such as Uganda and Nigeria, obesity is a growing concern (Popkin, 2007). This is conflated with urbanization and more sedentary lifestyles.

The change from traditional to transitional dietary patterns often means the addition of new foods to the diet rather than the replacement of traditional foods. Amongst urban adults in Benin, the transitional diet was more diversified but, at the same time, associated with a significantly higher intake of energy from sugar and fat (Sodjinou et al, 2009). In this study, dietary patterns (traditional or transitional) were significantly influenced by birthplace, urban or rural, but not by duration of urban life. This finding suggests that food habits acquired in early childhood have a long-lasting influence. Consequently, as more people in SSA are born in cities due to a rapid urbanization, children may acquire food habits that promote development of chronic diseases (Sodjinou et al, 2009).

In a study designed to understand how urbanization influences nutrition and health transition in South Africa, it was noticed that urban subjects consumed less maize porridge but more fruits, vegetables, animal-derived food, fats and oils than rural study participants. The nutrient intake of vitamin A, ascorbic acid, iron and calcium, all nutrients which can be derived from vegetables, increased with urbanization. At the same time, overweight and obesity rates increased with urbanization as well (Vorster et al, 2005). However, it was not clear if the increased vegetable consumption was from traditional or exotic types. Differences in dietary habits (i.e. consumption of green vegetables) were also found between urban, rural and pastoral populations in Tanzania. These variations were suggested as contributing significantly to the variation in cardiovascular disease risk pattern (Njelekela et al, 2003).

Contrary to these studies, urbanization is not generally believed by researchers to increase vegetable consumption uniformly for all urban dwellers. Only those who can afford it will have better access to a diverse and varied diet in urban rather than rural areas. At the same time, urbanization will distance more people from primary food production in the first place and will have a negative impact upon access to, and availability of, a nutritious diet containing enough vegetables, especially for the urban poor (WHO, 2003b).

In general, the shift between different dietary patterns is rapid and, in many developing countries, the shift from under-nutrition to over-nutrition has occurred in less than one generation (Popkin, 2007). However, reports of rising popularity of AIVs can be found – for example, in Kenya, where farmers even started to reduce the cultivated land under other crops to expand their production and marketing of traditional vegetables (Virchow et al, 2007); or in South Africa, where wild traditional vegetables are moving from the wild into home gardens, thereby becoming cash crops (Jansen van Rendsburg et al, 2007) and in such a way popular so that most consumers would like to eat them more often (van den Heever and du Plooy, 2007).

Nutrient content

Nutrient content of major food groups in African diets

Over the last decade (1995 to 2005), vegetable consumption in Africa has slightly increased, although it remains far below recommended levels. The increase was mainly due to the increased capita consumption of tomato (9.9g per day) and onion (3.4g per day) (FAO, 2007). In contrast, consumption of minor vegetable groups has decreased, which hints to less diversity in vegetable markets. Promotion of AIVs in order to diversify crops and diets is needed.

Major food groups consumed in African regions according to FAOSTAT 2005 data (FAO, 2007) are presented in Table 4.3. African populations consume cereal-based and/or starchy-based diets. Cereals and starchy foods are good and inexpensive sources of energy; however, they are low in micronutrients. The iron content of wheat, maize and sorghum is relatively high; but staple foods generally contain crude fibres, polyphenols and phytate, which

Table 4.3 *Consumption and nutrient content of major food groups in African regions*

Food group	*Consumption (g/capita/day)*					*Nutrient content (per 100g fresh weight)*			
	Central Africa	*East Africa*	*South Africa*	*West Africa*	*North Africa*	*Energy (kcal)*	*Protein (g)*	*Vitamin A (µg RAE)*	*Iron (mg)*
Total energy (kal/capita/day)	**1897**	**2042**	**2813**	**2802**	**3154**				
Cereals	**186**	**330**	**518**	**453**	**630**				
Wheat	36	59	127	51	392	339	13.7	0	3.88
Maize	70	140	303	67	88	365	9.4	0	2.71
Sorghum	26	35	10	92	45	339	11.3	0	4.40
Starchy roots	**629**	**335**	**100**	**510**	**90**				
Cassava (fresh and dried)	550	192	0	268	0	160	1.4	1	0.27
Potatoes	10	43	79	8	82	77	2.0	0	0.78
Sweet potatoes	26	60	3	23	4	86	1.6	709	0.61
Pulses	**18**	**34**	**11**	**23**	**20**				
Beans (including cowpeas), dry	15	18	8	20	1	343	23.9	2	9.95
Vegetables	**67**	**59**	**112**	**141**	**394**				
Tomatoes	17	6	22	27	160	18	0.9	42	0.27
Onions (including shallots)	4	5	18	16	34	40	1.1	0	0.21
Cabbages (white cabbage)	2	9	7	2	9	25	1.3	5	0.47
Pumpkins, squash and gourds	4	2	13	0	15	26	1.0	369	0.80
Chillies and peppers, green	0	1	0	10	15	40	2.0	59	1.20
Carrots	0	1	7	3	10	41	0.9	835	0.30
Other vegetables	33	33	35	77	30	23	2.9	469	2.71
Fruits	**140**	**151**	**147**	**155**	**199**				
Bananas	32	32	15	5	16	89	1.1	3	0.26
Oranges	5	2	19	4	36	47	0.9	11	0.10
Grapes	6	1	60	2	28	68	2.6	31	0.26
Lemon	0	0	2	0	6	20	1.2	2	0.70
Meat	**31**	**28**	**117**	**29**	**59**				
Bovine meat	10	15	36	9	21	276	15.0	0	5.67
Chicken meat	7	4	58	6	21	234	18.8	37	1.09
Milk, whole, fresh	**32**	**74**	**134**	**40**	**232**	60	3.2	28	0.03
Bird eggs (including hen eggs)	**1**	**2**	**14**	**6**	**10**	143	12.6	140	1.83

Note: RAE = retinol activity equivalent.

Source: consumption data retrieved from FAOSTAT, http://fao.stat.fao/org; nutrient data retrieved from USDA nutrient database (USDA, 2007)

significantly inhibits iron absorption by the human body and results in low bioavailability. Pulses are good sources of protein and energy, and generally are more abundant in the amounts and types of micronutrients compared to other staple foods. Increased pulse consumption could contribute significantly to better micronutrient supply in diets. Vegetables, in general, are micronutrient rich; and individual species may be nutritious in terms of certain types of essential nutrients. Commonly consumed vegetables in Africa, such as tomato, pepper and pumpkin, are rich in pro-vitamin A (β-carotene), but insufficient in iron. Green leafy vegetables are rich in pro-vitamin A, iron, folates and other micronutrients; but their consumption is insignificant compared to other food items. Meat is an excellent source of protein and micronutrients, and it provides bioavailable iron and high-quality protein complete in essential amino acids. Their nutritional contribution is limited due to low meat consumption in most African countries (Neumann et al, 2002).

Nutrient contents of important African indigenous vegetables

Table 4.4 lists group means and ranges of nutrient contents of priority AIVs categorized as exotic, adapted and indigenous according to PROTA (2005). Priority African vegetables by species and their nutrient contents are presented in Table 4.5. The nutrient values were retrieved from PROTA (Grubben and Denton, 2004), AVRDC (World Vegetable Center) data (AVRDC, 2002, 2004; unpublished data) and the US Department of Agriculture (USDA) nutrient database (www.nal.usda.gov/fnic/foodcomp/search/). The phytochemical composition of vegetables was retrieved from PROTA (Grubben and Denton, 2004) and searched from a phytochemical handbook (Harborne et al, 1999). Group means of exotic, adapted and indigenous vegetables are comparable for β-carotene and vitamin C contents. In Africa, about 80 per cent of dietary intake of vitamin A came from vegetable sources (SCN, 1993). However, AIVs followed by adapted vegetables have greater vitamin E, folate, calcium, iron, zinc and antioxidant activity mean values. Regarding anti-nutrient factors, indigenous vegetables are the highest in polyphenols, while exotic vegetables as a group are higher in oxalate content mainly due to the fact that 900mg/100g were found in New Zealand spinach (AVRDC data).

Nutrient densities vary from very low to very high among AIVs. On average, indigenous and adapted vegetables are more nutritious then exotic. In particular, values for antioxidant activity were high for indigenous vegetables (Yang et al, 2006a, 2007). Among species, AIVs show a wider content range for most micronutrients. The wide range of nutrient contents indicates that there is great potential to increase nutrient supply by incorporating nutrient-dense vegetables within diets. Among the AIVs currently consumed in East Africa (Weinberger and Swai, 2006) and West Africa (Grubben and Denton, 2004), cowpea leaves (*Vigna unguiculata*), baobab leaves (*Adansonia digitata*), amaranth (*Amaranthus viridis*), spider plant (*Cleome gynandra*), jute mallow (*Corchorus olitorius*), moringa leaves (*Moringa oleifera*) (see Plate 4.2), African nightshade (*Solanum scabrum*), cassava leaves (*Manihot esculenta*), pumpkin leaves (*Cucurbita* spp.) and sweet potato leaves (*Ipomoea batatas*) are nutritious and relatively high in β-carotene and iron content (see Table 4.5). These AIVs deserve greater promotion for household consumption.

Yang et al (2007) studied the wide variation that exists among the 120 vegetable species mostly native to tropical and subtropical zones for vitamins, minerals, polyphenols and oxalate. Values of the most nutritious African

Table 4.4 *Content range of nutrient and anti-nutrient components of three groups of vegetables used in Africa*

Vegetable groups		*DM* (%)	*A* (mg)	*C* (mg)	*E* (mg)	*F* (μg)	*Ca* (mg)	*Fe* (mg)	*Zn* (mg)	*AOA* (μgTE)	*TP* (g)	*Oxa* (mg)
Exotic	Minimum	3.9	0.01	3	0.27	5	18	0.20	0.00	388	80	196
(n = 13)	Maximum	23.2	14.00	242	1.40	159	358	6.50	1.28	685	128	915
	Median	7.8	0.96	27	0.75	31	67	0.99	0.17	560	112	244
	Mean	10.4	2.52	50	0.73	47	123	1.63	0.37	544	107	452
	Standard deviation	6.5	4.70	67	0.46	45	122	1.83	0.39	149	24	402
	N	13	8	13	5	11	12	13	11	3	3	3
Adapted	Minimum	4.2	0.00	3	0.00	3	9	0.21	0.00	276	26	21
(n = 33)	Maximum	41.4	8.29	221	4.52	140	589	8.90	1.48	3838	543	479
	Median	9.4	0.39	30	0.62	37	67	0.98	0.29	768	120	68
	Mean	11.0	1.88	46	1.03	47	135	1.63	0.46	1084	176	193
	Standard deviation	7.2	2.42	49	1.19	35	160	1.87	0.38	972	150	196
	N	31	31	30	20	23	31	31	20	13	13	9
Indigenous	Minimum	4.3	0.02	3	0.05	3	3	0.20	0.00	164	24	8
(n = 73)	Maximum	25.5	12.59	400	17.72	198	711	29.20	3.10	13,506	2140	353
	Median	13.0	2.14	40	1.40	70	172	1.81	0.46	847	204	33
	Mean	14.7	2.78	56	2.23	71	185	3.30	0.56	1683	345	67
	Standard deviation	11.9	2.99	68	3.72	56	165	4.96	0.72	2727	464	95
	N	54	30	39	21	15	54	45	17	22	21	14

Notes: Species included for calculation are listed in Table 4.5. Vegetable groups were classified according to PROTA. Content ranges (per 100g fresh weight) were calculated using available data listed in Table 4.2. Data for AOA, TP and Oxa were calculated from AVRDC data (unpublished data).

DM = dry matter; A = β-carotene; C = vitamin C; E = ·α-tocopherols; AOA = antioxidant activity; TE = tolox equivalent; TP = total phenolics; Oxa = oxalate.

Vegetable grouping are according to PROTA. Exotic = vegetables that have been recently introduced; adapted = vegetables that were introduced long ago and are now widely adapted; indigenous = vegetables that are native to Africa.

vegetables (see Box 4.1) are comparable to the maximum values of this data set. Palatability of these high-nutrient AIVs is a concern because of their high dry matter and crude fibre content. However, high dry matter could be an advantage to preserve the nutritional values of dried vegetables because shorter processing time is required, thus reducing nutrient losses.

Exceptionally high values for the iron content of AIVs have been reported (see Table 4.5) (Grubben and Denton, 2004). Iron contamination in African plant foods from soil and food processing facilities were also reported (Bothwell et al, 1979; Hallberg and Bjorn-Rasmussen, 1981; Abebe et al, 2007). Contaminant iron could significantly elevate total iron content of plant foods more than ten times higher than plant intrinsic iron. Calcium contamination from soil was also reported for vegetables containing exceptionally high calcium content (Abebe et al, 2007). The bioavailability of contaminant iron and calcium remains uncertain (Derman et al, 1982; Harvey et al, 2000; Hooda et al, 2004; Teucher et al, 2004). Studies confirming the iron and calcium contents of AIVs reported to contain high levels of these nutrients should be carried out under conditions that avoid contamination.

Box 4.1 Top three African indigenous vegetables high in nutrients and anti-nutrients

- *β-carotene: Moringa oleifera* (drumstick tree; adapted, leaf, 19.74mg); *Manihot esculenta* (cassava; indigenous, leaf, 16.87mg); and *Cnidoscolus aconitifolius* (chaya; exotic, leaf, 14mg).
- *Vitamin C: Moringa stenopetala* (cabbage tree; indigenous, leaf, 400mg); *Manihot esculenta* (cassava; indigenous, leaf, 387mg); and *Moringa oleifera* (drumstick tree; adapted, leaf, 274mg).
- *Vitamin E: Moringa stenopetala* (cabbage tree; indigenous, leaf, 17.72mg); *Moringa oleifera* (drumstick tree; adapted, leaf, 13.44mg); and *Manihot esculenta* (cassava, indigenous, leaf, 12.77mg).
- *Folates: Cleome gynandra* (spider plant; indigenous, leaf, 198μg); *Brassica rapa* (neep crops, caisin, turnip, Chinese cabbage, pak choi; exotic, leaf, 159μg); and *Celosia argentea* (celosia; indigenous, leaf, 159μg).
- *Calcium: Moringa stenopetala* (cabbage tree; indigenous, leaf, 711mg); *Senna obtusifolia* (sicklepod; adapted, leaf, 589mg); and *Moringa oleifera* (drumstick tree; adapted, leaf, 584mg).
- *Iron: Triumfetta annua* (burweed; indigenous, leaf, 29.2mg); *Hibiscus cannabinus* (kenaf; indigenous, leaf, 12.1mg); and *Cucumis africanus* (indigenous, leaf, 12mg).
- *Zinc: Pterocarpus mildbraedii* (indigenous, leaf, 3.10mg); *Moringa oleifera* (drumstick tree; adapted, leaf, 2.80mg); and *Manihot esculenta* (cassava; indigenous, leaf, 1.79mg).
- *Antioxidant activity: Adansonia digitata* (baobab; indigenous, leaf, 13,506 TE); *Rorippa madagascariensis* (watercress; indigenous, leaf, 12,839 TE); and *Manihot esculenta* (cassava; indigenous, leaf, 7527 TE).
- *Total phenolics: Rorippa madagascariensis* (watercress; indigenous, leaf, 2189mg), *Adansonia digitata* (baobab; indigenous, leaf, 2140mg); and *Manihot esculenta* (cassava; indigenous, leaf, 1863mg).
- *Oxalate: Tetragonia tetragonioides* (New Zealand spinach; exotic, leaf, 915mg); *Amaranthus dubius* (amaranth; adapted, leaf, 479mg); and *Amaranthus cruentus* (amaranth; adapted, leaf, 408mg).

Table 4.5 *Nutrient contents and bioactives of vegetables in tropical Africa*

Number	*Species*	*DS*	*DM (%)*	*Pro (g)*	*A (mg)*	*C (mg)*	*E (mg)*	*F (µg)*	*Ca (mg)*	*Fe (mg)*	*Zn (mg)*	*Phytochemicals/bioactives*
Well-known, widely distributed vegetables												
1	*Allium ampeloprasum* (leek, great-headed garlic, pearl onion,	3	17	1.5	1.0	12	0.9	64	59	2.1	0.1	Chrysoeriolmalonic acid, phenylpropanoids, sesquiterpenoids, saponins, propane-L-thiol; anti-tumour activity; antibacterial and antifungal activities; anti-hyperglycaemic and anti-asthmatic properties
2	*Allium cepa* (onion, shallot; a, b)	3	10.9	1.1	0.0	7	0.0	19	23	0.2	0.2	
3	*Allium fistulosum* (welsh onion; e, l)	3	9.5	1.9		27		16	18	1.2	0.5	
4	*Allium sativum* (garlic; a, b)	3	41.4	6.4	0.0	31	0.1	3	181	1.7	1.2	
5	*Apium graveolens* (celery; e, l)	3	4.6	0.7	0.3	3	0.3	36	40	0.2	0.1	
6	*Benincasa hispida* (wax gourd; e, l)	3	3.9	0.4		13		5	19	0.4	0.6	Anti-ulcer activity, histamine-release inhibitors alnusenol, multi-florenol
7	*Beta vulgaris* (garden beet; e, l)	2	6.3	1.5	0.9	13	0.8	31	46	0.4	0.3	Glucopyranosides, L-azetidine 2-carboxylic acid, flavonoids, phenylpropanoids, galactinol, hypox-anthine, octacosan-1-ol, triterpenoid saponins, skatole, betalain alkaloids
8	*Brassica oleracea* (headed cabbage; a, l)	1	9.9	1.7	0.4	49		75	52	0.7	0.3	Glucosinolates, cyanin, hexacosane-1-ol, indole-3-acetonitrile, nonacosane, sinapic acid, triacontan-1-ol
9	*Brassica oleracea* (kohlrabi; e, l)	1	8.3	1.6		43		82		0.3	0.1	
10	*Brassica oleracea* (leaf cabbage; a, l)	1	15.5	3.3	4.6	120		29	135	1.7	0.4	
11	*Brassica rapa* (neep crops, caisin, turnip, Chinese cabbage, pak choi; e, l)	3	7.8	2.2		130		159	210	1.5	0.2	Glucosinolates; antibacterial and antifungal activity
12	*Capsicum annuum* (capsicum; a, l)	2	11.8	4.4	2.5	93	0.4	53	188	2.1	0.8	Carotenoids and steroidal glycosides
13	*Cucumis anguria* (West Indian gherkin; i, f)	1	7	1.4	0.1	51			26	0.6		Cucurbitacins

14	*Cucumis melo* (melon; i, f)	2	6.3	0.9	0.0	14	0.1		9	0.7		
15	*Cucumis sativus* (cucumber; a, f)	3	4.8	0.7	0.0	3	0.0	7	16	0.3	0.2	Cucurbitacins, galactinol, nona-2,6-dienal, tryptamine
16	*Daucus carota* (carrot; a, r)	3	11.7	0.9	8.3	6	0.7	19	33	0.3	0.2	Sesquiterpenoids, anthocyanins, phenylpropanoids, falcarindiol, flavonoids, lycopene, miscellaneous phenolics, coumarins
17	*Lactuca sativa* (lettuce; a, l)	3	4.4	0.9	0.3	3	0.2	29	18	0.4	0.2	Chioric acid, dihydroconiferyl alcohol, lactucaxanthin, plastocyanin
18	*Lycopersicon esculentum* (tomato; a, f)	2	4.6	0.9	0.2	30	0.9	5	9	0.6	0.2	6-methyl-5-hepten-2-on, β-ionone, β-damascenone, geranylaceton, bergapten, α-γ-ζ-falcarindiol, fusaric acid, hentriacontane, carotenoids, prunin, serotonin, spermidine, steroidal alkaloids, tryptamine
19	*Phaseolus vulgaris* (French bean; a, f)	1	9.3	1.9	0.3	12		80	36	1.2	0.2	2-carboxyarabinitol-1-phosphate, 3′:5′-cyclic amp, glucosamine, nortriterpenoids, phaseolic acid, flavonoids, phaseoloside d, phaseolotoxin, phytohaemagglutinin, L-pipecolic acid, propane-1-thiol, 5-ribosyluracil
20	*Pisum sativum* (field pea, garden pea, sugar pea; i, l)	2	4.9	2.8	0.9	13			3	0.7		Cadaverine, L-noradrenaline, pinolidoxin, flavonoids, propane-L-thiol and anthocyanins
21	*Praecitrullus fistulosus* (tinda; e, f)	1	6.5	1.4	0.0	18			25	0.9		
22	*Raphanus sativus* (radish; a, l)	3	4.7	0.7	0.0	15	0.0	25	25	0.3	0.3	Glucosinolates, phenylpropanoids, methyl mercaptan; fungicidal, bactericidal, nematicidal, allelopathic properties and against cancer
23	*Rheum x hybridum* (rhubarb; e)	1	6.4	0.9	0.0	8		7	86	0.2	0.1	Oxalate, malate, anthocyanins and anthraquinone
24	*Solanum melongena* (eggplant; a, f)	2	7.5	1.5	0.0	13	0.0	5	9	0.9	0	Steroidal alkaloids, tropane alkaloids, flavonoids, anthocyanins, sesquiterpenoids, and melongosides; antioxidant activity and hypolipidemic action
25	*Talinum triangulare* (waterleaf; i, l)	2	6.2	1.9	3.3	3	0.9		41	1		Oxalate, hydrocyanic acid and saponins
26	*Telfairia occidentalis* (fluted pumpkin; i, l)	1	13.6	2.9								Cyanide and tannins
26	*Telfairia occidentalis* (fluted pumpkin; i, s)	1	93.8	20.5					84			

Table 4.5 *continued*

Number	Species	DS	DM (%)	Pro (g)	A (mg)	C (mg)	E (mg)	F (μg)	Ca (mg)	Fe (mg)	Zn (mg)	Phytochemicals/bioactives
27	*Vigna unguiculata* (cowpea, yard-long bean, catjang cowpea; i, l)	2	11.8	4.5	2.9	75	3.0	123	275	1.9	0.6	Diterpenoids, flavonoids, benzofurans
Species deserving more attention from extension and research												
28	*Abelmoschus caillei* (West African okra; i, f)	1	15	3.3	0.3	53			308	1		
29	*Abelmoschus esculentus* (common okra; i, f)	2	8.9	1.8	0.4	37	0.5	32	44	0.9	0	
30	*Acanthophoenix rubra* (barbel palm; i, l)											Catecholases
31	*Acanthosicyos horridus* (nara; i, s)	1	94.7	30.7					100	4	5.5	Cucurbitacins
31	*Acanthosicyos horridus* (nara; i, f)	1	16	1.4	0.1				21	0.5	0.6	
32	*Acanthosicyos naudinianus* (herero cucumber; i, f)	1	9.4	1.3		35			21	0.5		Cucurbitacins
33	*Acmella oleracea* (toothache plant; a)											Spilanthol, monoterpenoids and sesquiterpenoids; antibacterial activity and insecticide
34	*Adansonia digitata* (baobab; i, l)	2	18	3.8	2.7	69	3.5	35	264	2	0.6	Antiviral and antibacterial activities
35	*Amaranthus blitum* (amaranth; a, l)	1	11.1	3.5	1.7	42		85	270	3		Anthocyane, nitrate, oxalate, tannin, amasteroland and betalain alkaloids
36	*Amaranthus cruentus* (amaranth; a, l)	2	9.4	3.2	1.8	36	1.1	52	305	3.8	0.7	
37	*Amaranthus dubius* (amaranth; a, l)	2	11.2	3.5	3.1	78	2.1	82	582	3.4	1.5	
38	*Amaranthus tricolor* (amaranth; e, l)	2	11.7	3.9	1.8	62	0.3	31	358	2.4	0.8	

39	*Amaranthus viridis* (green amaranth; a, l)	1	16	4.6	5.7	64			410	8.9		
40	*Asystasia gangetica* (tropical primrose; i, l)	2	12	4.0	6.6	47	2.7	27	145	2.3	0	Analgesic and anti-asthmatic properties
41	*Basella alba* (ceylon spinach; a, l)	1	7	1.8	2.4	102		140	109	1.2		Triterpene oligoglycosides; antifungal properties
42	*Brassica carinata* (Ethiopian kale; i, l)	2	10.4	3.2	0.9	157	1.3	86	157	1.3	0.9	
43	*Brassica juncea* (brown mustard; a, l)	2	6.8	2.1	1.2	72	1.8	41	138	0.7	0.2	Glucosinolates
44	*Canavalia gladiata* (sword bean; a, f)	1	16.4	4.6	0.0	32			33	1.2		Toxic storage proteins
45	*Celosia argentea* (celosia; i, l)	2	10.4	3.7	4.1	35	1.4	159	162	2.3	0.3	Phytic acid and oxalic acid
46	*Celosia trigyna* (silver spinach; i, l)	1	11	2.7	1.9	10			154	5		Anthelmintic and acaricidal properties
47	*Ceratotheca sesamoides* (false sesame; i, l)	1	19	4.2		28			300	3.2		
48	*Citrullus lanatus* (egusi melon, cooking melon, watermelon; i, f)	3	8.6	0.6	0.3	8	0.1	3	7	0.2	0.1	Lycopene
49	*Cleome gynandra* (spider plant; i, l)	2	10.2	4.4	2.7	113	0.7	198	186	2.2	0.8	Glucosinolates; exhibit repellent and acaricidal properties, anti-HIV and antibacterial activities
50	*Cnidoscolus aconitifolius* (chaya; e, l)	1	22.5	6.0	14.0	242			320	6.5		Cyanogenic glycosides, flavonoids and proteolytic enzymes
51	*Coccinia grandis* (ivy gourd; i, l)	2	13	3.9	1.9	41	1.9	98	38	1.8	0	
52	*Coccinia sessilifolia* (wild cucumber; i, l)	1	17.7	2.1		25			38	0.2		
52	*Coccinia sessilifolia* (wild cucumber; i, r)	1	16	1.0		7			351	2.2		
53	*Colocasia esculenta* (taro; a, l)	2	6.5	0.3	0.2	5			79	0.4		
54	*Corchorus asplenifolius* (wild jute mallow; i, l)											

Table 4.5 *continued*

Number	*Species*	*DS*	*DM (%)*	*Pro (g)*	*A (mg)*	*C (mg)*	*E (mg)*	*F (µg)*	*Ca (mg)*	*Fe (mg)*	*Zn (mg)*	*Phytochemicals/bioactives*
55	*Corchorus olitorius* jute mallow; i, l)	2	15	4.8	4.7	105	3.6	92	259	4.5	0.4	Phenylpropanoids and ionone glucosides; inhibitory activity on histamine release
56	*Crassocephalum crepidioides* (ebolo; i, l)	1	20.1	3.2					260			Monoterpenoids and sesquiterpenoids; antimutagenic activity
57	*Crassocephalum rubens* (yuruban bologi; i, l)	1	20.1	3.2					260			
58	*Crotalaria brevidens* (slenderleaf; i, l)	1	25.5	8.8					222	0.8		Pyrrolizidine alkaloids, monoterpenoids and phenolic compounds
59	*Crotalaria ochroleuca* (rattlepod; i, l)	1	25.5	8.8					222	0.8		Pyrrolizidine alkaloids, monoterpenoids and phenolic compounds
		1		4.6	5.8	122			270	4		
60	*Cucumeropsis mannii* (egusi-itto; i, s)	1	91.7	26.2					86			
61	*Cucumis africanus* (i, l)	1	7.8	1.3		81			216	12		Cucurbitacins; anti-tumour, anti-inflammatory and analgesic activities
61	*Cucumis africanus* (i, f)	1	11.8	2.8		13			13	1.1		
62	*Cucumis metuliferus* (horned melon; i, f)	1	9	1.1		19			12	0.5		Cucurbitacins; anti-tumour and anti-inflammatory activities
63	*Cucurbita ficifolia* (fig-leaf gourd; a)											
64	*Cucurbita maxima* (pumpkin; a, f)	2	7.1	2.3	0.2	18	1.7	36	65	1.1	0.5	Spinasterol; proteinase inhibitors
65	*Cucurbita moschata* (musk pumpkin; a, f)	3	13.6	1.0	4.2	21	1.4	27	48	0.7	0.2	Cucurbitacins
66	*Digera muricata* (i)											
67	*Dypsis baronii* (sugarcane tree; i)											
68	*Eruca vesicaria* (garden rocket; a, l)	1	8.2	2.7	1.4				352	0.8		Glucosinolates
69	*Erythrococca kirkii* (i)											
70	*Gnetum africanum* (eru; i)											C-glycosylflavones

71	*Gnetum buchholzianum* (eru; i)											C-glycosylflavones
72	*Guizotia scabra* (i)											
73	*Hibiscus acetosella* (false roselle; i)											
74	*Hibiscus cannabinus* (kenaf; i, l)	1	21	5.5		75			484	12.1		
75	*Hibiscus sabdariffa* (roselle; i, l)	2	10.8	2.5	2.5	28	1.9		183	1.2		Antispasmodic and anthelminthic; bactericidal properties
76	*Hibiscus surattensis* (wild sour; i, l)											
77	*Ipomoea aquatica* (kangkong; i, l)	2	11.3	2.4	0.4	40	1.8	29	220	2.5	0.5	
78	*Justicia ladanoides* (justicia; i, l)	1	13.2	3.3					510			
79	*Kedrostis pseudogijef* (i)											
80	*Lablab purpureus* (lablab; i, p)	2	12.1	2.9	0.9	21	0.1	31	45	1	0.6	Cyanogenic glucosides, trypsine and flavonoids
81	*Lagenaria siceraria* (bottle lettuce; i, f)	2	5.1	0.6	0.0	5			7	0.3		Cucurbitacins
82	*Launaea cornuta* (bitter lettuce; i, l)	1	13.2	3.9		19			214	7.2		
83	*Launaea taraxacifolia* (yanrin; i, l)	1	15.7	3.2					326			
84	*Lemuropisum edule* (tara nut; i, l)											
85	*Leptadenia hastata* (i, l)	1	19	4.9	4.9	78			417	5.4		
86	*Luffa acutangula* (ridged gourd; a, l)	1	5.8	0.8	0.0	3		37	12	0.3		
86	*Luffa acutangula* (ridged gourd; a, f)	1	11	5.1	9.2	95			56	11.5		
87	*Micrococca mercurialis* (i)											
88	*Momordica charantia* (bitter gourd; i, f)	2	5.7	0.8	0.3	75	1.3		11	0.4		Cucurbitacins and lycopene
89	*Moringa oleifera* (drumstick tree; a, l)	2	24.9	8.6	19.7	274	13.4		584	10.7	2.8	Glucosinolates, oxalic acid and flavonoids; bactericide and fungicide

Table 4.5 *continued*

Number	*Species*	*DS*	*DM (%)*	*Pro (g)*	*A (mg)*	*C (mg)*	*E (mg)*	*F (μg)*	*Ca (mg)*	*Fe (mg)*	*Zn (mg)*	*Phytochemicals/bioactives*
90	*Moringa stenopetala* (cabbage tree; i, l)	2	23.8	5.8	12.6	400	17.7		711	5.4		Glucosinolates; lower blood glucose concentration
91	*Myrianthus arboreus* (giant yellow mulberry; i, l)	1	14.5	1.9					44	1.1		
92	*Portulaca oleracea* (purslane; i, l)	2	4.3	1.3	2.4	9	0.4		39	1.1		Oxalate, nitrate, alkaloid norepinephrine, L-noradrenaline and acylated betacyanins; antioxidant and anti-inflammatory effects
93	*Psophocarpus scandens* (African winged bean; i, l)	1	18	7.1					565			
93	*Psophocarpus scandens* (African winged bean; i, f)	1	13	3.6					297			
94	*Psophocarpus tetragonolobus* (winged bean; e, l)	3	23.2	5.9		45		16	224	4	1.3	
95	*Pterocarpus mildbraedii* (i, l)	1	15	3.8					72	4.7	3.1	Hydrogen cyanide and oxalate
96	*Rorippa madagascariensis* (watercress; i)	2	5.6	2.2	3.0	45	0.9		146	0.7		
97	*Rumex abyssinicus* (sorrel; i)	2	6.2	1.5	0.5	22	0.8	16	46	0.5	0.3	Oxalic acid and quinones; anti-inflammatory and anti-viral activities; promotes regeneration of epithelial cells
98	*Salicornia pachystachya* (glasswort; i)											Rich in salt and the ash was used for washing.
99	*Sechium edule* (chayote; a, f)	3	5.8	0.8	0.0	8	0.1	93	17	0.3	0.7	Anti-mutagenic activity
100	*Senna obtusifolia* (sicklepod; a, l)	1	20.3	5.6	7.9	113			589	5.9		Quinones and xanthones; laxative properties and herbicide
101	*Sesamum radiatum* (black benniseed; i)											Lignans; antioxidant activity, anti-inflammatory, anti-hypertensive, anti-tumour and insecticidal activities

102	*Solanecio biafrae* (worowo; i)											Sesquiterpene
103	*Solanum aethiopicum* (African eggplant; i, f)	2	8.2	1.1	0.5	10	0.2	10	9	1	0	Sitosterol glucosides, oxalate, steroidal alkaloids and steroid saponins; antifungal activity and glycocorticoid effects
104	*Solanum americanum* (glossy nightshade; i)											Steroidal alkaloids and steroid saponins; diarrhoea and cardiac arrest
105	*Solanum anguivi* (i)											Steroidal glycosides
106	*Solanum macrocarpon* (gboma; i, l)	1	14.4	4.6					391			Steroidal alkaloids; acaricidal activity and poisonous
106	*Solanum macrocarpon* (gboma; i, f)	1	11	1.4					13			
107	*Solanum scabrum* (African nightshade; i, l)	2	10.5	4.4	5.8	75	2.3	70	194	3	0.5	Steroidal alkaloids; diarrhoea, vomiting, dizziness, blindness and bitterness
108	*Solanum tarderemotum* (black nightshade; i)											
109	*Solanum torvum* (pea eggplant; a)	2	20.4	2.7	0.2	15	0.6		82	1		Steroidal alkaloids, isoflavonoid and steroidal glycosides; hepatic tumours, antimicrobial and antiviral activities
110	*Solanum villosum* (red-fruited nightshade; a, l)	2	11.1	4.2	2.9	79	2.1	61	175	3.3	0.8	Diosgenin and steroidal alkaloids
111	*Talinum paniculatum* (flameflower; a, l)	2	9.1	2.8	6.1	74	4.5		106	1.5		
112	*Tetragonia tetragonioides* (New Zealand spinach; e, l)	2	7.3	1.9	2.1	34	1.4	68	74	1	0	Oxalate and saponin; anti-ulcerogenic activity
113	*Trichosanthes cucumerina* (snake gourd; a, f)	2	4.2	0.6	0.1	8	0.0		18	0.4		
114	*Triumfetta annua* (burweed; i, l)	1	21.6	4.2					392	29.2		
115	*Vernonia amygdalina* (bitterleaf; i, l)	1	17.4	5.2		51			145	5		Sesquiterpene lactones and steroid glucosides; anti-aggregating properties, cytostatic action, anti-parasitic, anti-hepatotoxic, anti-leishmanial , antibacterial, antiviral and cytotoxic activities

Table 4.5 *continued*

Number	Species	DS	DM (%)	Pro (g)	A (mg)	C (mg)	E (mg)	F (µg)	Ca (mg)	Fe (mg)	Zn (mg)	Phytochemicals/bioactives
116	*Vernonia hymenolepis* (sweet bitterleaf; i, l)	2	13.6	4.2	4.4	58	1.6	74	257	2	0.4	Sesquiterpene lactones; anti-tumour activity and anti-aggregating properties
117	*Ipomoea batatas* (sweet potato; l)	2	12.3	2.3	3.0	40	3.1		272	3.9	0.5	
118	*Cucurbia* spp. (pumpkin; l)	2	19.8	4.4	3.5	77	8.6	84	246	3.6	1.3	
119	*Manihot esculenta* (cassava; l)	2	24.5	8.1	16.9	387	12.8	95	403	4.5	1.8	

Notes: Nutrient content values are based on 100g fresh weight of edible portion.

DS = data source (1 = PROTA, 2 = AVRDC and 3 = USDA databases); 0.0 = value below detective level; DM = dry matter; Pro = protein;

A = β-carotene; C = vitamin C; E = ·α-tocopherol;

F = folates.

Bioactive compounds were retrieved from PROTA (Grubben and Denton, 2004) and a phytochemical dictionary (Harborne et al, 1999).

Nutrients, organic acids and flavonoids are universal and thus were not included.

Species lists are set out according to the priority groups of PROTA (2005). Parts for analysis and vegetable groups were listed after each vegetable name: l = leaf, shoot or aerobic part; f = fruit; s = stem or tube; r = root; b = bulb. Vegetable groups: e = exotic; a = adapted; i = indigenous.

Vegetable items 117–119 were not included in PROTA (2005).

Health-promoting properties

Nutritional components in plants include 90 to 98 per cent macro-constituents such as protein, carbohydrate and oil, and 1 to 10 per cent micro-constituents, including at least 17 essential minerals, 13 essential vitamins and an abundance of secondary metabolites estimated at greater than 200,000 compounds (Wildman, 2001). Some phytochemicals have positive physiological effects on humans and act as protective agents for health maintenance and promotion, or possess medicinal properties. Beneficial phytochemicals include, for example, flavonoids, carotenoids, glucosinolates and allyl-sulphur compounds, which function as antioxidant, anti-cancer, anti-inflammatory or anti-microbial compounds or influence blood lipid profiles (Hasler, 2000).

Flavonoids, in particular, are abundant in plants and function to produce colour and act as antioxidants to protect plants against UV-radiation (Winkel-Shirley, 2002). Oxidative stress is an underlying factor in health and disease in humans (Halliwell, 1993; Aruma, 1994; Hughes, 2006). A proper balance of oxidants and antioxidants is important to maintain health and improve longevity. Altering the balance in favour of oxidants may result in pathological responses. Work that profiled the flavonoids of more than 100 plant species, including mostly Asian and African indigenous, found that species high in flavonoids were among those most underutilized (Yang et al, 2008). AIVs, including baobab tree leaves (*Adansonia digitata*: 44mg), cowpea leaves (Vigna unguiculata: 125mg), moringa leaves (*Moringa oleifera*: 129mg), jute mallow (*Corchorus olitorius*: 64mg), Ethiopian kale (*Brassica carinata*: 56mg) and sweet potato leaves (*Ipomoea batatas*: 125mg) are good sources of flavonoids, and are rich in other dietary antioxidants as well, including carotenoids, vitamin C and tocopherols (see Table 4.5). These vegetables could contribute to diets high in antioxidants and micronutrients recommended to supplement medicinal therapy in fighting HIV/AIDS (Friis, 2002).

Some AIVs showed hypoglycemia activity as anti-diabetic agents (Katerere and Eloff, 2006). The dietary approach for diabetes management has been practised using bitter gourd (*Momordica charantia*), cowpea (*Vigna unguiculata*) and its leaves, wild cabbage (*Brassica oleracea*), ivy gourd leaves (Coccinia grandis), bitter melon/gourd (*Momordica charantia*), French beans (*Phaseolus vulgaris*) and jute mallow (*Corchorus olitorius*). Supplementation with vitamin A and a mixture of micronutrients is recommended to potentiate host resistance to malaria through enhanced parasite clearance and inhibition of excessive pro-inflammatory response (Serghides and Kain, 2002). Furthermore, micronutrients as therapeutic tools in the management of infectious diseases (Friis et al, 2002; Tomkins, 2005) and sickle cell disease were practised and discussed elsewhere (Okochi and Okpuzor, 2005). Application of processing AIVs high in micronutrients as dietary or herbal supplements in infectious disease management merits more exploration.

Nutrient retention and bioavailability

Nutrient values of vegetables are affected by variety, crop management, processing, cooking methods, and bioavailability. Nutrient values could be improved by selecting species and varieties high in nutrient content. Crop management practices such as fertilizer types and amounts, water management and choice of season strongly influence nutrient content. AIVs are usually grown in home gardens without fertilizers or are harvested in the wild, so there is the potential to improve the nutrient content of AIVs through better management practices. In addition, nutrient losses can be minimized by improved post-harvest handling and modification of current food practices, such as reduced time of thermal treatment, improved drying process, avoiding the chopping of vegetables before washing, and adding vegetables to boiling water instead of cold water for cooking.

The potential contribution of plant foods to micronutrient status depends upon the retention of the nutrients after processing and cooking. It is important to determine the nutrient content of cooked/processed foods and to verify losses during cooking and processing. Nutrient data for cooked/processed foods commonly consumed in Africa are insufficient due to high costs of chemical analysis, and the variety of different food preparation methods and conditions. Data for a specific dish may not be applicable to another dish with the same ingredients. Development of algorithms to predict nutrient retention of African dishes based on knowledge of traditional food methods by locations or by at-risk populations, as well as nutrient data of raw ingredients, could be an alternative way to calculate nutrient supply in diets.

Bioavailability measures the amounts of nutrient absorbed by the body from a food, and is influenced by diet- and host-related factors (Gibson, 2007). The dietary factors affecting bioavailability include the nature of nutrients in the food matrix and interactions of nutrient–food components in plant foods and during food processing. Iron, calcium, zinc and vitamin carotenoids, folate and niacin are most affected by their chemical form. For example, haem iron in animal food is more bioavailable (25 per cent) than non-haem iron in plant foods (5 per cent). Natural folates in a polyglutamate form are less available than the synthetic monoglutamate form (Gibson, 2007).

The food matrix and food processing influence carotenoid bioavailability, which in vegetables is highest in squash, yams or sweet potato cooked with some oil and lowest in raw green leafy vegetables (IOM, 2000; Underwood, 2000; Faber and van Jaarsveld, 2007). Most AIVs are green leafy vegetables; so chopping them prior to cooking with some oil could release and dissolve β-carotene and other carotenoids and increase bioavailability.

Poor mineral bioavailability is particularly problematic in diets based on unrefined cereals, starchy roots and legumes with a high content of mineral inhibitors such as phytate, polyphenols and crude fibres. Oxalate in green leafy vegetables is also an inhibitor of minerals, particularly for calcium. The non-competitive interactions of minerals and these inhibitors form insoluble complexes during food preparation and the digestion process, thus rendering the dietary iron insoluble (Lee, 1982). Organic acids such as citric acid and ascorbic acid are enhancers for minerals and may form soluble ligands with minerals in the gastrointestinal tract. Ethiopian kale (*Brassica carinata*), chaya (*Cnidoscolus aconitifolius*), Ceylon/Malabar spinach (*Basella alba*) and particularly moringa leaves (*Moringa oleifera*) are high in ascorbic acid and minerals and low in oxalate content. These AIVs could be a relatively good source of bioavailable minerals. Ethiopian kale (*Brassica carinata*) was suggested to be a good calcium source for Sidama households in Ethiopia due to its lack of phytate and low oxalate content (Weaver et al, 1999; Bohn et al, 2004).

Traditional food preparation methods such as thermal processing, baking, boiling, germination and malting, village-based milling or home pounding and microbial fermentation to improve micronutrient bioavailability of African diets at household level are discussed elsewhere (Gibson and Hotz, 2001; Gibson et al, 2006; Hotz and Gibson, 2007). Community-based dietary interventions using traditional food methods and modified diets by incorporating food items with mineral enhancers were investigated to enhance the micronutrient adequacy of plant-based staples diets (Yeudall et al, 2005; Gibson et al, 1998; Gibson and Hotz, 2001). Previous work using tomato and moringa as enhancers to increase iron bioavailability of cereal-, rice- and legume-based meals was tested *in vitro* and showed increased accessible iron in legume-based meals (AVRDC, 2000). Tomato–mungbean recipes were designed accordingly (Subramanian and Yang, 1998; Bains et al, 2003) and proved to improve haemoglobin levels of school children (Vijayalakshmi et al, 2003). Diet modification by cooking staple food with tomato and selected AIVs that are low in mineral inhibitors and high in mineral enhancers and micronutrients could be further investigated.

Nutrition and health outcome

Status: Nutritional situation in rural and urban Africa

While protein-energy malnutrition (PEM) is suggested to be the most lethal form of malnutrition, micronutrient deficiency (MND) is the most serious threat to the health and development of populations the world over, particularly to pre-school children and pregnant women in low-income countries (WHO, 2004). Micronutrient deficiency, also called 'hidden hunger', was not addressed before the 1980s, when development practitioners began to be concerned with whether people were getting enough micronutrients – instead of being concerned only with whether people were getting enough energy (Fritschel, 2000). Although data on people affected by MND are scarce or even non-existent for some countries, available estimates are cause for alarm. Vitamin A deficiency (VAD) is the cause of night blindness as the first stage in a set of increasingly severe eye problems, as well as the cause of an impaired resistance to infections (Benson, 2004). An estimated 100 million to 250 million pre-school children alone are affected by severe VAD (Fritschel, 2000). In Africa, about 30 per cent of all pre-school children are affected and in single countries these figure can reach up to more than 60 per cent (see Figure 4.5).

Iron-deficiency anaemia (IDA) affects more than half of pregnant women and school-age children (Fritschel, 2000) and can cause complications in

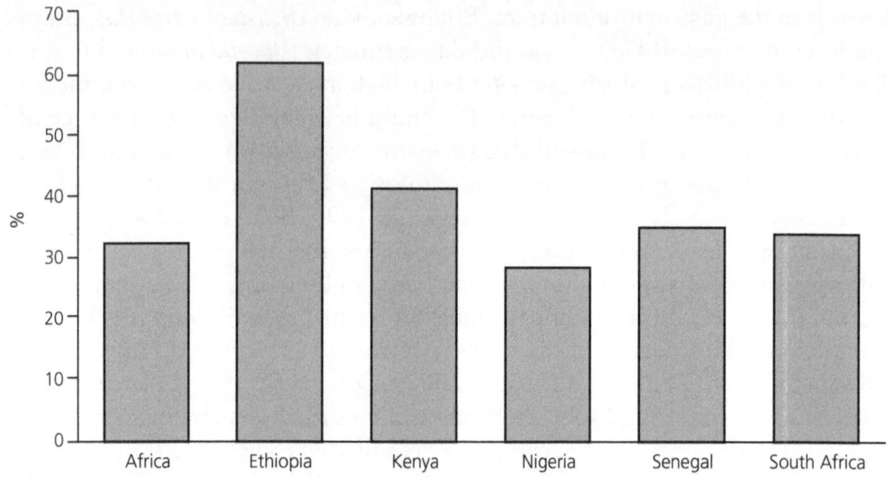

Source: adapted from SCN (2004)

Figure 4.5 *Prevalence of pre-school children (<5 years) with vitamin A deficiency (VAD) in Africa and selected African countries*

pregnancy, maternal mortality, premature birth and low birth weight (Benson, 2004). If mild and moderate anaemia is taken into account, in some African countries more than 80 per cent of pre-school children are affected (see Figure 4.6). In general, one third of the diseases in Africa caused by under-nutrition can be attributed not to insufficient calories or protein, but to deficiencies in micronutrients – mainly iron and vitamin A, as well as iodine and zinc (Benson, 2004).

Vegetables are an efficient source of micronutrients, both with respect to unit cost of production and per unit of land (Ali and Tsou, 1997). Therefore, the incorporation of vegetables as a food source rich in micronutrients provides one strategy to sustainably improve the micronutrient status in the human body. While this is known and acknowledged by the research community, this message has not yet reached consumers, especially in developing countries. While several promotion activities, such as the 5 a Day programme to enhance vegetable and fruit consumption, are already established in developed countries, fruit and vegetable promotion initiatives are only about to be established in some developing countries (WHO, 2003a).

For example, when knowledge of dietary and behaviour-related determinants of non-communicable diseases was assessed among urban Senegalese women, the protective effect of fruit and vegetables was least understood by study participants (Holdsworth et al, 2006). In Nigeria, the high prevalence of micronutrient deficiencies was found to be due to seasonal variation in vegetable production, as well as inadequate processing and preservation of vegetables. Furthermore, culture was an important factor that may limit adequate consumption of leafy vegetables even if they are available in abundance (Hart et al, 2005).

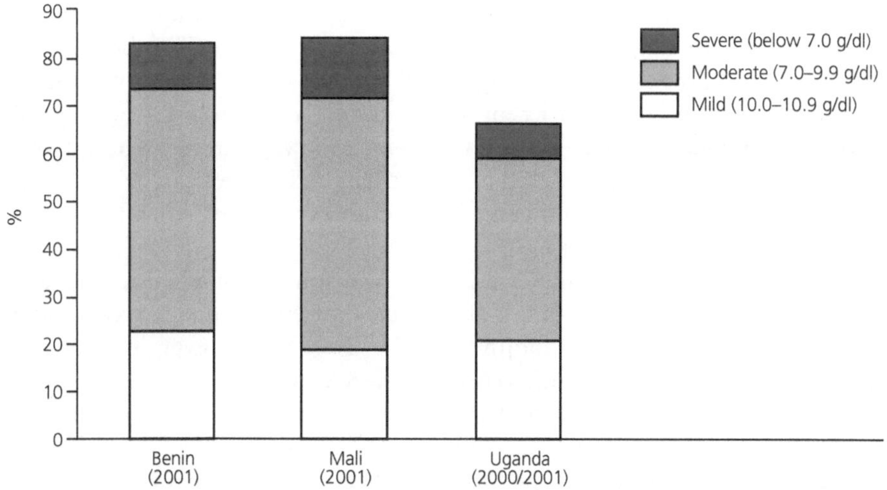

Source: adapted from SCN (2004)

Figure 4.6 *Prevalence of anaemia in pre-school children (aged 6 to 59 months) in selected African countries*

Contribution: Nutrient intake from vegetables

Traditional leafy vegetables are among the most nutritive vegetables both on a fresh weight basis and when prepared, and provide significant amounts of provitamins A and C and several minerals (Abukutsa-Onyango, 2003; Bosland, 2003; IPGRI, 2003; Weinberger and Msuya, 2004). Moreover, leafy vegetables grow rapidly and allow several harvests per season and are, therefore, among the world's most productive plants in terms of nutritional value per unit. For example, 1.5 cups of cooked spinach (*Spinacia oleracea*), which is comparable to amaranth (*Amaranthus* spp.) in its constituents (only the content of vitamin A is higher in spinach), has 40 calories and provides 70 per cent of the recommended daily allowance (RDA) for adults of vitamin A, 25 per cent of the RDA of vitamin C, and 20 per cent of the RDA of iron (Bosland, 2003). In terms of iron, traditional leafy vegetables such as amaranth and African nightshade (*Solanum* spp.), with as high as 37mg iron per 100g edible portion, are an even better source than spinach, with less than 2mg (Weinberger and Msuya, 2004).

In Yaoundé (Cameroon) it was found that the nutritional contribution of AIVs was extremely high. This was relative to other ingredients that are typically used to prepare the sauces for accompanying starchy staples and especially so among the poor. Since traditional leafy vegetables were the most important sources of calcium, vitamin B1 and B2, and iron, it was concluded that promoting traditional leafy vegetable consumption among the poor will help to address nutritional deficits (Gockowski et al, 2003).

Similarly, in Tanzania, it was revealed that mainly poor households rely on the consumption of AIVs to fulfil their daily requirements of micronutrients. This especially concerns vitamin A and iron; in fact, about 50 per cent of all vitamin A requirements and slightly less than one third of iron requirements were consumed through indigenous vegetables in Tanzania (Weinberger and Msuya, 2004). In the same study it appeared that the share of exotic vegetables was much lower and only 3 per cent of all vitamin A and only 1.5 per cent of all iron requirements were consumed through exotic vegetables. Hence, it was concluded that part of the observed micronutrient deficiencies can be related to a fading importance of indigenous vegetables in the Tanzanian diet (Weinberger and Msuya, 2004).

In a study among urban adults in Benin it was found that a higher intake of vegetables was associated with a higher micronutrient adequacy score and a higher healthfulness score (Sodjinou et al, 2009). However, it was not mentioned if traditional or exotic vegetables were consumed.

Most traditional vegetables in Africa do not contribute to the diet of people only but have further usages, especially as medicines for a range of different illnesses, such as stomach pain, peptic ulcers, headache, anaemia or scabies. In Tanzania, wild vegetables such as jute mallow, wild simsim, false sesame, bitter lettuce and African spider plant, in particular, are attributed to many different medicinal usages and obviously contain more useful ingredients in terms of medicinal value than cultivated vegetables (Keller, 2004). A key element in the selection of plants by humans comprises allelo-chemicals, which serve a variety of functions, such as repellents of herbivores and competitors, or attractants of pollinators or seed dispersers. It is assumed that these compounds have been eliminated from most domesticated plants by human action (Etkin, 1994), which would explain why wild traditional vegetables have more different medicinal applications.

Changes: Health outcomes from various consumption patterns

Despite different dietary patterns, low vegetable consumption is a persistent phenomenon in many regions of the developing world (WHO, 2003b). When vegetable consumption patterns were analysed in Nigeria, it was ascertained that the main influencing factors were season and culture. Other important factors for vegetable consumption were availability and price, while taste and nutritional knowledge were of minor importance (Hart et al, 2005). In contrast, taste and cultural reasons were the main determinants of traditional vegetable consumption in South Africa, followed by health and medicinal properties and affordability, while season was not mentioned (van den Heever and du Plooy, 2007).

In a study on dietary changes in response to currency devaluation in urban households in Dakar (Senegal) and Brazzaville (Congo), it was found that food was less affordable, resulting in reduced dietary diversity. Theses changes, if lasting, would pose a health risk. However, it was also noticed that due to high food prices (e.g. for the preparation of sauces), exotic vegetables such as tomatoes, carrots and cabbages were the first to be reduced. At the same time, traditional vegetables such as cassava leaves (*Manihot esculenta*) or cowpeas (*Vigna unguiculata*) replaced meat dishes due to high meat prices. The authors argue that nutrient value has diminished as the quantity of food has either decreased or been substituted with other ingredients of lower quality (Fouéré et al, 2000).

In general, data on food consumption patterns and trends, especially in Africa, are scarce and usually come from food balance sheets rather than from food consumption surveys and should be titled 'Apparent food consumption patterns' (WHO, 2003b). While data exist for the last 40 years on *per capita* food consumption in terms of energy from various sources, and the amount of fat or livestock products, data on consumption of vegetables are usually deduced from supply data. Nonetheless, these data can give an overall idea for Africa, where vegetable consumption patterns have appeared to change as the supply of vegetables rose from 45.4kg per capita per year in 1979 to 52.1kg in 2000 (WHO, 2003b). The role that traditional and exotic vegetables play has obviously increased; however, it is unclear if the trend has been similar for rural and urban areas.

Future work

It is generally believed that the introduction of exotic vegetable varieties in Africa contributed to the decline in the production and consumption of African indigenous vegetables (Smith and Eyzaguirre, 2007). However, some literature states that declining use of traditional vegetables is due to declining availability (Adedoyin and Taylor, 2000; Okeno et al, 2003), while others argue that traditional vegetables are readily available, especially during the rainy season, but remain among the least consumed foods (Maziya-Dixon et al, 2004). Thus, conserving traditional plants themselves is only one part of the story; while it is important to maintain traditional knowledge of the advantages of these crops, it is also important to use, prepare, store and consume them – in short, conservation through usage.

When looking especially at urban vegetable consumption, it must be kept in mind that it is not possible to refer to an 'average' urban consumer. Rather, different urban consumer groups should be identified who have various cultural backgrounds, socio-economic status and lifestyles resulting in different consumer needs and behaviours (Delisle, 1990). Again, the urban poor, in particular, should be considered in terms of vegetable consumption as they cannot afford a diverse and varied diet that is otherwise available in urban areas.

All studies on vegetable consumption can only provide an idea of consumption patterns of traditional vegetables in Africa, and as information is very limited, it should be interpreted with caution. Consequently, there is an enormous information gap on the *per capita* consumption of traditional vegetables in sub-Saharan Africa which needs to be overcome in order to evaluate the effectiveness of interventions, as well as to develop further vegetable consumption promotion strategies (Smith and Eyzaguirre, 2007).

When looking at changing consumption patterns, the issue on where food consumption takes place becomes increasingly important. Eating meals prepared outside – so-called 'street food' – is a typical feature of urban lifestyles. While street food provides cheap food, it also offers work and income, notably to women (Delisle, 1990). If food vendors, in turn, prepare local foods and traditional dishes, including indigenous vegetables, they contribute to the promotion and conservation of traditional plants, even in urban areas.

Highly nutritious AIVs consumed today in East and West Africa are cowpea leaves, baobab leaves, amaranth, spider plant, jute mallow, moringa leaves, African nightshade, cassava leaves, pumpkin leaves and sweet potato leaves, all found with relatively high β-carotene and iron content. These AIVs should be promoted and grown for household consumption. The potential contribution of plant foods to micronutrient status depends upon the retention of the nutrients after processing and cooking. Data measured for a specific dish may not be applicable to another dish with the same ingredients. Development of algorithms based on knowledge of nutrient retention of traditional food methods and nutrient data of raw ingredients could be an alternative way of predicting the nutrient content of cooked foods.

Nutrition research and interventions have been and still are focusing strongly on single nutrients such as protein, vitamin A, iron, iodine or zinc, which are deficient in certain population groups (DeMaeyer, 1989; Waterlow, 1992; UNICEF and WHO, 1999) while less is known about the health outcomes from a combination of foods and food patterns. However, it is increasingly acknowledged that single nutrients alone are not the key to solving nutritional problems and to preventing chronic diseases; but whole foods or food groups in the right combinations and, thus, a variety of vegetables and fruit as part of a plant food-based diet are desirable. In this way, the interaction between nutrients plays an important role, as well as complementary and overlapping mechanisms of actions of different phytochemicals. Vegetables were especially found to be rich sources of a variety of nutrients and biologically active compounds (Lampe, 1999) and, in addition to providing several nutrients for a healthy and balanced diet, protect against a variety of diseases, such as cardiovasular diseases and cancers. This protection is attributed mainly to antioxidants and dietary fibre (Eastwood, 1999). The contribution of traditional and wild vegetables to a healthy diet and the impact that they could have upon diet and, consequently, upon the health of urban dwellers still needs to be studied in more detail in Africa.

References

Abebe, Y., Bogale, A., Hambidge, K. M., Stoecker, B. J., Bailey, K. and Gibson, R. S. (2007) 'Phytate, zinc, iron and calcium content of selected raw and prepared foods consumed in rural Sidama, southern Ethiopia, and implications for bioavailability', *Journal of Food Composition Analysis*, vol 20, pp161–168

Abukutsa-Onyango, M. O. (2003) 'Unexploited potential of African indigenous vegetables in western Kenya', *Maseno Journal of Education, Arts and Science*, vol 4, pp103–122

Adedoyin, S. A. and Taylor, O. A. (2000) 'Homestead nutrition garden project: An extension communication strategy for eliminating vitamin A deficiency (VAD) among poor households in Nigeria', *Nigerian Journal of Nutritional Science*, vol 21, pp34–40

Ali, M. and Tsou, S. C. S. (1997) 'Combating micronutrient deficiencies through vegetables: A neglected food frontier in Asia', *Food Policy*, vol 22, no 1, pp17–38

Aruoma, O. I. (1994) 'Nutrition and health aspects of free radicals and antioxidants', *Food Chemistry and Toxicology*, vol 32, pp671–683

AVRDC (World Vegetable Center) (2000) 'Bioavailability of vegetables', in *AVRDC Report 1999*, AVRDC – World Vegetable Center, Tainan, Taiwan, pp70–73

AVRDC (2002) 'Nutritional potential of indigenous vegetables', in *AVRDC Report 2001*, AVRDC – World Vegetable Center, Tainan, Taiwan, pp189–792

AVRDC (2004) 'Nutrient values of vegetables harvested from the AVRDC indigenous vegetable display garden', in *AVRDC Report 2003*, AVRDC – World Vegetable Center, Tainan, Taiwan, pp120–123

Bains, K., Yang, R. Y. and Sundanmugasundaram, S. (2003) *High-Iron Mungbean Recipes for North India*, AVRDC – World Vegetable Center, Tainan, Taiwan

Benhura, M. A. and Chitsaku, I. C. (1992) 'Seasonal variation in the food consumption patterns of the people of Mutambara district of Zimbabwe', *Central African Journal of Medicine*, vol 38, no 1, pp8–13

Benson, T. (2004) *Africa's Food and Nutrition Security Situation: Where Are We and How Did We Get Here?*, 2020 Discussion Paper 37, International Food Policy Research Institute (IFPRI), Washington, DC

Bentley, G. R., Aunger, R., Harrigan, A. M., Jenike, M., Bailey, R. C. and Ellison, P. T. (1999) 'Women's strategy to alleviate nutritional stress in a rural African society', *Social Science Medicine*, vol 48, no 2, pp149–162

Bohn, T., Davidson, L., Walczyk, T. and Hurell, R. F. (2004) 'Fractional magnesium absorption is significantly lower in human subjects from a meal with an oxalate-rich vegetable, spinach, as compared with a meal served with Kale, a vegetable with a low oxalate content', *British Journal of Nutrition*, vol 91, pp601–606

Bosland, P. W. (2003) 'Leaf vegetables', in Katz, S. H. and Weaver, W. W. (eds) *Encyclopaedia of Food and Culture*, vol 2, Charles Scribner's Sons, New York, NY

Bothwell, T. H., Charlton, R. W. and Cook, J. D. (1979) *Iron Metabolism in Man*, Blackwell Scientific Publications, Oxford, UK

Burgess, A. and Glasauer, P. (2004) *Family Nutrition Guide*, FAO, Rome, Italy

Chigumira, N. F. and Mvere, B. (1999) 'Zimbabwe', in Chweya, J. A. and Eyzaguirre, P. B. (eds) *The Biodiversity of Traditional Leafy Vegetables*, International Plant Genetic Resources Institute, Rome, Italy

Chweya, J. A. and Eyzaguirre, P. B. (1999) *The Biodiversity of Traditional Leafy Vegetables*, IPGRI, Rome, Italy

Cocks, M. (2006) *Wild Resources and Cultural Practices in Rural and Urban Households in South Africa: Implications for Bio-Cultural Diversity Conservation*, PhD thesis, Wageningen Universiteit, Wageningen, The Netherlands

Delisle, H. (1990) Patterns *of Urban Food Consumption in Developing Countries: Perspectives from the 1980s*, FAO, Rome, Italy

deMaeyer, E. M. (1989) *Preventing and Controlling Iron Deficiency Anaemia through Primary Health Care: A Guide for Health Administrators and Programme Managers*, WHO, Geneva, Switzerland

den Hartog, A. P. (2003) 'Taboos', in Katz, S. H. and Weaver, W. W. (eds) *Encyclopaedia of Food and Culture*, vol 3, Charles Scribner's Sons, New York, NY

Derman, D. P, Bothwell, T. H., Torrance, J. D., MacPhail, A. P., Bezwoda, W. R., Charlton, R. W. and Mayet, F. G. (1982) 'Iron absorption from ferritin and ferric hydroxide', *Scandinavian Journal of Haematology*, vol 29, pp18–24

Dovie, D. B. K., Shackleton, C. M. and Witkowski, E. T. F. (2007) 'Conceptualising the human use of wild edible herbs for conservation in South African communal lands', *Journal of Environmental Management*, vol 84, pp146–156

Eastwood, M. A. (1999) 'Interaction of dietary antioxidants in vivo: How fruit and vegetables prevent disease?', *The Quarterly Journal of Medicine*, vol 92, pp527–530

Engle, L. E., Shanmugasundaram, S. and Hanson, P. M. (2003) 'Evaluation and utilization of vegetable genetic resources for the tropics', *Acta Horticultureae*, vol 623, pp263–274

Etkin, N. L. (1994) *Eating on the Wild Side: The Pharmacologic, Ecologic and Social Implications of using Noncultigens*, Arizona Series in Human Ecology, University of Arizona Press, Tucson, AZ

Ezzati, F., Lopez, A. D., Rodgers, A., Hoorn, S. V. and Murray, C. J. L. (2002) 'Selected major risk factors and global and regional burden of disease', *Lancet*, vol 360, no 9343, pp1347–1360

Faber, M. and van Jaarsveld, P. J. (2007) 'The production of provitamin A-rich vegetables in home-gardens as a means of addressing vitamin A deficiency in rural African communities', *Journal of Science, Food and Agriculture*, vol 87, pp366–377

FAO (United Nations Food and Agriculture Organization) (2007) *FAOSTAT Database*, http://faostat.fao.org/default.aspx, assessed November 2007

Fouéré, T., Maire, B., Delpeuch, F., Martin-Prével, Y., Tchibindal, F. and Adoua-Oyila, G. (2000) 'Dietary changes in African urban households in response to currency devaluation: Foreseeable risks for health and nutrition', *Public Health and Nutrition*, vol 3, no 3, pp293–301

Friis, H. (2002) 'Micronutrients and infections: An introduction', in Friis, H. (ed) *Micronutrients and HIV Infection*, CRC, Boca Raton, FL, pp2–21

Friis, H., Gomo, E. and Michaelsen, K. F. (2002) 'Micronutrient interventions and the HIV pandemic', in Friis, H. (ed) *Micronutrients and HIV Infection*, CRC, Florida, pp220–245

Fritschel, H. (2000) 'Fighting hidden hunger', *2020 News and Views*, no 20, International Food Policy Research Institute (IFPRI), Washington, DC

Gibson, R. S. (2007) 'The role of diet- and host-related factors in nutrient bioavailability and, thus, in nutrient-based dietary requirement estimates', *Food and Nutrition Bulletin*, vol 28, no 1, ppS77–100

Gibson, R. S. and Hotz, C. (2001) 'Dietary diversification/modification strategies to enhance micronutrient content and bioavailability of diets in developing countries', *British Journal of Nutrition*, vol 85, Supplement 2, ppS159–S166

Gibson, R. S., Yeudall, F., Drost, N., Mtitimuni, B. and Cullinan, T. (1998) 'Dietary interventions to prevent zinc deficiency', *American Journal of Clinical Nutrition*, vol 68, pp484S–487S

Gibson, R. S., Perlas, L. and Hotz, C. (2006) 'Improving the bioavailability of nutrients in plant foods at the household level', *Proceedings of the Nutrition Society*, vol 65, pp160–168

Gockowski, J., Mbazo'o, J., Mbah, G. and Moulende, T. F. (2003) 'African traditional leafy vegetables and the urban and peri-urban poor', *Food Policy*, vol 28, no 3, pp221–235

Grubben, G. J. H. and Denton, O. A. (eds) (2004) *Plant Resources of Tropical Africa 2: Vegetables*, PROTA Foundation, Wageningen, Netherlands/Backhuys Publishers, Leiden, Netherlands/CTA, Wageningen, Netherlands

Hallberg, L. and Bjorn-Rasmussen, E. (1981) 'Measurement of iron absorption from meals contaminated with iron', *American Journal of Clinical Nutrition*, vol 34, pp2808–2815

Halliwell, B. (1993) 'The role of oxygen radicals in human disease, with particular reference to the vascular system', *Haemostasis*, Supplement 1, pp118–126

Harborne, J. B., Baxter, H. and Moss, G. P. (1999) *Phytochemical Dictionary: A Handbook of Bioactive Compounds from Plants*, 2nd edition, Taylor and Francis, Philadelphia, US

Hart, A. D., Azubuike, C. U., Barimalaa, I. S. and Achinewhu, S. C. (2005) 'Vegetable consumption pattern of households in selected areas of the old Rivers State in Nigeria', *African Journal of Food, Agriculture and Nutritional Development,* vol 5, no 1, www.ajfand.net/Vol5no1.htm

Harvey, P. W. J., Bexter, P. B. and Darnton-Hill, I. (2000) 'The impact of consuming iron from non-food sources on status in developing countries', *Public Health and Nutrition*, vol 3, pp375–383

Hasler, C. M. (2000) 'Plants as medicine: the role of phytochemcials in optimal health', in Shahidi, F. and Ho, C. T. (eds) *Phytochemcials and Phytopharmaceuticals*, AOCS Press, Champaign, IL, pp517–528

Holdsworth, M., Delpeuch, F., Landais, E., Gartner, A., Eymard-Duvernay, S. and Marie, B. (2006) 'Knowledge of dietary and behaviour-related determinants of non-communicable disease in urban Senegalese women', *Public Health and Nutrition*, vol 9, no 8, pp975–981

Hooda, P. S., Henry, C. J. K., Seyoum, T. A., Armstrong, L. D. M. and Fowler, M. B. (2004) 'The potential impact of soil ingestion on human mineral metabolism', *Science of the Total Environment*, vol 333, pp75–87

Hotz, C. and Gibson, R. S. (2007) 'Traditional food-processing and preparation practices to enhance the bioavailability of micronutrients in plant-based diets', *Journal of Nutrition*, vol 137, no 4, pp1097–1100

Hughes, D. A. (2006) 'Antioxidant vitamins and immune function', in Calder, P. C., Field, C. J. and Gill, H. S. (eds) *Nutrition and Immune Function*, CABI, Wallingford, pp171–192

IFPRI (International Food Policy Research Institute) (2001) *Empowering Women to Achieve Food Security: Vision 2020*, Focus No 6, IFPRI, Washington, DC

IOM (Institute of Medicine) (2000) *Dietary Reference Intakes for Vitamin C, Vitamin E, Selenium, and Carotenoids*, National Academy Press, Washington, DC

IPGRI (2003) *Rediscovering a Forgotten Treasure*, IPGRI Public Awareness, Rome, Italy, www.biodiversityinternational.org/news_and_events/news/news/article/rediscovering-a-forgotten-treasure.html, accessed May 2009

Jansen van Rendsburg, W. S., Vorster, H. J. and Ntombela, S. P. (2007) 'Morogo (traditional leafy vegetable) is not a weed or poor man's crop anymore', Poster presentation at the 5th International Symposium on New Crops and Uses: Their Role in a Rapidly Changing World, 3–4 September, Southampton, UK

Katerere, D. R. and Eloff, J. N. (2006) 'Management of diabetes in African traditional medicine', in Soumyanath, A. (ed) *Traditional Medicines for Modern Times: Antidiabetic Plants*, CRC, Boca Raton, FL, pp203–220

Keding, G. B., Virchow, D. and Krawinkel, M. B. (2007a) 'Impact of vegetable diversity in the field and on the plate on iron-deficiency anaemia of women in rural Tanzania', Poster presented at the 1st FANUS meeting on Nutrition, Health and Human Development in Africa: Breaking the Downward Trend, 7–9 May 2007, Ouarzazate, Morocco

Keding, G. B., Swai, I. S. and Virchow, D. (2007b) 'Traditional versus exotic vegetables in Tanzania: Producing, collecting and purchasing during dry and rainy season', Paper presented at the 5th International Symposium on New Crops and Uses: Their Role in a Rapidly Changing World, 3–4 September 2007, Southampton, UK

Keller, G. B. (2004) *African Nightshade, Eggplant, Spiderflower et al – Production and Consumption Aspects of Traditional Vegetables in Tanzania from the Farmers' Point of View*, MSc thesis, Georg-August University, Goettingen, Germany

Lampe, J. W. (1999) 'Health effects of vegetables and fruit: Assessing mechanisms of action in human experimental studies', *American Journal of Clinical Nutrition*, vol 70, pp475S–490S

Lee, K. (1982) 'Iron chemistry and bioavailability in food processing', in Kies, C. (ed) *Nutritional Bioavailability of Iron*, American Chemical Society, Washington, DC, pp27–54

Legay, C. (2004) 'Shell-type solar driers for home use', www.alin.or.ke/tech-note/data/solar.html, accessed 1 April 2004

Lock, K., Pomerleau, J., Causer, L., Altmann, D. R. and McKee, M. (2005) 'The global burden of disease attributable to low consumption of fruit and vegetables: Implications for the global strategy on diet', *Bulletin of the World Health Organization*, vol 83, no 2, pp100–108

Martorell, R. (2002) 'Obesity in the developing world', in Popkin, B. M. and Caballero, B. (eds) *The Nutrition Transition: Diet and Disease in the Developing World*, Academic Press, London, UK

Maundu, P. M. (1997) 'The status of traditional vegetable utilization in Kenya', in Guarino, L. (ed) *Traditional African Vegetables: Promoting the Conservation and Use of Underutilised and Neglected Crops*, Proceeding of the IPGRI International Workshop on Genetic Resources of Traditional Vegetables in Africa: Conservation and Use, 29–31 August 1995, Nairobi, Kenya, IPGRI, Rome, Italy

Maundu, P. M. and Imbumi, M. (2003) 'East Africa', in Katz, S. H. and Weaver, W. W. (eds) *Encyclopaedia of Food and Culture*, vol 1, Charles Scribner's Sons, New York, NY

Maziya-Dixon, B., Akinyele, I. O., Oguntona, E. B., Nokoe, S., Sanusi, R. A. and Harris, E. (2004) *Nigeria Food Consumption and Nutrition Survey (2001–2003)*, International Institute for Tropical Agriculture, Ibadan, Nigeria

Mgoba, C. M, Temalilwa, C. R., Mulokozi, G., Wandema, F. and Ndossi, G. D. (1993) 'Development of simple solar driers for vitamin A rich foods', *Tanzania Food and Nutrition Newsletter*, vol 001, pp20–23

Nekesa, P. and Meso, B. (1997) 'Traditional African vegetables in Kenya: Production, marketing and utilization', in Guarino, L. (ed) *Traditional African Vegetables: Promoting the Conservation and Use of Underutilised and Neglected Crops*, Proceeding of the IPGRI International Workshop on Genetic Resources of Traditional Vegetables in Africa: Conservation and Use, 29–31 August 1995, Nairobi, Kenya, IPGRI, Rome, Italy

Neumann, C., Harris, D. M. and Rogers, L. M. (2002) 'Contribution of animal source foods in improving diet quality and function in children in the developing world', *Nutrition Research*, vol 22, pp193–220

Njelekela, M., Sato, T., Nara, Y., Miki, T., Kuga, S., Noguchi, T., Kanda, T., Yamori, M. Ntogwisangu, J., Masesa, Z., Mashalla, Y., Mtabaji, J. and Yamori, Y. (2003) 'Nutritional variation and cardiovascular risk factors in Tanzania: Rural–urban difference', *South African Medical Journal*, vol 93, no 4, pp295–299

Okeno, J. A., Chebet, D. K. and Mathenge, P. W. (2003) 'Status of indigenous vegetables in Kenya', *Acta Horticulturae*, vol 621, pp95–100

Okochi, V. I. and Okpuzor, J. (2005) 'Micronutrients as therapeutic tools in the management of sickle cell disease, malaria and diabetes', *African Journal of Biotechnology*, vol 4, no 13, pp1568–1579

Oomen, H. A. P. C. and Grubben, G. J. H. (1978) *Tropical Leaf Vegetables in Human Nutrition*, Orphan Publishing Company, Amsterdam, The Netherlands

Popkin, B. M. (2002) 'An overview on the nutrition transition and its health implications: The Bellagio meeting', *Public Health and Nutrition*, vol 5, no 1A, pp93–103

Popkin, B. M. (2007) 'The world is fat', *Scientific American*, September, pp60–67

PROTA (2005) 'Comparative data on 275 vegetables', in Bosch, C. H., Borus, D. J. and Siemonsma, J. S. (eds) *Vegetables of Tropical Africa: Conclusions and Recommendations based on PROTA 2: 'Vegetables'*, PROTA Foundation, Wageningen, The Netherlands

Raschke, V., Oltersdorf, U., Elmadfa, I., Wahlqvist, M. L., Cheema, B. S. and Kouris-Blazos, A. (2007) 'Content of a novel online collection of traditional east African food habits (1930s–1960s): Data collected by the Max-Planck-Nutrition Research Unit, Bumbuli, Tanzania', *Asia Pacific Journal of Clinical Nutrition*, vol 16, pp140–151

Ruel, M. T., Minot, N. and Smith, L. (2005) 'Patterns and determinants of fruit and vegetable consumption in sub-Saharan Africa: A multicountry comparison', Background paper for the Joint FAO/WHO Workshop on Fruit and Vegetables for Health, 1–3 September 2004, Kobe, Japan, World Health Organization, Geneva, Switzerland

SCN (Standing Committee on Nutrition) (1988) *First Report on the World Nutrition Situation*, SCN News, no 2, United Nations System, Standing Committee on Nutrition, Geneva, Switzerland

SCN (1993) *Focus on Micronutrients: A Periodic Review of Developments in International Nutrition*, Administrative Committee on Coordination and Sub-Committee on Nutrition, Washington, DC

SCN (2004) *Fifth Report on the World Nutrition Situation: Nutrition for Improved Development Outcomes*, United Nations System, Standing Committee on Nutrition, Geneva, Switzerland

Seck, A., Sow, I. and Niass, M. (1999) 'Senegal', in Chweya, J. A. and Eyzaguirre, P. B. (eds) *The Biodiversity of Traditional Leafy Vegetables*, International Plant Genetic Resources Institute, Rome, Italy

Serghides, L. and Kain, K. C. (2002) 'Mechanism of protection induced by vitamin A in falciparum malaria', *The Lancet*, vol 359, pp1404–1406

Shackleton, C. M. and Shackleton, S. E. (2006) 'Household wealth status and natural resource use in the Kat River Valley, South Africa', *Ecological Economics*, vol 57, pp306–317

Shackleton, S. E., Shackleton, C. M., Dzerefos, C. M. and Mathabela, F. R. (1998) 'Use and trading of wild edible herbs in the central lowveld savanna region, South Africa', *Economic Botany*, vol 52, pp251–259

Siemonsma, J. S. and Kasem, P. (1994) *Prosea: Plant Resources of South-East Asia, 8 – Vegetables*, 2nd edition, Prosea Foundation, Indonesia.

Smith, F. I. and Eyzaguirre, P. (2007) 'African leafy vegetables: Their role in the world health organization's global fruit and vegetables initiative', *African Journal of Food, Agriculture, Nutrition and Development*, vol 7, no 3, www.ajfand.net/Vol7No3.html

Sodjinou, R., Agueh, V., Fayomi, B. and Delisle, H. (2009) 'Dietary patterns of urban adults in Benin: Relationship with overall diet quality and socio-demographic characteristics', *European Journal of Clinical Nutrition*, vol 63, pp222–228

Steinmetz, K. A. and Potter, J. D. (1996) 'Vegetables, fruit and cancer prevention: A review', *Journal of the American Dieticians Association*, vol 96, pp1027–1039

Subramanian, A. M. and Yang, R. Y. (1998) *High-Iron Mungbean Recipes from South Asia*, AVRDC publication no 98-480, AVRDC – World Vegetable Center, Tainan, Taiwan

Teucher, B., Olivares, M. and Cori, H. (2004) 'Enhancers of iron absorption: Ascorbic acid and other organic acids', *International Journal of Vital Nutritional Research*, vol 74, pp403–419

Tomkins A. (2005) 'Nutrition, infection and immunity: Public health implications', in Calder, P. C., Field C. J. and Gill, H. S. (eds) *Nutrition and Immune Function*, CABI, Massachusetts, US, pp375–412

Underwood, B. A. (2000) 'Dietary approaches to the control of vitamin A deficiency: An introduction and overview', *Food and Nutrition Bulletin*, vol 21, pp117–123

UNICEF and WHO (1999) *Prevention and Control of Iron Deficiency Anaemia in Women and Children*, Report of the UNICEF/WHO Regional Consultation, 3–5 February 1999, Geneva, Switzerland

UNU (United Nations University) (1985) 'Workshop on agricultural research and human nutrition', *Food and Nutrition Bulletin*, vol 7, no 1, United Nations University Press

USDA (United States Department of Agriculture) (2007) *Nutrient Database*, www.ars.usda.gov/nutrientdata, accessed November 2007

van den Heever, E. and du Plooy, C. P. (2007) 'Status and market potential of traditional leafy vegetables in South Africa', Paper presented to the Fifth International Symposium on New Crops and Uses: Their Role in a Rapidly Changing World, 3–4 September, Southampton, UK

Vijayalakshmi, P., Amirthaveni, S., Devadas, R. P., Weinberger, K. Tsou, S. C. S. and Shanmugasundaram, S. (2003) *Enhancing Bioavailability of Iron from Mungbeans and its Effects on Health of Schoolchildren*, Technical Bulletin no 30, AVRDC Publication 03-559, AVRDC – World Vegetable Center, Tainan, Taiwan

Virchow, D., Oluoch, M. and Kimathi, M. (2007) 'Indigenous vegetables in East Africa: Sorted out, forgotten, revitalised and successful', Paper presented to the Fifth International Symposium on New Crops and Uses: Their Role in a Rapidly Changing World, 3–4 September, Southampton, UK

Vorster, H. H., Venter, C. S., Wissing, M. P. and Margetts, B. M. (2005) 'The nutrition and health transition in the North West Province of South Africa: A review of the THUSA (Transition and Health during Urbanisation of South Africans) study', *Public Health Nutrition,* vol 8, no 5, pp480–490

Vorster, H. J. I., Jansen van Rensburg, W., van Zujl, J. J. B. and Venter, S. L. (2007) 'The importance of traditional leafy vegetables in South Africa', *African Journal of Food, Agriculture, Nutrition and Development*, vol 7, no 4, www.ajfand.net/Vol7No4.html

Waterlow, J. C. (1992) *Protein Energy Malnutrition*, Edward Arnold, London, UK

Weaver, C. M., Proulx, W. R. and Heaney, R. (1999) 'Choices for achieving adequate dietary calcium with a vegetarian diet', *American Journal of Clinical Nutrition*, vol 70, pp543S–548S

Weinberger, K. and Msuya, J. (2004) *Indigenous Vegetables in Tanzania: Significance and Prospects*, Technical Bulletin no 31, AVRDC – World Vegetable Center, Tainan, Taiwan

Weinberger, K. and Swai, I. S. (2006) 'Consumption of traditional vegetables in central and north-eastern Tanzania', *Ecology of Food and Nutrition*, vol 45, no 2, pp87–103

WHO (World Health Organization) (2003a) *Fruit and Vegetable Promotion Initiative*, Report of a meeting of 25–27 August 2003, World Health Organization, Geneva, Switzerland

WHO (2003b) *Diet, Nutrition and the Prevention of Chronic Diseases*, WHO Technical Report Series 916, Report of a joint WHO/FAO expert consultation, World Health Organization, Geneva, Switzerland

WHO (2004) *Micronutrient Deficiencies: Battling Iron Deficiency Anaemia*, Geneva, Switzerland, www.who.int/nut/ida.htm, accessed 10 May 2004

WHO (2007) *Prevention of Chronic Diseases: A Vital Investment*, www.who.int/chp/chronicdiseasereport, accessed November 2007

Wildman, R. E. C. (2001) 'Nutraceuticals: A brief review of historical and teleological aspects', in Wildman, R. E. C. (ed) *Handbook of Nutraceuticals and Functional Foods*, CRC, FL, pp1–12

Winkel-Shirley, B. (2002) 'Biosynthesis of flavonoids and effects of stress', *Current Opinions in Plant Biology*, vol 5, no 3, pp218–223

Yang, R. Y., Tsou, S. C. S., Lee, T. C., Wu, W. J., Hanson, P. M., Kuo, G., Engle, L. M. and Lai, P. Y. (2006a) 'Distribution of 127 edible plant species for antioxidant activities by two assays', *Journal of the Science of Food and Agriculture*, vol 86, pp2395–2403

Yang, R. Y., Tsou, S. C. S., Lee, T. C., Chang, L. C., Kuo, G. and Lai, P. Y. (2006b) 'Moringa, a novel plant rich in antioxidants, bioavailable iron, and nutrients', in Ho, C. T. (ed) *Challenges in Chemistry and Biology of Herbs*, American Chemical Society, Washington, DC, pp224–239

Yang, R. Y., Wu, J., Chen, D., Lin, M., Shiao, R. and Kuo, G. (2007) 'Distribution of edible plants for nutraceutical values', *Acta Hortucultureae*, vol 752, pp115–121

Yang R. Y., Lin, S. and Kuo, G. (2008) 'Content and distribution of flavonoids among 91 edible plant species', *Asian Pacific Journal of Clinical Nutrition*, vol 17, no 1, pp275–279

Yeudall, F., Gibson, R. S., Cullinan, T. R. and Mtimuni, B. (2005) 'Efficacy of a community-based dietary intervention to enhance micronutrient adequacy of high-phytate maize-based diets of rural Malawian children', *Public Health and Nutrition*, vol 8, no 7, pp826–836

5

Production and Harvesting Systems for African Indigenous Vegetables

Mel O. Oluoch, Germain N. Pichop, Drissa Silué, Mary O. Abukutsa-Onyango, Meissa Diouf and Charlie M. Shackleton

Introduction

African indigenous vegetables (AIVs) have been collected from the wild and as 'weeds' in agricultural and disturbed spaces for millennia (Jansen van Rensburg et al, 2007). As such, they are a significant component of the diet of households throughout sub-Saharan Africa (SSA) (FAO, 1988; Grubben and Denton, 2004). Hundreds, if not thousands, of species are used (see Chapter 3), many as daily components of the diet. For example, Vainio-Mattila (2000) found that the Sambaa people in Tanzania consume 73 species of wild plant foods, most of which are ruderals growing by the roadsides or in arable lands. Ogle and Grivetti (1985a) identified more than 200 wild edible species used by the Swazi of Swaziland, most of which were collected on a daily basis, by women, just before the preparation of meals. These were sourced from agricultural fields (56 per cent) as well as from other disturbed environments that include 'near home' (18 per cent), cattle or goat kraals (2 per cent), or from household gardens (1 per cent). Many others are used less frequently, and some only in times of drought or as famine foods (e.g. Zinyama et al, 1990). Regardless, there is little doubt that these species and varieties are extremely important for food security, nutrition, culture and poverty alleviation throughout Africa. Some species have a wide distribution (e.g. *Amaranthus* spp., *Solanum* spp. and *Cleome* spp., to mention a few) and are used wherever they are found; other species or varieties are extremely localized and are hardly known by anyone other than the few communities who make use of them (see Chapter 3). Collectively, they represent an immense store of species and genetic diversity. However, this resource is under threat due to changing lifestyles, land transformation and inadequate knowledge by scientific and conservation communities. This is also a consequence of the fact that the number of species and varieties involved is large, and many are used only locally.

These species are not only collected from the wild. There is long history of them being tolerated and nurtured in agricultural lands and homestead plots (Vainia-Mattila, 2000; Keller, 2004; Jansen van Rensburg et al, 2004, 2007). While in many instances not actively planted, the association of many of these species with cultivated areas is well known and appreciated by local communities. Many households resist conventional agricultural extension advice to remove them as they potentially compete with the planted crop and thereby reduce yields. Thus, when tending their field or garden, they leave these species in place and do not weed them out as extension officers frequently urge. For example, High and Shackleton (2000) reported that just over one third of the total value of plants in homestead plots in the Busbuckridge region of South Africa was from non-planted species. In such an arrangement, the AIVs benefit from whatever care and inputs the farmer provides to the staple crop, such as watering, fertilizer, manure or pest control. However, few agricultural models have factored them in as a productive component of the field or bothered to measure the combined yield (i.e. of both the planted crop and the accompanying AIV). Indeed, yield estimates for AIVs, whether planted in monoculture or as weeds in association with other crops, have hardly been measured. There is a dire shortage of information on AIV yields and what factors influence them, as discussed later in this chapter. The ratio of collected AIVs to cultivated ones varies enormously from place to place, the reasons for which currently remain unknown. Thus, for some households or communities, all of the AIVs are wild collected; but for others, all the AIVs come from fields and gardens. There can be any ratio in between. For example, recent results in Tanzania show that 17.2 per cent of the AIVs consumed are collected from the wild, while 59.6 per cent are produced (Weinberger and Msuya, 2004); but in South Africa over 90 per cent are wild collected (see Chapters 6 and 7). This possibly relates to the availability and abundance of AIVs from wild places. Interestingly, some species may be both cultivated but also occasionally collected in the wild (Keller, 2004). Cultivation, however, does not necessarily mean that species have been fully domesticated. There is also an indication that management of this class of food plants is predominantly undertaken by women, including collection, conservation, and transplanting in fields and gardens (Ogle and Grivetti, 1985b; Leimar Price, 1997; Moreno-Black and Somnasang, 2000; Keller, 2004).

Moving on from nurturing of self-seeded plants in agricultural fields, there is also active sowing and cultivation of AIVs in different parts of SSA (see Chapter 7). In the simplest form, farmers collect seeds from preferred AIVs, store them and then at the onset of the rains, scatter the seeds in their fields, along the peripheries or against fences (for climbing species such as

Momordica). For species with vegetative propagation material, such as yams or taro, tubers may be collected from wild populations and transplanted into the garden (e.g. Hildebrand, 2003). The resultant crop of AIVs is once again tended alongside the staple crop with which it has been sown. Typically, these are used for home consumption; but if the field is large, some may be collected and sold, or the owner may allow kin, neighbours or other residents to access the fields and collect AIVs. In some parts of South Africa, it has been reported that large-scale commercial farmers enter into agreements with small-scale vendors of AIVs, in which the vendors collect AIVs from the fruit orchards or fields in the mornings, which the vendors then sell at neighbouring city markets in the afternoons (Mavimbela, 2004, Vorster et al, 2007). Both benefit from the arrangement; the commercial farmers have their orchards weeded for free, and the vendors get access to a large concentration of AIVs for free.

In a more controlled and managed model, some farmers actively plant and grow AIVs as their primary crop in home gardens or arable fields, rather than as something additional to or amongst other staple crops (e.g. Diouf et al, 2007). This is the primary focus of this chapter. However, unlike exotic vegetables, there is relatively little seed stock and information available to support such enterprises. But international and national agencies in Africa are beginning to appreciate the potential of these crops; therefore, several research and conservation programmes have been initiated during the last few decades and are increasing. For example, in order to combat the risk of genetic erosion of valuable AIVs, common germplasm collection is being carried out for *ex situ* conservation in gene banks (e.g. Diouf et al, 2007). In the Southern African Development Community (SADC) region (comprising South Africa, Namibia, Zimbabwe, Mozambique, Botswana, Swaziland, Malawi, Angola, Zambia, Lesotho, Tanzania and Mauritius), over 1200 accessions of local traditional vegetables were collected through the International Board for Plant Genetic Resources (IBPGR) collection missions during the late 1970s to mid 1980s (Nkhoma et al, 1997). In the recent past, Biodiversity International (formerly International Plant Genetic Resources Institute or IPGRI), the AVRDC – World Vegetable Center, the Regional Centre for Africa (AVRDC–RCA) and other agricultural organizations and universities (e.g. Maseno University in Kenya) have been actively engaged in the collection, maintenance and genetic improvement of AIVs, along with promotional activities for their cultivation and consumption. Since 1999, over 1250 accessions of many different species of AIVs have been collected. These collections are being stored at the AVRDC – World Vegetable Center and at national gene banks in Africa. These collections have been followed up by production trials, both on farm and at research stations, both in rural and urban and peri-urban settings (e.g. Nairobi, Kenya, and Arusha/Dar es Salaam, Tanzania). However, collection from the wild still remains widespread in rural areas and in countries such as South Africa, Zambia, Malawi, Mozambique, etc. Commercial production is still deficient for many common species due to the lack of seeds and market demand for a resource that can be freely collected. Promotional activities for AIV consumption are therefore required to boost demand and, consequently, large-scale production. However, for promotion to be successful, data on production technologies, nutrient content, medicinal value, soil fertility improvement, and pest and disease control must be gathered and made available to producers and consumers. Efforts have been made, especially by different organizations such as the AVRDC – World Vegetable Center, several universities and national agricultural organizations to develop production packages addressing leaf, fruit and pod yield, seed production and processing technologies. However, it is only recently that these initiatives are becoming networked, sharing information between countries and agencies, thereby accelerating the rate of knowledge accumulation and decreasing duplication. This chapter synthesizes a variety of information in this regard, focusing on research work at each of the authors' respective institutions, albeit not restricted to such.

A key finding to date is that the yields obtained on research trials are significantly greater than is commonly measured on farm. Thus, in many places where intensive and intended production does occur (whether for commercial sale or home garden for own consumption), the yields and production of AIVs are frequently well below their potential, despite the fact that they have several agronomic advantages over exotic vegetables, such as being adapted to the local climate and soils, and lower pest incidence. For example, the majority of farmers in the East African region grow AIVs mainly at a subsistence rain-fed production level (Nekesa and Meso, 1997) with leaf yields of 1 to 2 tonnes per hectare (t/ha) against a documented potential of 40t/ha (Abukutsa-Onyango, 2003). Schippers (2002) reported that the total leaf yield of *Brassica carinata* varies considerably between cultivars, and one of the best yields obtained by farmers has been 30t/ha against a potential of 55t/ha. Currently, this discrepancy is attributed to low-quality seed and lack of information provided to farmers, extension personnel and researchers on optimum production strategies, right from the field preparation stage through to harvesting and storage.

Production systems

As with all crops, indigenous or exotic, in Africa there are many different production systems. The main differentiating attributes include location, size, proximity to homestead, commercial or subsistence, nature and quantity of inputs, and whether the crops are planted in mixed or pure stands (see Chapter 1). Different combinations of these attributes result in a wide array of production systems, both within countries and between them. For example, Table 5.1 shows considerable regional differences in AIV production systems between two cities in Uganda.

In comparison, in northern Tanzania, approximately one third of AIVs are intercropped, while two-thirds are cultivated in pure stands, with 67 per cent of all plots using sowing rather than broadcasting (Weinberger and Msuya, 2004). Intercropping of AIVs with field crops such as maize, cassava and sugar cane has multiple uses. The ever-changing climatic situation globally has left

Table 5.1 *Range of AIV production systems in and around two cities in Uganda*

AIV	*Kampala (n = 94)*					*Mbale (n = 74)*			
	Field	*Home garden*	*Inter-cropped*	*Pure stand*	*Mix-cropped*	*Field*	*Home garden*	*Inter-cropped*	*Pure stand*
African eggplant	66.2	10.8	13.8	4.3	2.1	44.0	25.6	2.7	74.3
Ethiopian mustards	55.4	9.5	10.6	9.6	1.1	4.4	24.4	6.7	55.4
Amaranthus	29.7	23.0	46.8	17.0	2.1	54.4	30.0	2.7	50.0
Cowpea	21.6	5.4	4.36	1.1	–	48.9	23.3	2.7	28.4
African nightshade	8.1	4.1	–	–	–	–	–	–	12.2
Spider plant	8.1	5.4	15.1	7.5	7.5	7.8	30.0	–	13.5

Source: IndigenoVeg survey data, 2006

Africa, in particular, very vulnerable to the unpredictability of weather. Indigenous crops are better placed to withstand drastic changes in natural systems. Intercropping of adaptable species could be used by farmers as an insurance against crop failure. The most commercially viable crops are also produced in farms as opposed to kitchen gardens.

Even within urban areas, AIV production systems are not uniform. Gockowski et al (2003) identified three distinctive styles of production across the urban/peri-urban landscape in Yaoundé:

1 an intensive urban (IU) system located within the city limits, characterized by mono-cropping, often on raised beds in inland valleys using high levels of inputs;
2 a semi-intensive peri-urban (SIPU) style extending approximately 30km outside the city limits that also mono-crops on raised beds in inland valleys but using fewer inputs than intensive urban producers; and
3 an extensive peri-urban (EPU) style within an approximate radius of 30km of the city limits that produces AIVs in mixed associations with staple crops and no purchased inputs.

In rural areas, subsistence AIV cultivation generally follows an extensive cropping pattern in association with staples or tree crops. AIVs are planted between and around other staple crops such as maize, cassava, etc. In rural areas, production and marketing is undertaken mainly by women. However, as soon as cash generation potential of the crop increases, men become more involved (Jansen van Rensburg et al, 2007) (as is the case with many other natural resources). This may be one of the reasons why more men are involved in production activities in urban and peri-urban areas, while marketing is still left to women. In addition, producers are younger in urban areas, where commercial production is more labour intensive and often necessitates hired labour, which is mainly offered by young men. Some young people who migrated from rural and resource-poor regions in search of improved living standards convert to the production of AIVs when they are unable to secure the jobs they hoped for, which provides them with a source of income and food for subsistence. The further away one moves from the urban centre, women increasingly engage in production activities, while men are called upon for tasks that require greater physical strength. Activities such as ploughing, irrigation, fertilizer and pesticide application have always been considered men's activities, while sowing, harvesting and trading are considered women's activities (Gockowski et al, 2003).

Although there are a great variety of production systems, the two most common ones, typically differentiated by size and location, are arable fields and home gardens. There are many variations of these, as described in Chapter 1, but these are the two most common.

Home gardens

Home gardens refer to traditional land-use practices near the homestead where different plant species (predominantly vegetables combined with fruits, fodders, medicinal herbs and ornamental plants) are maintained by members of the household with the products intended primarily for household consumption (Gautam et al, 2007). They are common throughout SSA (as well as other developing regions), such that many different types have evolved to make the best of local conditions. Well-known systems include the *kibanja* or banana-based agroforestry home gardens in northern Tanzania (Baijukya and Piters, 1998), the *chagga* systems around Mount Kilimanjaro (e.g. Soini, 2005) and the *dambo* gardens in Zimbabwe and Zambia (e.g. Bell and Hotchkiss, 1989). Home gardens are so widespread that most authors examining household livelihoods or agricultural practices in rural areas throughout the continent report that all or over 90 per cent of households maintain home gardens, whether in East (e.g. Musotsi, 2004; Soini, 2005), West (Dabi and Anderson, 1999) or Southern Africa (e.g. High and Shackleton, 2000; Campbell et al, 2002). Home gardens are typically much smaller than arable fields, and range from a few square metres to perhaps half a hectare. In extreme cases, where land or space is limited, home gardens can be established on verandas or rooftops using growing containers (e.g. plastic pots, plastic bags, clay pots or other convenient containers) (see Chapter 1). Climbing plants can be trained on fences or roofs. Water-loving plants can be grown in drainage ditches. The diversity of vegetables and availability of water from the homestead activities often permits year-round production and forms the basis of a more balanced diet meant to fulfil the daily dietary requirements of a household. Home gardens exist in different parts of the world under different names that include kitchen garden, dooryard garden and backyard garden. Vegetables that can tolerate some extent of shading can be grown under taller plants. Vegetables grown in home gardens are those that are eaten often by the family and can be easily grown without too much attention or expense. The home gardens can sometimes be located away from home depending upon the availability of land, water and other facilities.

The essential characteristics of a home garden are:

- location near to the home to make it convenient for working at odd hours;
- location near a water source;
- generally a small plot size;
- production primarily for household consumption (surplus is often donated to kin or neighbours, or some might be sold);
- management by members of the household (wife, husband, children);
- use of low-cost inputs;
- cultivation of different varieties of vegetables;
- cultivation throughout the year or for a longer duration than arable fields; and
- greater incorporation of indigenous varieties of vegetables and fruits.

Home gardening is likely to be more successful if local technologies are used. It requires some management, including land/pit preparation, sowing/planting, weeding and composting. The amount of time spent in the garden will depend upon its size, although continuous and hard labour is not required. Inputs (such as seeds, seedlings, fertilizers and pesticides) are important elements in AIV production. It is difficult for landless and marginal families to afford such high-cost inputs. A good vegetable garden can be maintained by using low-cost inputs and technology such as compost, local fencing materials or live fencing and biological/mechanical/cultural methods of pest control. Home gardens offer numerous benefits to the household, including:

- a fresh supply of vegetables at the family's doorstep;
- the provision of safe and good-quality AIVs for family consumption;
- cost savings;
- a place of pride for the family (visitors and neighbours can appreciate a home garden);
- a sense of cooperation among family members while working together in the garden;
- making use of a family's spare time;
- occasional extra income from selling excess vegetables;
- a training ground for farming, especially for children;
- a place where the home waste or city waste can be used as a compost for vegetable growing;
- the possibility of alleviating malnutrition through the consumption of fresh and nutritious vegetables;
- an increase in food production, ensuring food security;
- protection for the environment; and
- genetic diversity and conservation.

Difficulties encountered in establishing home gardens include year-round availability of water, especially in cities; availability of inputs such as seeds and chemicals; the availability of know-how for families; and damage caused by animals (especially chickens), which is a serious problem for home gardeners.

Home garden plans and yields

To make the best possible use of the land throughout the year an advance planting plan of the garden should be prepared. In view of seasonal differences, the planting dates will have to be adjusted according to local conditions. Where rainfall is likely to be heavy at times, in order to obtain optimum drainage, raised beds should be constructed, particularly during the rainy season. When not in use, the ground should be manured, dug up and left fallow until needed.

Land around houses is frequently limited, so it is wise to make the most of any available area. Several vegetables can be grown on the same ground throughout the year by planting another vegetable as soon as the first has been harvested. Enough time should be allowed between each crop to spade, pulverize and properly prepare the soil for the next crop. Crops that mature quickly, such as amaranth, may be planted between the rows of tomato, eggplant and other widely spaced crops, which occupy the space throughout the entire season. Vine crops may be sown near the walls or the trees so that they can trail upon them. The sowing of crops should be systematic. Perennial crops such as asparagus, globe artichoke, bitter leaf, moringa, etc., and small fruits should be located on one side of the garden where they will not interfere with ploughing. All crops, which occupy the land for about the same period, should be planted side by side. Studies by the AVRDC – World Vegetable Center in Arusha (Tanzania) have revealed some common cropping and sowing patterns using healthy diet gardening kits and where early-, medium- and long-duration crops are mixed in a pattern that provides a household with their daily vegetable requirements year round (Chadha and Oluoch, 2007). Healthy diet gardening kits containing up to 14 different types of vegetables have been distributed to thousands of farmers in Tanzania, Zambia, Sudan, Mozambique, Uganda, Kenya, Rwanda and Malawi, spawning the development of vegetable home gardens and the popularization of improved varieties of some AIVs in these countries. Through these healthy diet gardening kits, the AVRDC has developed model home gardens, as shown in Figure 5.1. Different kinds of vegetables were planted in a repeated block design to allow for maximum utilization of land in an intercropped or mono-crop system depending upon the combination of species.

From the home garden models described above, yield results are shown in Table 5.2, indicating the overall mean harvests in one season. The result of the cropping season revealed that the average yield per hectare varied from 5t/ha to 82t/ha. Leafy vegetables can be divided into two groups, one with the highest yield (*Amaranthus* with 82t/ha), while the second one ranges between 12t/ha for jute mallow and 16t/ha for spider plants. Fruit vegetables also have a wide range of yield with okra yielding the lowest (5t/ha), while tomato – tengeru 97 performed better in fruit vegetable with 39t/ha.

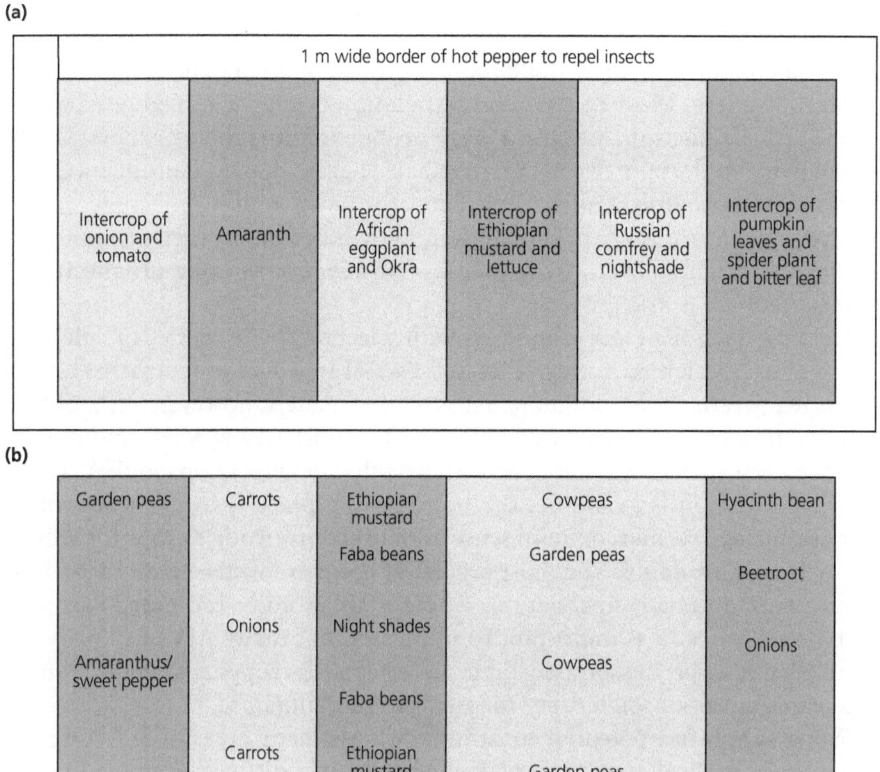

Note: A = okra and African eggplant; B = tomato and jute mallow; C = spider plant; D = onion; E = tomato and Amaranthus.
Source: AVRDC-RCA, unpublished information

Figure 5.1 *(a) Home garden layout 1 (each bed is 1m wide by 2m long); (b) home garden layout 2 (each bed is 1m wide and 2m long except for cowpeas, which is 2m wide); (c) home garden layout 3 (each bed is 1m wide by 1m long)*

Arable fields

Africa is a continent of almost 800 million people, of whom about 60 per cent are rural dwellers. Most of the rural population is engaged in some form of agricultural production, and for a large proportion it is their primary livelihood activity. In the poorer nations, over half of gross domestic product (GDP) derives from agricultural initiatives. Thus, for most rural dwellers, access to land for agriculture is crucial, including knowledge and information about optimal and sustainable cropping options. However, with rapid urbanization, urban agriculture is also growing.

Field production of any crop starts with selection of the species or cultivars to be planted, which is, in turn, dictated by soil type and bioclimatic factors such as temperature and rainfall patterns. Unfortunately, however, optimal soil and climatic conditions are undocumented for most AIV species, and only a few have been researched and even then at only a few locations in SSA. Until better information is available, the best starting place is to use traditional practices and agronomic common sense established from other crops for which information is available. The most crucial of these are watering and fertilizer requirements, especially in situations where AIVs are to be promoted in intensive monocultures. It is important to point out that many AIVs can provide multiple harvests per season (see Table 5.2); but such cropping systems require good management of soil fertility for yields to be maintained.

Water supply is a potential constraint because many areas of SSA are arid or semi-arid. In rural areas, this typically results in a distinct cropping season during the rains and a non-cropping season during dry periods. However, in urban areas, there are often supplementary sources of water that may permit a longer cropping season or cropping throughout the entire year. In any case, the greater drought tolerance of AIVs compared to exotic species means that they can be cropped for longer periods of the year and are better able to withstand intra-seasonal periods of low rainfall (Dzerefos et al, 1995). This vital characteristic is of primary interest in plant breeding and selection trials (e.g. Slabbert et al, 2004).

Table 5.2 *Mean yields (one season) of different vegetable crops from a home garden model shown in Figure 5.1*

Crop	*Number of harvests*	*Leaves or fruits*	*Overall mean yields (t/ha)*
Amaranthus	3	Leaves	82.8
Okra	2	Fruits	5.0
Onion	1	Fruits	74.0
Spider plant	2	Leaves	15.9
Jute mallow	3	Leaves	12.0
Tomato – tanya	4	Fruits	37.8
Tomato – tengeru 97	4	Fruits	39.1
African eggplant	2	Fruits	15.2

Source: AVRDC-RCA, unpublished information

With respect to planting material, AIVs are mainly propagated by seed or tubers that are normally sourced from seed material produced by the previous season's crop, neighbours or local markets. Over the past few years there has been a growing, albeit still limited, supply of species such as *Solanum scabrum* and *S. villosum*, *Cleome gynandra*, *Amaranthus dubius* and *A. blitum*, *Crotalaria ochroleuca*, *Corchorus olitorius*, *Brassica carinata* and *Vigna unguiculata* via research agencies and commercial outlets (e.g. the Kenya Seed Company, Maseno University, KARI/GBK, KARI-KISII in Kenya; the Alpha Seed Company, Incofin, AVRDC-RCA and Hort-Tengeru in Tanzania; the ARC in South Africa; and the ISRA in Senegal); as a result, supply of quality seed is seen as a major constraint to scaling up production of AIVs in both rural and urban contexts (Diouf et al, 2007). Several species have low seed germination rates due to dormancy and improper seed processing. Seed treatments to remove dormancy include soaking in water overnight, soaking in acetone, soaking in warm water at 45°C, scarification for 20 minutes and vernalization (AVRDC, 1992). Seed germination of *Solanum scabrum* has been reported to be problematic due to low vigour caused by improper seed extraction and therefore inadequate removal of sugars and germination inhibitors present in the fruit (Mwai and Schippers, 2004). In order to ensure uniform germination, it is recommended that seeds are treated before sowing. Seed dormancy has been reported for *Corchorus olitorius* (jute mallow), and a general solution to seed dormancy in this species has been to parboil the seed by placing them in simmering water for five to ten seconds, which ensures uniform seed germination (Schippers, 2002). In comparison, *Crotalaria brevidens* (slenderleaf) has minimal problems of germination (Abukutsa-Onyango, 2004), maintaining a germination percentage of over 90 per cent regardless of the source (see Table 5.3). Oseko (2007) reported a germination percentage of 25 to 37 per cent for *Cleome gynandra* (spider plant) collected from different research organizations in Kenya compared to 15 per cent from farmers' fields (see Table 5.3); for *Crotalaria brevidens*, it was over 90 per cent. It has also been reported that seeds of *Crotalaria brevidens* stored for a period of five years maintained a germination percentage of 90 per cent and above according to Abukutsa-Onyango (2005).

Table 5.3 *Germination test on seeds before planting*

African indigenous vegetable (AIV)	*Germination (%)*	
	Farmers' seed	*KARI seed*
Cowpeas	51	100
Spider plant	15	65
Slenderleaf	92	92
Jute mallow	51	75
Vegetable nightshades	5	50
Amaranthus	70	80

Source: Onyango (1992); Onyango et al (1999a); KARI- Kenya Agricultural Research Institute, unpublished data

Several species, such as African nightshades, can be propagated via stem cuttings. Vegetative propagation is also useful for those species that do not produce viable seed such as *eru* (*Gnetum africanum*) in Cameroon, Congo and Nigeria, or those that have seeds that show poor germination and vigour, such as *Solanum scabrum* in Nigeria (Schippers, 2002). The stem cutting must have a single or two pairs of leaves; if kept sufficiently moist, rooting should occur within a month. The advantage of this method is that it takes a considerably shorter time to reach harvest period compared to direct or transplanted nightshades. Tuberous species, such as sweet potato, are easily propagated provided that there is an adequate supply of quality tubers. Traditionally, taro (*Colocasia esculenta*) is stored in shallow soil pits (30cm to 50cm deep). The deeper the pit and the larger the corm stored, the better the initial growth after planting out (Modi, 2004, 2007).

Quality seed is one of the main factors that determine the success of a crop (AVRDC, 1992; Diouf et al, 2007). Seed production at the farm level by the farmer is still the most common source of seeds of AIVs. Very often the seeds produced are of a poor quality. In most cases, no positive selection is conducted, and the right varietal or species isolation distance is frequently not followed to prevent cross-pollination. The plant population then becomes genetically mixed in succeeding plantings. For the farmer to produce quality seed, the best plants in terms of growth and yield are selected for seed production; the best piece of land should also be used, while the seed plot should be isolated from other plots of the same species using the isolation distance outlined in Table 5.4.

Table 5.4 *Harvesting time, seed maturity and isolation distance in the production of selected AIVs*

Species	*Pollination*	*Vegetable maturity*	*Seed maturity*	*Isolation distance (m)*
Amaranthus hypochondriacus	Self-pollinated	21–28 days	90–100 days	200
Amarathus cruentus	Self-pollinated	35–42 days	100–120 days	200
Amaranthus dubius	Self-pollinated	42–48 days	130–150 days	200
Nightshades	Highly self-pollinated	21–35 days	120–150 days	100
African eggplant	Partially cross-pollinated	Green mature	Red ripe mature	200–700
African kale	Highly cross-pollinated	28–35 days	75% of pods yellowish grey	1000
Cowpeas	Self-pollinated	28–35 days	70% of pods are yellowish brown	100–200
Okra	Partially cross-pollinated	Young green pods	Pods turn brown	200–400

Source: AVDRC (2002); Schippers (2002); Grubben and Denton (2004)

The seed crops of AIVs can be grouped according to the state of the seed at harvest – namely, dry seeds or wet seeds. Dry seed refers to seed that is harvested when the fruits ripen and are left to dry on the plant (e.g. slenderleaf and *Amaranthus*). Proper harvesting, preparation and storage are essential for good-quality seed. Harvesting of seed is normally done by cutting off the whole plant. The harvested plant is dried for two to five days to ensure that pods are dry. Threshing and winnowing is done once the seeds are dry. The cleaned seed is dried for a further two to five days to ensure a moisture content of 7 to 9 per cent. Packaging occurs in airtight containers and in appropriate quantities, and products are stored in a dry cool place. Wet seeds are those of fleshy fruits or a mucilaginous layer such as African nightshade fruits, for which mature berries are harvested and then soaked in clean water overnight to enable fermentation to take place. Seeds can also be crushed by hand immediately after harvesting. Berries are squeezed in water to separate the seeds from the fruit coat. The empty fruit coats are sieved out first, followed by the seed. Seeds are dried for four to five days, initially under shade to avoid baking the seed. Packaging should be in airtight containers. Once dry and sealed, seeds are easy to store. These can be kept for a year or more with few problems. However, once the container is opened and not resealed properly, leakage occurs and the remaining seeds start to reabsorb water moisture from the atmosphere and will deteriorate (AVRDC, 1992). It must be noted that storability of different vegetable seeds differs even under similar conditions. Abukutsa-Onyango (2005) reported a germination percentage of over 90 per cent in a seed lot of slenderleaf (*Crotalaria brevidens*) stored for a period of five years under room temperature, while the seed of most AIVs can only be stored for no more than one year (Schippers, 2002).

Arable field management and yields

Research has been carried out at the AVRDC – World Vegetable Center on the harvesting systems of different AIVs in order to improve productivity on farm. The following section provides examples for different species. A key facet of field production systems is the spacing of plants in order to optimize yields and quality, while minimizing costs. As a rule, crops tend to increase yield per unit area as the plant density increases up to a maximum beyond which the yield may not increase further due to interplant competition (AVRDC, 1992). The optimal plant density depends upon site factors, but, more importantly, also upon the size of the mature crop plant and the part to be harvested – namely, tubers, leaves or fruits/pods (see Table 5.5). Typically, large plants require wider spacing.

However, plant spacing depends upon whether or not the crop is grown in pure stands or intercropped with other species. Long growing seasons, small landholdings and high labour–land ratios make multi-crop production systems advantageous in many parts of tropical Africa (AVRDC, 1992). Intercropping within the field production unit is the norm rather than the exception throughout the continent (e.g. Hoffman et al, 2001; Büttner and Hauser, 2003).

Table 5.5 *Recommended spacing for selected AIVs*

AIV	*Plant height (cm)*	*Spacing (cm)*	
		Between rows	*Within rows*
African nightshades	80	20–40	20–60
Spider plant	80	30–50	15–20
Slenderleaf	75	30	15–20
Ethiopian kale	200	50–75	30–35
Cowpeas	70	75	15–30
Jute mallow	100	50	15–30
Amaranthus	100	10–20	10–20
Pumpkins	40	60	60

Source: Schippers (2002); Abukutsa-Onyango (2004); Mwai and Schippers (2004)

Multiple cropping systems can optimize new technologies to exploit natural resources such as space, moisture and radiation for productivity and sustained production. Due to their diversity and relatively short maturity, AIVs can be easily incorporated within many cropping systems. They are usually grown in association with other crops in Eastern, Western and Southern Africa (Gockowski and Ndumbe, 1997; Heever, 1997; Abukutsa-Onyango, 2007). The advantages of intercropping include the following (Ofori and Stern, 1987; AVRDC, 1992; Trenbath, 1993; Sekamatte et al, 2003, Aladesanwa and Adigun, 2007):

- increased total yields;
- yield stability and risk aversion;
- diversified production;
- reduced disease incidence;
- reduced rates of fertility depletion;
- reduced weed growth (and, hence, labour required in weeding);
- greater diversity and abundance of soil fauna; and
- the provision of nitrogen to other crops if one of the crops is a nitrogen-fixing legume.

Perhaps the key aspect is the total productivity of the unit of land. Unfortunately, many early studies examining intercropping considered the yields of only one of the crops. For example, Ikeorgu (1991) reported that yields of *egusi* melon and okra significantly decreased when intercropped with either maize or cassava or both, but did not report the total yield of all crops together. The term land equivalent ratio (LER) is commonly used to indicate biological efficiency of an intercropping system and is defined as the relative land area required by single crops to produce the same yield as in intercropping and is calculated as follows (Willey, 1979):

LER = (Yield of intercrop 1/yield of sole crop 1) + (yield of intercrop 2/yield of sole crop 2).

An LER greater than 1 means that intercropping is more efficient than monocropping of the component crops. The yield advantage can be explained by the fact that there is better use of light, nutrients and water by allowing the intercrops to utilize resources that would otherwise be wasted. A crop that covers the soil rapidly without competing excessively with the associated crop can reduce the growth of weeds and reduce their competition. A study conducted to investigate the compatibility of African nightshades (*Solanum scabrum*) and maize (*Zea mays*) indicated that although leaf yield was reduced by 40 per cent, the land equivalent ratio was greater than 1 (Chweya, 1997). The same applied with intercropping maize and spider plant (*Cleome gynandra*). Results of intercropping slenderleaf (*Crotalaria brevidens*) with finger millet (*Eleusine corocana*) did not significantly affect their leaf and grain yield (Chweya, 1997). Interestingly, Ikeorgu (1990) investigated the LER when intercropping three AIVs (*Amaranthus hybridus*, *Celosia argentea* and *Corchorus olitorius*) and found that the LER was best with all three together (3.8) than with any pairing of any two (2 to 2.5). Other work by Ikeorgu et al (1989) reported an LER of 1.6 for a cassava–maize system intercropped with *egusi* melon and okra in Nigeria. The LER is focused on usable yield. However, the other above-listed advantages may mean that it remains beneficial even if the LER is less than 1. For example, Aladesanwa and Adigun (2007) report on the intercropping of maize with sweet potatoes, which successfully suppressed weeds and improved yields of maize. In this instance the sweet potato was not used as a crop. Although, traditionally, many of the AIVs have been grown in intercrops, the suitability or efficiency of intercropping requires further investigation across a variety of species and conditions.

Yields of selected African indigenous vegetables

With growing interest in African indigenous vegetables, several institutions in Africa have initiated research programmes into various dimensions of AIV propagation, care and yields. The yield aspects are particularly important as many farmers wish to optimize yields and profits for their efforts. If AIVs do not perform better than exotic vegetables, then some farmers will not be willing to cultivate them for market purposes. The following subsections present recent findings on a number of trails of several AIV species that are in high demand. These trials are ongoing.

*Spider plant (*Cleome gynandra*)*

Preliminary investigations have compared different harvesting systems. These included harvesting:

- the whole plant at the base (above the first four nodes);
- the tender stems and leaves only (leaving the flowers intact);
- the tender stems, leaves and flowers;
- the leaves only (and leaving the stems and flowers intact); and
- uprooting the entire plant four to five weeks after sowing.

Table 5.6 *Determining harvesting practices for improved productivity of the spider plant, Arusha, Tanzania (July–December 2006 and October 2007–March 2008)*

Type of harvesting	*Biological yield*		*Economic yield*		
	2007–2008 (g/plant)	*2007–2008 (t/ha)*	*2006 (g/plant)*	*2007–2008 (g/plant)*	*Mean (g/plant)*
Whole plant at the base (above the first four nodes)	337.2[a]	149.9a	138.7[b]	84.3[a]	111.5
Tender stems and leaves only (leaving the flowers intact)	70.0[c]	31.1[c]	185.8[ab]	37.1[b]	111.4
Tender stems, leaves and flowers	239.0[b]	106.3[b]	218.8[a]	80.3[ab]	149.5
Stripping the leaves only and leaving the stems and flowers intact	–	–	142.5[b]	93.9[a]	118.2
Uprooting the entire plant four to five weeks after sowing (result of two harvests)	171.5[b]	76.2[b]	17.9[c]	115.0[a]	66.4
F-test	***	***	**	*	
Least significant difference (LSD)	87.1	38.7	66.1	45.3	
Coefficient of variation (CV) (%)	28.7	28.3	25.0	29.3	

Notes: * = significant ($P<0.05$); ** = high significance ($P<0.01$); *** = highly significant ($P<0.001$).
'Biological yield' means the entire plant, including stems and roots; 'economic yield' means parts used for consumption (leaves, tender stems and flowers).
Means within the same column followed by the same letter(s) are not significantly different at 5% probability levels according to Duncan's Multiple Range Test (DMRT). If a mean has two letters, for example, ab, it does not differ significantly either from the means marked with a or with means marked with b.
The experiments were laid out in randomized complete block design (RCBD), with three replications. AVRDC Line purple stem was used for the study. The plots were fertilized with nitrogen, phosphorous and potassium (NPK) (20-10-10) and urea. The data were then subjected to an analysis of variance (ANOVA), using COSTAT software.
Source: AVRDC-RCA, unpublished information

As indicated in Table 5.6, harvesting the tender stems, leaves and flowers gave the highest economic leaf yields per plant with 218.8g, although harvesting the whole plant at the base gave overall significantly higher mean economic leaf yields (149.5g per plant) compared to the other harvesting methods. It appears that leaving the flowers intact acts as a sink for assimilates, resulting in significantly reduced biological and economic yields. Since most communities who consume spider plants harvest tender stems, leaves and flowers for consumption, this study confirms this practice as the most productive one which should be promoted.

*Vegetable cowpea (*Vigna unguiculata*)*

Cowpea is indigenous to Africa and has been cultivated for centuries. However, in arable production systems, there is a lack of knowledge on the optimum time to begin harvesting leaves for consumption in order to achieve high yields. Recent studies to determine the effect of harvesting time and frequency on yield characteristics of vegetable cowpea varieties have shown that, in general, the later the harvesting was started, the higher the leaf yield, with 12 weeks after sowing giving optimum leaf yields (AVRDC, 2002). An additional study on three cowpea lines, however, showed that harvesting at

Table 5.7 *Leaf yield of three cowpea varieties as affected by harvest times; the trial was carried out at the Regional Centre for Africa in Arusha, Tanzania, from July to December 2005*

Variety	*Number of leaves per plant*			*Leaf size*		*Total yield*		*Marketable leaf yield (g/plant)*
	Total	*Marketable*	*Non-marketable*	*Length (cm)*	*Width (cm)*	*g/plant*	*t/ha*	
Tumaini	69.6^{a}	51.6^{a}	31.6^{a}	26.0^{ab}	15.2^{a}	86.5^{a}	28.8^{a}	57.0^{a}
Dakawa	41.0^{b}	30.8^{b}	16.2^{b}	27.5^{a}	15.4^{a}	52.4^{b}	17.5^{b}	36.4^{b}
Fahari	62.0^{a}	51.2^{a}	28.3^{a}	24.9^{b}	15.4^{a}	82.1^{a}	27.4^{a}	56.1^{a}
F-test	***	***	***	*	ns	***	***	***
LSD0.05	11.47	9.64	7.83	2.28	1.29	10.89	3.59	7.43
Time to start harvesting (weeks after sowing)								
5	55.3^{cd}	39.1^{cd}	25.1^{b}	31.8^{b}	17.7^{a}	64.3^{cd}	21.4^{cd}	41.9^{e}
6	56.6^{bcd}	47.4^{abc}	24.7^{b}	35.8^{a}	18.1^{a}	73.9^{bc}	24.6^{bc}	48.4^{cd}
7	66.1^{abc}	58.2^{ab}	24.8^{b}	36.3^{a}	17.1^{a}	85.1^{ab}	28.4^{ab}	64.0^{b}
8	76.2^{ab}	63.8^{a}	23.2^{b}	27.4^{c}	16.4^{ab}	100.9^{a}	33.6^{a}	77.7^{a}
9	83.6^{a}	56.7^{ab}	39.0^{a}	20.0^{d}	14.1^{cd}	91.4^{ab}	30.5^{ab}	60.6^{bc}
10	49.9^{cde}	42.7^{bc}	25.3^{b}	19.1^{d}	12.1^{d}	83.6^{ab}	27.9^{ab}	55.0^{bc}
11	37.5^{de}	24.3^{d}	20.0^{b}	20.0^{d}	12.8^{cd}	51.7^{de}	17.2^{de}	29.7^{e}
12	35.0^{e}	24.1^{d}	21.4^{b}	18.7^{d}	14.4^{bc}	38.3^{e}	12.8^{e}	21.5^{e}
F-test	***	***	*	***	***	***	***	***
LSD0.05	18.73	15.74	12.79	3.73	2.11	17.64	5.88	12.14
CV (%)	34.31	37.26	53.04	15.02	14.51	25.24	25.24	25.67

Notes: * = significant ($P<0.05$); ** = high significance ($P<0.01$); *** = highly significant ($P<0.001$); ns = not significant.

The experiments were laid out in RCBD, with three replications. The seeds were sown directly on both sides of the ridges at a spacing of 10cm between plants and 30cm between ridges. No fertilizer was applied and harvesting was carried out by hand-picking the leaves every week for eight consecutive weeks. Data were analysed using COSTAT software. Means within the same column followed by the same letter(s) are not significantly different at 5% probability levels according to DMRT. If a mean has two letters, for example, ab, it does not differ significantly either from the means marked with a or with means marked with b.

Source: AVRDC-RCA, unpublished information

eight weeks after sowing gave the optimum total leaf yield (100.9g per plant; 33.6t/ha) and marketable leaf yield (30.9t/ha) that can maximize productivity (see Table 5.7).

Another study on the influence of density, leaf harvesting and plant genotype on cowpea leaf and seed yields showed that a plant spacing of 10cm × 30cm gave the highest leaf yield (30.5t/ha) and seed yield (2.5t/ha) due to the higher number of plants per plot. However, not unexpectedly, the widest spacing (30cm × 30cm) gave the highest yield per plant – leaves (111.1g per plant), pods (11.6g per plant) and seeds (9.5g per plant) – and leaf yields. When the effect of harvesting leaves on seed yield was studied, the results show that unharvested plots had significantly longer (16.2cm) and wider (0.7cm) pods, as well as a greater numbers of pods (14.7), seeds per pod (11.5) and larger seeds (see Table 5.8). Farmers in most communities in SSA produce cowpea seeds from plants where the leaves have been harvested for consumption. However, results of this study show that this practice can reduce seed yields threefold and may consequently not be recommended in communities

Table 5.8 *Leaf and seed yield characteristics of cowpea as affected by spacing, leaf harvesting and plant genotype; the study was conducted at the Regional Centre for Africa in Arusha, Tanzania, from July to December 2006*

	Pod length (cm)	*Pod width (cm)*	*Leaf yield (g/plant)*	*Leaf yield (t/ha)*	*Number of pods/ plant*	*Seed yield (g/plant)*	*Seed yield (t/ha)*	*Seed size (g/seed)*	*Number of seeds/ pod*
Variety									
Tumaini	16.2	0.7[ab]	92.3	19.1[ab]	11.2	8.5	1.80	1.43[ab]	10.0
Dakawa	15.6	0.7[a]	85.3	17.3[b]	10.3	7.7	1.59	1.34[b]	9.7
Fahari	16.2	0.7[b]	105.0	21.2[a]	10.3	8.9	1.71	1.50[a]	10.0
F-test	ns	*	ns	*	ns	ns	ns	*	ns
LSD0.05	1.01	0.04	18.97	3.17	2.51	1.78	386.55	0.14	1.39
Spacing									
10cm x 30cm	15.8[ab]	0.7	91.5 ab	30.5a	10.0	7.6	2.52a	1.42	9.3
15cm x 30cm	15.6[ab]	0.7	89.4 ab	19.9b	10.5	8.5	1.90b	1.43	10.1
20cm x 30cm	15.1[b]	0.7	84.2 b	14.0c	10.3	7.9	1.31c	1.48	10.3
30cm x 30cm	15.5[a]	0.7	111.7 a	12.4c	11.7	9.5	1.06c	1.37	9.8
F-test	*	ns	*	***	ns	ns	***	ns	ns
LSD0.05	1.16	0.04	21.91	3.66	2.89	2.06	446.35	0.16	1.61
Leaf harvesting									
Harvested	15.3[b]	0.7[b]	-	-	6.5[b]	4.1[b]	0.87[b]	1.20[b]	8.3
Unharvested	16.2[a]	0.7[a]	-	-	14.8[a]	12.6[a]	2.53[a]	1.65[a]	11.5
F-test	*	***	-	-	***	***	***	***	***
LSD0.05	0.82	0.03	-	-	2.05	1.46	315.61	0.11	1.14
CV (%)	11.00	9.17	23.79	19.48	40.68	36.62	39.16	16.66	24.32

Notes: * = significant (P<0.05); ** = high significance (P<0.01); *** = highly significant (P<0.001); ns = not significant.
The experiments were laid out in RCBD, with three replications. No fertilizer was applied.
Means within the same column followed by the same letter(s) are not significantly different at 5% probability levels according to DMRT. If a mean has two letters, for example, ab, it does not differ significantly either from the means marked with a or with means marked with b.
Source: AVRDC-RCA, unpublished information

where cowpea seeds are preferred over leaves for consumption. However, a spacing of 10cm × 30cm may be recommended for higher seed yields per unit of land utilized.

Amaranth (Amaranthus *spp.)*

Vegetable amaranth (*Amaranthus* spp.) is widely grown in the tropics and is one of the most important leafy vegetables in Africa, Asia and South America. Amaranth is an annual fast-growing plant, and it is easily cultivated in gardens and fields. However, several species are still collected from the wild in some countries. Depending upon the species and accessions used, amaranth is ready for harvest 20 to 45 days after sowing. Plants may be harvested once or several times. When direct seeding is used, seeds are either broadcasted or sown in rows. Different communities use different harvesting techniques and there is a need to optimize the harvesting methods to determine the best practice needed to improve productivity. A study carried out at the World Vegetable Center evaluated the effects of several harvesting practices on leaf yield. The harvesting practices studied included:

Table 5.9 *Effects of harvesting techniques on leaf yield characteristics of three amaranth accessions evaluated in Arusha, Tanzania, from July to December 2006*

	Economic yield		*Biological yield*		*Leaf size*		*Plant height*
					Length	*Width*	*at maturity*
	(g/plant)	*(t/ha)*	*(g/plant)*	*(t/ha)*	*(cm)*	*(cm)*	*(cm)*
Accessions							
AM-Zim (*Amaranthus cruentus*)	299.6^{a}	29.4^{ab}	725.1	55.5	12.9^{a}	3.1^{c}	28.5^{bc}
AM-TL (*A. hypochondriacus*)	311.9^{a}	32.4^{a}	513.1	43.8	12.0^{ab}	7.0^{ab}	42.3^{a}
AM-NL (*A. cruentus*)	205.8^{b}	19.5^{c}	716.0	52.4	9.7^{c}	7.1^{a}	23.9^{c}
AM-Ex mombo (*A. dubius*)	202.2^{b}	24.9^{bc}	629.1	51.2	11.0^{b}	6.1^{b}	35.0^{ab}
F-test	**	***	ns	ns	***	***	***
LSD0.05	62.21	5.61	500.61	33.31	1.19	1.0	8.22
Harvesting method							
Uprooting whole plant (two harvests)	16.8^{c}	29.8^{a}	13.6^{b}	23.9^{b}	8.2^{b}	4.8^{b}	–
Continuous, with topping	480.5^{a}	32.0^{a}	881.8^{a}	58.8^{a}	12.9^{a}	6.4^{a}	45.6^{a}
Continuous, without topping	267.4^{b}	17.8^{b}	1042.1^{a}	69.5^{a}	13.1^{a}	6.3^{a}	51.6^{a}
F-test	***	***	***	**	***	**	***
LSD0.05	53.88	4.86	433.54	28.85	1.03	0.89	7.12
CV (%)	25.0	21.6	79.3	67.2	10.6	17.9	25.9

Notes: * = significant (P<0.05); ** = high significance (P<0.01); *** = highly significant (P<0.001); ns = not significant.

The experiments were laid out in RCBD, with three replications. Three harvesting practices were evaluated (uprooting the whole plant after five to six weeks and resowing, and continuous harvesting with or without topping). Plot size was 1m x 6m. For uprooted treatments, seeds were broadcasted on the beds. For continuous harvesting treatments, seeds were drilled (after mixing with sand) on two ridges/plot spaced at 60cm, then later thinned to 25cm between plants. Two harvests were carried out for uprooted treatments.

Means within the same column followed by the same letter(s) are not significantly different at 5% probability levels according to DMRT. If a mean has two letters, for example, ab, it does not differ significantly either from the means marked with a or with means marked with b.

Source: AVRDC-RCS, unpublished information

- uprooting the whole plant and resowing again;
- continuous harvesting with topping (removal of flowers); and
- continuous harvesting without topping.

Evaluation of harvesting techniques showed that continuous harvesting with topping gave the highest economic leaf yield (32.0t/ha), while continuous harvesting without topping gave the lowest (17.8t/ha). Uprooting the whole plant (after two plantings and two harvests) was the second highest yielding method with 29.8t/ha, and plants harvested with this method had the smallest leaf size but provided better marketable leaf quality (see Table 5.9).

*Nightshade (*Solanum scabrum, S. americanum *and* S. villosum*)*

Vegetable nightshade is among the most popular and nutritious African indigenous vegetables (see Chapter 7) and has the potential to improve the nutritional and economic status of the rural resource poor and small-scale farmers, especially women. Nightshades are particularly rich in several essential nutrients, including iron, vitamins, fibre and calcium, among others (Grubben and Denton, 2004). However, yields are still relatively low due to poor agronomic practices, including proper harvesting technologies. Another study carried out at the World Vegetable Center was to assess the effects of different harvesting technologies on leaf yield. The harvesting techniques evaluated were:

- cutting the whole plant from the base (above the first three nodes);
- harvesting lateral tender stems and leaves only;
- harvesting vertical tender stems and leaves only; and
- harvesting tender leaves only.

Results obtained (see Table 5.10) show that harvesting lateral tender stems and leaves only gave the highest mean leaf yields, probably due to faster regeneration. This practice may be recommended to farmers.

As mentioned previously, maintaining optimal yields of AIVs, in common with any other crop, requires sound management of soil fertility and structure. Once again, there is a dearth of literature on the fertilizer needs of AIVs across different soil types and agro-ecological conditions. Several species have received attention in research conditions at one or two locations, such as *Amaranthus* species (e.g. Akanbi and Togun, 2002; Materechera and Medupe, 2006; Alonge et al, 2007), *Solanum nigrum* (e.g. Murage et al, 1996) and *Vigna unguiculata* (e.g. Abayomi et al, 2008); but there is need for a comprehensive review and more programmatic approaches across species and agro-ecological conditions. Overall, farmers adopt a number of strategies to keep soil fertility as high as possible; but at times labour and/or cash constraints limit such efforts (Vanlauwe and Giller, 2006). The majority of farmers do add organic or inorganic fertilizers or manure to their fields (e.g. Chibudu et al, 2001; Diouf et al, 2007); but quantities may be below recommended levels. Fertility levels in home gardens or fields closest to the homestead are typically much higher than those further away, reflecting the higher inputs and attention that they receive (e.g. Chibudu et al, 2001).

Abukutsa-Onyango (2003) reported that, not unsurprisingly, both organic and inorganic sources of fertilizer significantly increased yields of edible portions of cowpeas (*Vigna unguiculata*) and spider plant (*Cleome gynandra*) (see Figure 5.2). It was further shown that combinations of organic and inorganic fertilizers can be used for AIVs such as cowpeas and spider plants. The growth of pot- and field-grown broad-leafed African nightshade (*Solanum scabrum*) was significantly increased by nitrogen levels of up to 25mg N/kg and 40kg N per hectare, after which any further increase in N levels either led to a stagnation or decrease of growth (Abukutsa-Onyango and Karimi, 2007). Omotoso and Shittu (2007) recommended that okra yield was optimized at intermediate levels of nitrogen addition (approximately 150kg/ha) rather than high additions.

Table 5.10 *Effects of different leaf harvesting techniques on the productivity of nightshade: The experiment was carried out at the Regional Centre for Africa in Arusha, Tanzania, from July to November 2005 and August 2007 to January 2008*

	Economic yield (fresh leaf weight; g/plant)			*Economic yield (fresh leaf weight; t/ha)*		
	2005	*2007*	*Mean*	*2005*	*2007*	*Mean*
Accession/line						
SS52 (*Solanum scabrum*)	311.2^{b}	383.4^{a}	347.3	13.0^{b}	16.0^{a}	14.5
SS49 (*S. scabrum*)	365.3^{a}	293.2^{b}	329.3	15.2^{a}	12.2^{b}	13.7
SA (*S. americanum*)	182.9^{c}	229.0^{c}	206.0	7.6^{c}	9.5^{c}	8.6
SV (*S. villosum*)	217.0^{c}	92.9^{d}	154.9	9.0^{c}	3.9^{d}	6.5
MW04 (*S. tarderomotum*)	183.1^{c}	96.6^{d}	139.8	7.6^{c}	4.0^{d}	5.8
LSD0.05	46.89	39.93		1.95	1.66	
F-test	***	***		***	***	
Harvesting method						
Cutting the whole plant from the base (above the first three nodes)	279.8^{a}	191.7^{b}	235.8	11.7^{a}	8.0^{b}	9.8
Lateral tender stems and leaves	300.8^{a}	251.2^{a}	276.0	12.53^{a}	10.5^{a}	11.5
Vertical tender stems and leaves	222.2^{b}	207.9^{b}	215.0	9.25^{b}	8.7^{b}	9.0
Stripping the leaves only and leaving the stems intact	204.9^{b}	225.3^{ab}	215.1	8.5^{b}	9.4^{ab}	8.9
LSD0.05	41.94	35.71		1.74	1.49	
F-test	***	*		***	*	
CV (%)	22.52	22.06		22.46	22.06	

Notes: * = significant (P<0.05); ** = high significance (P<0.01); *** = highly significant (P<0.001). Mean separation in columns was done by Student Newman Keuls test (for year 2005) and Duncan's Multiple Range Test (DMRT) for year 2007 at P<0.05.

The experiments were laid out in RCBD, with three replications. Five nightshade accessions (representing genetic diversity) SS52, SS49, SA, SV and MW04, and four harvesting methods were evaluated. The plot size was 6m x 1.2m, with plant spacing of 60cm between rows and 40cm between plants. Each treatment was planted in two rows. Fertilization, weeding, irrigation and pest and disease control were carried out as per need. Data collected were subjected to ANOVA analysis using COSTAT software. Means within the same column followed by the same letter(s) are not significantly different at 5% probability levels according to DMRT. If a mean has two letters, for example, ab, it does not differ significantly either from the means marked with a or with means marked with b.

Source: AVRDC-RCA, unpublished information

The positive effects of nitrogen addition are widely reported for most vegetables – exotic and indigenous, leafy and non-leafy. However, Abayomi et al (2008) found that while nitrogen addition resulted in greater biomass of cowpeas, if did not increase the number of flowers per plant, and also decreased the production of nodules. Dzerefos et al (1995) demonstrated that yields of *Corchorus tridens* were best promoted by the addition of nitrate nitrogen as opposed to ammonium nitrate or a mixture of both. Importantly, nitrogen fertilization can also increase the crude protein contents of the AIVs,

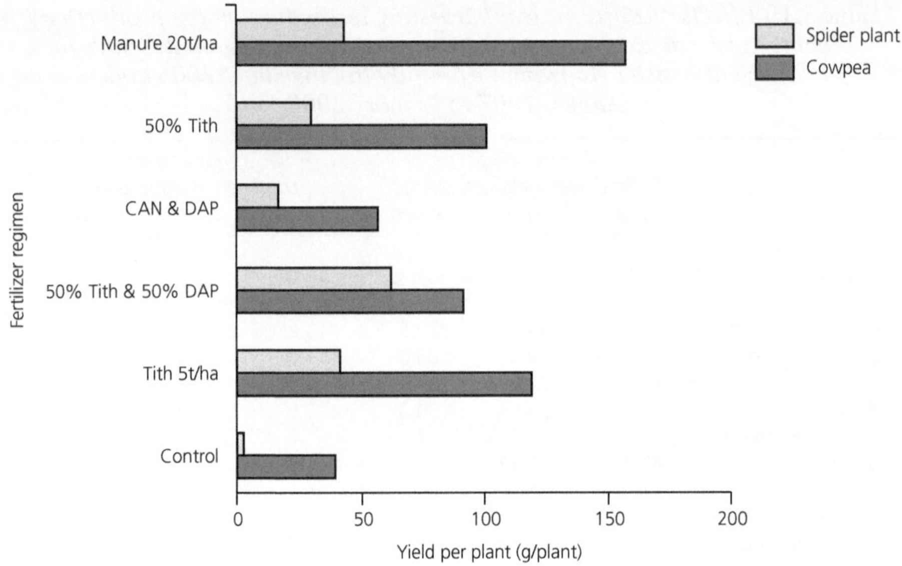

Note: CAN is calcium ammonium nitrate; DAP is diammonium phosphate; Tith is *Tithonia diversifolia*.
Source: Abukutsa-Onyango (2003)

Figure 5.2 *Comparison between responses of cowpea and spider plant to fertilizer sources at eight weeks*

as reported by Murage et al (1996) for black nightshade. As a general guide Abukutsa-Onyango (2003) recommends the applications set out in Table 5.11.

Yield is also related to a number of other management factors, such as pest management. The need for pest management systems is likely to increase as AIVs become more commercialized, with the associated trend towards monocultural systems. The most destructive insect pests of leafy AIVs affect yield by feeding on the leaves, sucking the cell sap and, in some cases, carrying disease-causing organisms that are transferred to other plants. Recent studies in Kenya by

Table 5.11 *Recommended fertilizer levels for commercial production of selected AIVs*

Organic sources	*Inorganic sources*
Manures: 10–40t/ha	African nightshades: 20-10-10 NPK and urea (200 kg/ha)
Amaranthus: 30–40t/ha	African eggplant: 15-15-15 NPK (250kg/ha)
Spider plant: 20–30t/ha	*Amaranthus*: 10-10-20 NPK (200 kg/ha)
Jute mallow: 10–20t/ha	Spider plant: 40 and 120 N kg/ha
Slenderleaf : 10–20t/ha	Ethiopian kale: CAN (200 kg/ha)
Leaf biomass: 2.5–5t dry tith leaves/ha	Slenderleaf : CAN (200 kg/ha)
Cowpeas, spider plant, African kale and slenderleaf	

Source: Abukutsa-Onyango (2004); Abukutsa-Onyango and Karimi (2007); Onyango et al (1999a, 1999b); Onyango (2005).

Sithanantham et al (2005) have provided on-station estimates of avoidable yield loss due to insect pests on *Cleome gynandra*, *Amaranthus lividus*, *Corchorus olitorius* and *Crotalaria brevidens* (see Figure 5.3). A total of ten insect species were found and identified from African nightshade (*Solanum villosum*) and *Cleome gynandra*, grown in Keiyo District, Kenya (Kipgosgei, 2004). These included cutworms (*Agrotis* spp.), black aphids (*Aphis fabae*), cotton aphids (*Aphis gossypi*), leaf miner (*Lyriomyza* spp.), red spider mite (*Tetranchus* spp.), nematodes (*Meloidogyne* spp.), flea beetles (*Chrysomelidae* spp.), epilachna beetles (ladybird), diamond back moth (*Plutella xylostella*) and short-horned grasshoppers. There are several methods used to control insect pests satisfactorily, but the first step is to identify the insect pests for the different AIVs.

Integrated pest management (IPM) requires three areas of competence that normally include prevention, observation and intervention (Heerden, 2000). Prevention is an indirect measure and observation aims to determine when and what action to take. Intervention aims at reducing the effects of economically damaging pest populations to acceptable levels. Integrated pest control and management is where several methods of controlling pests are used together (Dobson et al, 2002). Control methods that are normally integrated include cultural and physical, biological and chemical pest control methods, and the main difficulty is that some of the methods interfere with each other. Considering that the fresh produce from the AIVs may also be consumed raw, it is important to develop pest control options that are safe, besides being cheap and simple to adopt. A pesticide applied to control a pest may also interfere with beneficial wasps that are natural enemies of some pests. Cropping systems will dictate the type of pest management system to be adopted, and

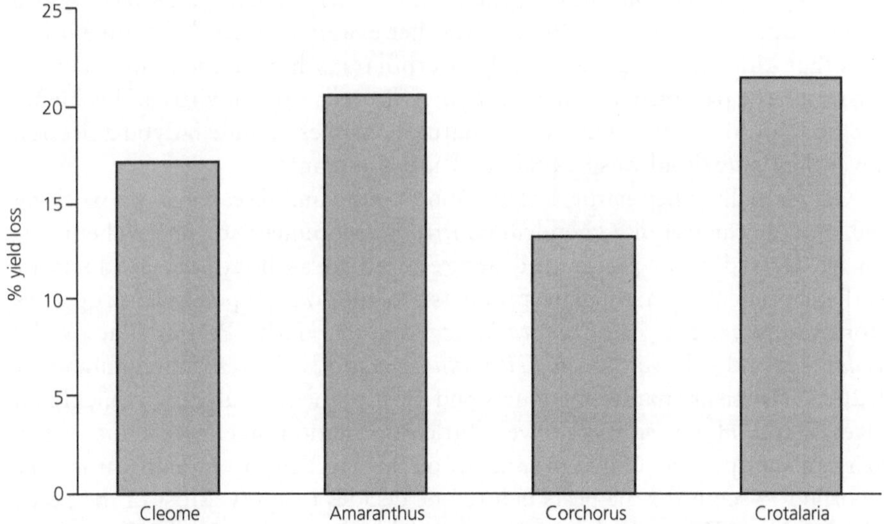

Source: Sithanantham et al (2005)

Figure 5.3 *Estimates of overall avoidable losses due to insect pests on four indigenous vegetables in Kenya (1996–1998)*

this is an area where further research is required. There is a pressing need for more research into IPM technologies and into ways of combining them in IPM packages that are cheap, safe and effective for small-scale vegetable farmers. The methods that are normally combined in IPM include cultural, biological and chemical control methods.

Cultural control is one of the oldest pest control methods and is based on changing the way in which the crop is grown or altering its habitat to reduce the likelihood of damage from pests and diseases. Cultural control includes using resistant and tolerant varieties, improving soil, avoiding pests and diseases, using crop rotation, destroying the source of infection, preventing disease with hot water seed treatment, water management, mulching, mechanical, mixed cropping and trap or catch crops (Sithanantham et al, 2005).

Biological or bio-control implies controlling pests by using living organisms. Beneficial living organisms that reduce pests are normally present in any crop unless broad spectrum pesticides have been used. These natural enemies can be conserved by taking care with farming practices so that their killing is minimized. Once natural enemies have been identified, it is possible to release additional beneficial organisms of the same type – a process normally called augmentation or inundation. Ensuring effective use of biological control and the conservation of natural enemies is imperative. Using insect pathogens in pest control spray has been found effective. Naturally occurring pathogens that kill insects can be obtained from diseased insects and incorporated within sprays and applied to control pests. Pathogen-based sprays or microbial biopesticides are not yet widely available for vegetable pests except for *Bacillus thuringiensis* (*Bt*), a bacterium that kills the larvae of moths and butterflies. *Bt* products available from agro-input retailers include *Dipel*, *Thuricide* and *Xentari* (Sithanantham et al, 2005). Another example under development is a virus that kills diamond back moth caterpillars, which are a serious pest in *Brassicas*. The pathogen is called *Plutella xyllostella* granulovirus and is highly specific (Dobson et al, 2002). Other natural enemies include ladybird beetles, Hover flies, parasitoid wasps, predatory mites and ants.

Occasionally when natural regulation of pests and diseases is not working well enough, the use of a chemical control is recommended. These chemicals can be derived from plants and are referred to as botanical products or synthetic pesticides. Many plant products are said to have pesticidal properties – for example, neem (*Azadirachta indica*) and its relative Persian lilac (*Melia azedarach*) and wild sunflower (*Tithonia diversifolia*) leaves (Sithanantham et al, 2005). These are natural products and most of them break down quickly on leaves or soil. However, there is very little information on effective dose rates, their impact upon beneficial organisms or their toxicity to humans. It is very important to scout for the pests before one decides to spray. Most of the pesticides used by AIV growers are either fungicides applied to control or prevent fungal diseases or insecticides to kill pests. Herbicides, which control weeds, are not widely used by small-scale AIV farmers in Central and Southern Africa. If labour for weeding becomes more scarce or expensive, herbicides may become more popular. In using pesticides, it is important to choose the right pesticide and to ensure that post-harvest interval is considered. Pesticides that require only a short period for farmers to wait between spraying and harvesting are recommended and are more practical.

Contribution to livelihood and food security

Improving health and nutrition

Quite a large number of AIVs have long been known and reported to have health-protecting properties and uses. Several of these species continue to be used for prophylactic and therapeutic purposes by rural communities. This indigenous knowledge of the health-promoting and protecting attributes of AIVs is clearly linked to their nutritional and non-nutrient bioactive properties. AIVs have long been, and continue to be, reported to significantly contribute to the dietary vitamin and mineral intakes of local populations.

Vegetables, especially AIVs, are excellent sources of vitamins A and C, and iron, as well as protein, minerals and fibre (see Chapter 4). Previous studies have shown that traditional leafy vegetables are richer in vitamins, mineral elements and crude fibres than European vegetables, and have appreciable amounts of crude protein, fat and oil, energy, vitamins and minerals. They have also been known to make food more palatable and digestible (Adebooye and Opabode, 2004). A diet rich in micronutrients can bolster the body's immune system and slow the progression of AIDS. Many indigenous vegetables can increase the bioavailability of micronutrients in staple foods when consumed. Because they are affordable, vegetables are also an important source of nutrition for poor urban farmers.

Income generation and food security

Indigenous leafy vegetables and fruits play a key role in income generation and subsistence. For example, Adebooye and Opabode (2004) reported that *Solanecio biafrae*, an indigenous leaf vegetable in south-west Nigeria, is several times more expensive than the routinely cultivated species, especially during the dry season. Experience has also shown that other indigenous leaf vegetables such as *Telfairia occidentalis*, *Celosia argentea*, *Amaranthus cruentus* and *Solanum macrocarpon* are also sold at high prices during the dry season in south-west Nigeria. In Kenya, Abukutsa-Onyago (2003) showed that AIVs offer a significant opportunity for the poor people in western Kenya to earn a living because indigenous leaf vegetable production can be done with little capital investment. The IndigenoVeg survey results (see Chapter 6) show that urban and peri-urban AIV production is perpetuated by vulnerable groups, often migrants who came to cities in search of jobs in the secondary or tertiary sector, but had engaged in AIV production in the absence of better options. AIV production thus provides employment opportunities and income for those who are outside the formal sector. The average revenue from the three most important AIVs were found to exceed US$540 for the Abidjan wholesalers and

US$200 for Kampala farmers. They are suitable for the resource poor since they are easy to grow and require minimal external inputs to do well (MATF, 2006).

In summary, improving AIVs requires the development of good farming systems practised for seed and leaf production. The wide variety and availability of AIVs makes them ideal candidate resources for food security and poverty alleviation in SSA. Non-availability of high-quality seeds, poor control of pests and diseases, lack of improved cultural practices and lack of effective technologies for post-harvest preservation and processing of leaves were found to be the main constraints to the successful production of AIVs.

Conclusions and recommendations

This chapter demonstrates that much rich information exists regarding the farming production and harvesting systems of AIVs. They are widely cropped in home gardens and arable fields throughout SSA. However, relative to exotic vegetables, they have received little research or extension support from national and international agencies. This in itself suggests the need to refocus national farming, agronomic and horticultural efforts to those species that are widely used and appreciated, and in doing so, respond to farmer and consumer needs, rather than externally driven research agendas. Farming research needs to be vested in the needs of the farmers if it is to be adopted and make an impact upon people's livelihoods.

This chapter has demonstrated the variety of systems, practices and species involved. Optimization of production strategies on the ground requires these aspects to be matched to local agro-ecological and market conditions. Thus, there is much simple, repetitive and basic, but useful, research that needs to be done, such as which species and varieties are best suited to which conditions and how production can be stimulated with the level of resource endowments available to typical farmers and gardeners. A key constraint commented upon by many authors is the lack of quality seed or propagation material. Initiatives are being developing in this regard; but they are still too few and far between. Farmers produce their own seed and keep some from one year to the next. However, increasing the availability of good-quality seed will play an important part in increasing production of AIVs, thereby benefiting the farmers, the consumers, and regional and national economies. Production, harvesting and storage protocols exist for only a few species. Key areas for investigation are water-use efficiencies, optimal intercropping combinations, identification of pests of AIVs, post-harvest technologies to minimize losses, processing and value addition.

Although site-specific trials and guidelines need to be developed, the sheer volume of potential combinations of site condition, climate, species, inputs and practices make the task daunting. No single agency can hope to make any impression at all. Consequently, it is imperative that research and extension agencies network across regions and the continent in order to promote joint learning and to eliminate duplication of efforts and research. Additionally, information and support packages for farmers developed in one country need to be shared regionally as it is often quicker to adapt existing ones to local conditions, languages and knowledge levels than to start from scratch.

References

Abayomi, Y. A., Ajibade, T. V., Sammuel, O. F. and Sa'adudeen, B. F. (2008) 'Growth and yield responses of cowpea (*Vigna unguiculata* (L.) Walp) genotypes to nitrogen fertiliser (NPK) application in the southern Guinea Savanna zone of Nigeria', *Asian Journal of Plant Sciences*, vol 7, pp170–176

Abukutsa-Onyango, M. O. (2003) 'The role of African indigenous vegetables in poverty alleviation in Kenya', in *Proceedings of the First PROTA International Workshop*, 23–25 September 2002, Nairobi, Kenya, pp269–270

Abukutsa-Onyango, M. O. (2004) '*Crotalaria brevidens*', in Grubben, G. J. H. and Denton, O. A. (eds) *Plant Resources of Tropical Africa 2: Vegetables*, PROTA Foundation, Wageningen, The Netherlands/ Backhuys Publishers, Leiden, The Netherlands, pp229–231

Abukutsa-Onyango, M. O. (2005) 'Effect of nitrogen rates on germination of stored seed of slenderleaf (*Crotalaria brevidens* benth) seeds', in *Proceedings of the Fourth Horticulture Workshop on Sustainable Horticultural Production in the Tropics*, 24–27 November 2004, Moi University, Eldoret, Kenya, pp188–195

Abukutsa-Onyango, M. O. (2007) 'The diversity of cultivated African leafy vegetables in three communities in western Kenya', *African Journal of Food Agriculture, Nutrition and Development* online journa, vol 7, no 3, www.ajfand.net

Abukutsa-Onyango, M. O. and Karimi, J. (2007) 'Effects of nitrogen levels on growth and yield of broad leafed African nightshade (*Solanum scabrum*)', *Acta Horticulturae,* vol 745, pp379–386

Adebooye, O. C. and Opabode, J. T. (2004) 'Status of conservation of the indigenous leaf vegetables and fruits of Africa', *African Journal of Biotechnology*, vol 3, pp700–705

Akanbi, W. B. and Togun, A. O. (2002) 'The influence of maize-stover compost and nitrogen fertiliser on growth, yield and nutrient uptake of amaranth', *Scientia Horticulturae*, vol 93, pp1–8

Aladesanwa, R. D. and Adigun, A. W. (2007) 'Evaluation of sweet potato (*Ipomoea batatas*) live mulch at different spacings for weed suppression and yield response of maize (*Zea mays* L.) in southwestern Nigeria', *Crop Protection*, vol 27, pp968–975

Alonge, S. O., Alonge, F. O., Bako, S. P., Olarewaju, J. D. and Adenji, O. B. (2007) 'Effects of rates and split application of compound NPK fertiliser on the growth and yield of three *Amaranthus* species in Nigeria Guinea Savanna', *Asian Journal of Plant Sciences*, vol 6, pp906–912

AVRDC (1992) *Vegetable Production Training Manual*, Asian Vegetable Research and Development Centre, Shanhua, Taiwan

AVRDC (2002) *AVRDC Report 2002*, AVRDC publication no 03-563, AVRDC – World Vegetable Center, Shanhua, Taiwan

Baijukya, F. P. and Piters, B. de S. (1998) 'Nutrient balances and their consequences in the banana-based land use systems of Bukoba district, northwest Tanzania', *Agriculture, Ecosystems and Environment*, vol 71, pp147–158

Bell, M. and Hotchkiss, P. (1989) 'Political interventions in environmental resource use', *Land Use Policy*, vol 6, pp315–323

Büttner, U. and Hauser, S. (2003) 'Farmers' nutrient management practices in indigenous cropping systems in southern Cameroon', *Agriculture, Ecosystems and Environment*, vol 100, pp103–110

Campbell, B. M., Jeffry, S., Kozanayi, W., Luckert, M., Mutamba, M. and Zindi, C. (2002) *Household Livelihoods in Semi-Arid Regions: Options and Constraints*, CIFOR, Bogor

Chadha, M. L. and Oluoch, M. (2007) 'Healthy diet gardening kit – for better health and income', *Acta Horticulturae*, vol 752, pp581–584

Chibudu, C., Chiota, C., Kandiros, E., Mavedzenge, B., Mombeshora, B., Mudhara, M., Mrimbarimba, F., Nasasara, A. and Scoones, I. (2001) 'Soils, livelihoods and agricultural change: The management of soil fertility in the communal lands of Zimbabwe', in Scoones, I. (ed) *Dynamics and Diversity: Soil Fertility and Farming Livelihoods in Africa*, Earthscan, London, pp116–163

Chweya, J. A. (1997) 'Genetic enhancement of indigenous vegetables in Kenya', in Guarino, L. (ed) *Traditional African Vegetables: Promoting the Conservation and Use of Underutilised and Neglected crops – 16*, Proceedings of the IPGRI International Workshop on Genetic Resources of Traditional Vegetables in Africa: Conservation and Use, 29–31 August 1995, ICRAF-HQ, Nairobi, Kenya, Institute of Plant, Genetic and Crop Plant Research, Gatersleben/International Plant Genetic Resources Institute, Rome, Italy

Dabi, D. D. and Anderson, W. P. (1999) 'Water use for commodity production in Katarko village, northern Nigeria', *Applied Geography*, vol 19, pp105–122

Diouf, M., Gueye, M., Faye, B., Dieme, O. and Lo, C. (2007) 'The commodity systems of four indigenous leafy vegetables in Senegal', *Water SA*, vol 33, pp343–348

Dobson, H., Cooper, J. Manyangarirwa, W., Karuma, J. and Chiimba, W. (2002) *Integrated Vegetable Pest Management*, Natural Resources Institute, University of Greenwich, Chatham, UK

Dzerefos, C. M., Shackleton, C. M. and Scholes, M. C. (1995) 'Seed germination, nitrogen nutrition and water requirements of the edible herb *Corchorus tridens* L.', *Economic Botany*, vol 49, pp380–386.

FAO (United Nations Food and Agriculture Organization) (1988) *Traditional Food Plants*, FAO, Rome, Italy

Gautam, R., Subedi, A., Shrestha, P. K. and Sthapit, B. R. (2007) 'Home gardens: On-farm agro-biodiversity conservation unit and source of dietary diversity of rural farmers of Nepal', *Acta Horticulturae*, vol 752, pp509–512

Gockowski, J. and Ndumbe, N. (1997) 'The transformation of leafy vegetable cropping systems along Cameroon's humid forest margins', in Schippers, R. and Budd, L. (eds) *African Indigenous Vegetables*, Proceedings of NRI/IPGRI/CPRO workshop, 13–18 January 1997, Limbe, Cameroon, ODA, London, pp46–51

Gockowski, J., Mbazo'o, J., Mbah, G. and Moulende, T. F. (2003) 'African traditional leafy vegetables and the urban and peri-urban poor', *Food Policy*, vol 28, pp221–235

Grubben, G. J. H. and Denton, O. A. (eds) (2004) *Plant Resources of Tropical Africa: 2: Vegetables*, PROTA Foundation, Wageningen, The Netherlands

Heerden, D. (2000) *Integrated Pest Management: A Practical Guide to Hydroponic Vegetable Production*, ARC-Roodeplaat, Pretoria, South Africa

Heever, E. (1997) 'The use and conservation of indigenous leafy vegetables in South Africa', in Guarino, L. (ed) *Traditional African Vegetables: Promoting the Conservation and Use of Underutilised and Neglected crops – 16*, Proceedings of the IPGRI International Workshop on Genetic Resources of Traditional Vegetables in Africa: Conservation and Use, 29–31 August 1995, ICRAF-HQ, Nairobi, Kenya,

Institute of Plant, Genetic and Crop Plant Research, Gatersleben/International Plant Genetic Resources Institute, Rome, Italy

High, C. and Shackleton, C. M. (2000) 'The comparative value of wild and domestic plants in home gardens of a south African rural village', *Agroforestry Systems*, vol 48, no 2, pp141–156

Hildebrand, E. A. (2003) 'Motives and opportunities for domestication: An ethnoarchaeological study in southwest Ethiopia', *Journal of Applied Archaeology*, vol 22, pp358–375

Hoffmann, I., Gerling, D., Kyiogwom, U. B. and Mané-Bielfeldt, A. (2001) 'Farmers' management strategies to maintain soil fertility in a remote area in northwest Nigeria', *Agriculture, Ecosystems and Environment*, vol 86, pp263–275

Ikeorgu, J. E. G. (1990) 'Glasshouse performance of three leafy vegetables grown in mixtures in Nigeria', *Scientia Horticulturae*, vol 43, pp181–188

Ikeorgu, J. E. G. (1991) 'Effects of maize and cassava on the performance of intercropped *egusi* melon (*Citrullus lanatus* (L.) Thunb.) and okra *(Abelmoschus esculentus* (L.) Moench.) in Nigeria, *Scientia Horticulturae*, vol 48, pp261–268

Ikeorgu, J. E. G., Ezumah, H. C. and Wahua, T. A. T. (1989) 'Productivity of species in cassava/maize/okra/*egusi* melon complex mixtures in Nigeria', *Field Crops Research*, vol 21, pp1–7

Jansen van Rensburg, W. S., Netshiluvhi, T. R., Venter, S. L., van den Heever, E., Vorster, H. J. and de Ronde, J. A. (2004) 'Role of indigenous leafy vegetables in combating hunger and malnutrition', *South African Journal of Botany*, vol 70, pp52–59

Jansen van Rensburg, W. S., van Averbeke, W., Slabbert, R., Faber, M., van Jaarsveld, P., van Heerden, I., Wenhold, F. and Oelofse, A. (2007) 'African leafy vegetables in South Africa', *Water SA*, vol 33, pp317–326

Keller, G. (2004) *African Nightshade, Eggplant, Spiderflower et al – Production and Consumption of Traditional Vegetables in Tanzania from the Farmers' Point of View*, MSc thesis, Georg-August Universität, Göttingen, Germany

Kipgosgei, K. L. (2004) 'Response of African nightshades (*Solanum villosum)* and spiderplant (*Cleome gynandra)* to fertilizer application, cooking and pest infestation in Keiyo District, Kenya', MSc thesis, University of Nairobi, Kenya

Leimar, D. and Price, L. (1997) 'Wild plant food in agricultural environments: A study of occurrence, management and gathering rights in Northeast Thailand', *Human Organization*, vol 56, no 2, pp209–221

Materechera, S. A. and Medupe, M. L. (2006) 'Effects of cutting frequency and nitrogen form fertiliser and cattle manure on growth and yield of leaf amaranth (*Amarathus hybridus*) in a South African semi-arid environment', *Biological Agriculture and Horticulture*, vol 23, pp251–262

MATF (2006) 'African indigenous vegetables: Spotlight on AVRDC', *MATF Newsletter*, no 4, June, pp11–12

Mavimbela, S. (2004) 'Local level trade in wild spinaches in Nelspruit town, Mpumalanga Province of South Africa: opportunities and constraints', Honours dissertation, Rhodes University, Grahamstown, South Africa

Modi, A. T. (2004) 'Short-term preservation of maize landrace seed and taro propagules using indigenous storage method', *South African Journal of Botany*, vol 70, pp16–23

Modi, A. T. (2007) 'Effect of indigenous storage method on performance of taro (*Colocasia esculenta* (L.) Schott) under field conditions in a warm subtropical area', *South African Journal of Plant and Soil*, vol 24, pp214–219

Moreno-Black, G. and Somnasang, P. (2000) 'In times of plenty and times of scarcity: Non-domesticated food in northeastern Thailand', *Ecology of Food and Nutrition*, vol 38, pp563–586

Murage, E. N., Chweya, J. A. and Imungi, J. K. (1996) 'Changes in leaf yield and nutritive quality of the black nightshade (*Solanum nigrum*) as influenced by nitrogen application', *Ecology of Food and Nutrition*, vol 35, pp149–157

Musotsi, A. A. (2004) *The Role of Home Gardening in Household Food Security in Butere Division, Western Kenya*, MSc thesis, Maseno University, Maseno, Kenya

Mwai, G. N. and Schippers, R. R. (2004) '*Solanum scabrum*', in Grubben, G. J. H. and Denton, O. A. (eds) *Plant Resources of Tropical Africa 2: Vegetables*, PROTA Foundation, Wageningen, The Netherlands/Backhuys Publishers, Leiden, The Netherlands, pp493–501

Nekesa, P. and Meso, B. (1997) 'Traditional African vegetables in Kenya: Production, marketing and utilization', in Guarino, L. (ed) *Traditional African Vegetables: Promoting the Conservation and Use of Underutilised and Neglected crops – 16*, Proceedings of the IPGRI International Workshop on Genetic Resources of Traditional Vegetables in Africa: Conservation and Use, 29–31 August 1995, ICRAF-HQ, Nairobi, Kenya, Institute of Plant, Genetic and Crop Plant Research, Gatersleben/International Plant Genetic Resources Institute, Rome, Italy

Nkhoma, C. N., Mkamanga, G. Y. and Ruredzo, T. J. (1997) 'Conservation of traditional vegetable germplasm in the SADC Region', in Guarino, L. (ed) *Traditional African Vegetables in Africa: Proceedings of the IPGRI International Workshop on Genetic Resources of Traditional Vegetables in Africa: Conservation and Use*, 29–31 August 1995, ICRAF, Nairobi, Kenya

Ofori, F. and Stern, W. R. (1987) 'Cereal–legume intercropping systems', *Advances in Agronomy*, vol 41, pp41–90

Ogle, B. M and Grivetti, L. E. (1985a) 'Legacy of the chameleon: Edible wild plants in the kingdom of Swaziland, Southern Africa – A cultural ecological, nutritional study. Part III – cultural ecological analysis', *Ecology of Food and Nutrition*, vol 17, no 3, pp1–40

Ogle, B. M and Grivetti, L. E. (1985b) 'Legacy of the chameleon: Edible wild plants in the Kingdom of Swaziland, Southern Africa, Part II – demographics, species availability and dietary use, analysis by ecological zone', *Ecology of Food and Nutrition*, vol 17, pp1–30

Omotoso, S. O. and Shittu, O. S. (2007) 'Effect of NPK fertiliser rates and method of application on growth and yield of okra (*Abelmoschus esculentus* (L.) Moench) at Ado-Ekiti Southwestern, Nigeria', *International Journal of Agricultural Research*, vol 2, pp614–619

Onyango, M. O. A. (1992) *Viability, Growth and Seed Production of Six Indigenous Vegetables*, Research report, JKUAT, Nairobi, Kenya

Onyango, M. O. A. (2005) 'Studies on horticultural practices of six African indigenous vegetables at Maseno University', in *Proceedings of the Third Horticulture Workshop on Sustainable Horticultural Production in the Tropics*, 26–29 November 2003, Maseno University MSU, Maseno, Kenya, pp13–18

Onyango, M. O. A., Onyango, J. C. and Jama, B. (1999a) *Traditional Crops in Western Kenya: Adaptive Technologies for Increased Production*, IRPS, Maseno University, Kenya, Reprint Series no 3, pp14–23

Onyango, M. O. A., Onyango, J. C., Bashir, J., Niang', A. and Obiero, H. M. (1999b) 'Effect of organic and inorganic sources of fertilizer on growth and yield of five priority vegetable crops in Western Kenya', in *Proceedings of the 2nd National Horticulture Seminar*, JKUAT, Nairobi Kenya, pp20–29

Oseko, J. K. (2007) 'Response of African kale (*Brassica carinata)* to intercropping with four African indigenous vegetables', MSc thesis, Maseno University, Kenya

Schippers, R. R. (2002) *African Indigenous Vegetables: An Overview of the Cultivated Species*, revised edition, Natural Resources Institute/ACP-EU Technical Centre for Agriculture and Rural Cooperation, Chatham, UK

Sekamatte, B. M., Ogenga-Latigo, M. and Russell-Smith, A. (2003) 'Effects of maize–legume intercrops on termite damage to maize, activity of predatory ants and maize yields in Uganda', *Crop Protection*, vol 22, pp87–93

Sithanantham, S., Matoka, C. M., Maundu, M., Jakari, M. and Agong, S. G. (2005) 'Integrated crop protection research for sustainable production of indigenous vegetable crops in Eastern Africa', in *Proceedings of the Third Horticulture Workshop on Sustainable Horticultural Production in the Tropics*, 26–29 November 2003, Maseno University, Maseno, Kenya, pp37–42

Slabbert, R., Spreeth, M. and Krüger, G. H. J. (2004) 'Drought tolerance, traditional crops and biotechnology: Breeding towards sustainable development', *South African Journal of Botany*, vol 70, pp116–123

Soini, E. (2005) 'Land use change patterns and livelihood dynamics on the slopes of Mt Kilimanjaro, Tanzania', *Agricultural Systems*, vol 85, pp306–323

Trenbath, B. R. (1993) 'Intercropping for the management of pests and diseases', *Field Crop Research*, vol 34, pp381–405

Vainia-Mattila, K. (2000) 'Wild vegetables used by the Sambaa in the Usambara Mountains NE Tanzania', *Annales Botanici Fenici,* vol 37, pp57–67

Vanlauwe, B. and Giller, K. E. (2006) 'Popular myths around soil fertility management in sub-Saharan Africa', *Agriculture, Ecosystems and Environment*, vol 116, pp34–46

Vorster, I. H. J., Jansen van Rensburg, W., van Zijl, J. J. B. and Venter, S. (2007) 'The importance of traditional leafy vegetables in South Africa', *African Journal of Food Agriculture, Nutrition and Development*, vol 7, no 4, www.ajfand.net

Weinberger, K. and Msuya, J. (2004) *Indigenous Vegetables in Tanzania: Significance and Prospects*, Technical Bulletin no 31, AVRDC, Shanhua

Willey, R. W. (1979) 'Intercropping: Its importance and research need. Part 1: Competition and yield advantages – field crop abstracts', *Commonwealth Bureau of Pastures and Field Crops*, vol 32, pp1–10

Zinyama, L. M., Matiza, T. and Campbell, D. J. (1990) 'The use of wild food plants during periods of food shortage in rural Zimbabwe', *Ecology of Food and Nutrition*, vol 24, pp251–265

6

The Production of African Indigenous Vegetables in Urban and Peri-Urban Agriculture: A Comparative Analysis of Case Studies from Benin, Kenya and South Africa

Margaret W. Pasquini, Françoise Assogba-Komlan, Ineke Vorster, Charlie M. Shackleton and Mary O. Abukutsa-Onyango

Introduction

African indigenous vegetables (AIVs) comprise an enormous range of species that are consumed throughout the continent (FAO, 1988; Grubben and Denton, 2004). A number of studies have recorded an impressive diversity of species with known vegetable use in various countries. For example, across Kenya there is knowledge on the use of 210 species (Maundu, 1996), and in Benin surveys from an ongoing project involving two of the authors of this chapter have recorded 250 species so far (unpublished data).

However, various studies have been highlighting how food habits across Africa are shifting towards increasingly monotonous diets, with a decline in the consumption of traditional foods, including AIVs. This is leading to a loss of indigenous knowledge on the use and preparation of many traditional food resources (Maundu et al, 1999; Vorster et al, 2007a; Akpavi et al, 2008).

Until very recently, AIVs received very little attention from formal research and development agencies (Pasquini and Young, 2006), and it is only now that their contribution to local diets and economies is starting to be investigated and quantified, albeit still in a very patchy way.

Although it is generally perceived that there has been a widespread overall trend towards the reduction in the diversity of AIVs being used by communities, there is encouraging evidence that specific AIVs are still widely used or even instances where their popularity is increasing. In Yaoundé in Cameroon, species such *Amaranthus cruentus*, *Solanum scabrum*, *Corchorus olitorius* and the leaves of *Manihot esculenta* are widely produced and consumed; and contrary to patterns observed in other African cities, where exotic *Brassica* species were increasingly replacing indigenous species, in Yaoundé the consumption of exotic *Brassica* spp. was low (accounting for only 7 per cent of total household expenditures on leafy vegetables) (Gockowski et al, 2003).

In Mali, interviews with village-based key informants conducted in the course of a survey in 2006 (unpublished data from an ongoing project)[1] indicated that while the consumption of certain AIVs was waning, other AIVs had become very popular. The use of *Corchorus* species, in particular, appeared to be a relatively new development, and a couple of respondents indicated that because of migration from rural to urban areas, young women had been exposed to new ideas and recipes, and had transmitted the information back to their villages of origin. Demand is so high that although cultivation of *Corchorus olitorius* and the wild collection of *Corchorus tridens* appear to be widespread, there is also significant importing from neighbouring countries. As found in a study by Abasse et al (2007), the Tillabery region in Niger exported large quantities of *Corchorus* spp. to Mali so that producers of *Corchorus* had annual revenues higher than US$1000. The study found that overall in this region, leafy AIVs could contribute more than 50 per cent of the annual income of the farmers, though in the regions of Maradi and Dosso, the contribution was about 20 to 30 per cent.

In Nairobi, Kenya, prior to 2000, AIVs had been traded in a localized way, mainly in areas inhabited or regularly visited by people originally from major growing areas. Since 2000, trade has increased both in terms of volume and number of species on offer, thanks to a combination of consumer promotion initiatives, stimulation of production in peri-urban and rural areas, linking of small-scale producers with supermarkets and strengthening of producer groups (Irungu et al, 2006).

With few exceptions, most work on AIVs has been carried out in rural areas. Few studies have explicitly attempted to collect information on the specific features of AIV production systems in urban and peri-urban areas and to determine the relative importance of AIVs within urban and peri-urban agriculture (UPA).

The IndigenoVeg project (see www.indigenoveg.org) sought to address this gap by developing a survey to collect baseline data from two cities in seven African countries (Benin, Côte d'Ivoire, Senegal, Kenya, Uganda, Tanzania and

South Africa) on the characteristics of AIV production systems in UPA, and to study the scope and value of economic activities related to the marketing of AIVs by documenting different supply chains, the range of different actors involved, the volumes moved along the chain, as well as the value of produce. The outcomes of the survey in relation to the marketing aspects are presented and discussed in Chapter 7, whereas this chapter focuses on the production aspects of AIVs in urban and peri-urban agricultural systems.

The main body of this chapter is devoted to the presentation and discussion of data on the main features of AIV production systems for two cities from three of the seven survey countries – namely, Benin, Kenya and South Africa – thereby having contributions from West, East and Southern Africa. The chapter includes data (disaggregated, where possible, by urban and peri-urban location) on producer characteristics, access to and use of land, AIVs produced, place and form of production of main crops, use of inputs, and degree of integration of AIVs within the wider farming systems. The results section is followed by an overview of the most salient features of the three countries' AIV production systems, and an assessment of the probable role played by, and relative importance of, AIVs within UPA systems in each city. In the light of the specific features exhibited in the three countries, city and urban/peri-urban location-specific differences, the concluding section identifies the key opportunities and constraints for future research and development interventions to promote AIVs in UPA.

Survey methodology

Data in the three countries were collected using a standardized questionnaire survey schedule developed through the IndigenoVeg project. In Benin, the questionnaire survey was administered to 59 and 28 farmers, respectively, in the urban and peri-urban areas of Cotonou, and 47 and 51 farmers in the urban and peri-urban zones of Lokossa. Unstructured interviews were also carried out in the course of the survey in order to verify and complement certain items of information collected by the enumerators. In Kenya, 9 and 55 farmers in the urban and peri-urban areas of Nairobi were interviewed, and 38 and 39 in the urban and peri-urban zones of Kisumu. In Soshanguve and Durban (in South Africa), 166 and 165 farmers were interviewed, respectively. It should be noted that in the case of one farmer in Durban and 18 farmers in Soshanguve, there was a problem with the data collection and the gender and other personal characteristics of the respondent were not recorded. In most cases, the analysis excludes these respondents.

In Benin, vegetable production sites located within the area of Cotonou city itself were classified as urban locations. Peri-urban vegetable production sites were located within the neighbouring communes of Abomey-Calavi, Ouidah, Torri-Bossito and Sèmè-Kpodji. In the case of Lokossa, sites located within a 3km radius of the city centre were considered urban, whereas the rest of the town and the communes of Athiémè, Dogbo, Bopa and Houéyogbé were classified as peri-urban. As yet, there is no consensus on how to situate farming areas along the rural–urban continuum. Iaquinta and Drescher (2000) have proposed a very interesting peri-urban typology model derived from underlying socio-demographic processes, especially migration. However, as the present study was extremely limited in terms of resources, it was not possible to collect all of the information that would have been required to classify the farming locations into the different types proposed by the model. Furthermore, the model is based on the assumption that respondents live and farm in the same neighbourhood; but this is not always the case. The vegetable production site that is situated on airport land in Cotonou is a non-residential area, and farmers are likely to be drawn from several different neighbourhoods. Survey partners examined and classified target survey locations mainly according to physical criteria (street patterns and housing density), administrative criteria, distance to city limits and homogeneity of the population. When comparing and interpreting the results for the different countries, the reader should bear in mind that as partners worked on the classification of each area on an individual country-specific basis, they may not have always come to the same decisions regarding the placement of the boundary line between urban and peri-urban.

In Kisumu, the main urban production areas were Nyamasaria-Dunga, Kanyakwar, Kibos-Chiga and Mamboleo-Wathorego, whereas the main peri-urban production areas were Kiboswa, Ahero, Kisian-Kombewa and Maseno. In Nairobi, indigenous vegetable farmers were mostly located in the peri-urban environs, which comprised Wangige, Kikuyu, Uthiru, Kinoo, Kiambu-Githunguri and Ngong-Kiserian. Only nine respondents from urban locations were interviewed and they were located in Dandora, Mathare, Kawangare, Ridgeways and Ngei.

All farms around Soshanguve were classified as peri-urban locations. In Durban, only around 10 per cent of the respondents were located in urban areas, and for this reason no distinction was made in the analysis between urban and peri-urban location.

There were no sampling frames available for any of the survey locations. The institutions responsible for the work carried out reconnaissance surveys of vegetable production sites to identify areas with reasonably widespread indigenous vegetable production activities. A shortlist of areas was drawn up and enumerators sampled from these locations, approaching all farmers who were available on their farms at the time of the visit. In Soshanguve, representatives from projects in the area and the Tshwane Metropolitan Council development agents were approached and asked for advice on suitable locations. In these locations, farmers were also asked to identify other farming areas or farmers. All individual farmers were interviewed, and at least half the farmers who formed part of communal areas or groups were interviewed due to their small number. In Nairobi, few indigenous vegetable farmers were encountered in urban locations, and for this reason, most respondents were drawn from peri-urban locations.

Interviews were conducted, where possible, in the local language preferred by the respondent. All farmers were informed of the nature of the project, the confidentially of their responses, and their right to withdraw from the interview at any stage or to pass over any specific question.

Many farmers did not keep records and, hence, the accuracy of many answers was hard to capture and, indeed, often not available to the researchers. Nearly all farmers in Kenya and South Africa produced other crops alongside AIVs, and did not consider inputs separately between AIVs and the other crops, nor between different species of AIVs, a problem also noted by Gockowski et al (2003).

Survey locations

Benin

Benin is a West African country that lies between the Equator and the Tropic of Cancer. It is bordered to the south by the Atlantic Ocean, to the east by Nigeria, to the west by Togo, and to the north by Niger and Burkina Faso. The country extends approximately 700km from the coast to the northern boundary. It therefore covers three agro-ecological zones, Guinean, Sudano-Guinean and Sudanian. In the south, where the survey cities are located, the climate is characterized by two rainfall seasons (April to July, and October to November).

Benin's economy is chiefly based on agriculture, with cotton being a key crop for export. Poverty levels are high. In 2007, the per capita income was US$570 (World Bank, 2008). The Human Poverty Index is 47.6, and the country therefore ranks 100th among 108 developing countries for which the index has been calculated (2004 statistics). The Human Development Index is 0.437, which gives the country a rank of 163rd out of 177 assessed countries (2005 statistics) (UNDP, 2008a).

The subequatorial maritime climate that characterizes the south of the country facilitates vegetable production throughout the year, particularly in the low-lying floodplains, where even in the dry season farmers have good access to irrigation water. Although neglected by the research and development sector, AIVs occupy an important place in local food traditions, and cultivated species such as *Amaranthus hibridus*, *Solanum macrocarpon*, *S. aethiopicum*, *Celosia argentea*, *Corchorus olitorius*, *Vernonia amygdalina* and *Basella alba* can be found in urban markets throughout the year (Colin and Heyd, 1991; Guidi, 2007). Wild-collected species are also widely used in the country. A recent survey carried out in the course of an ongoing project identified over 250 species used as vegetables and approximately 60 per cent of these were wild (unpublished data). Examples of popular wild resources are *Vitex doniana*, *Talinum triangulare* and *Moringa oleifera* leaves.

The popularity of AIVs in urban areas in Benin has been remarked upon in a consumer survey from the above-mentioned ongoing project in Benin. In Dassa-Zoumé (located in the centre of Benin), over 85 per cent of respondents reported that they only used AIVs (respondents were referring to their consumption of vegetables other than tomato and onion), while in Parakou (in the north of the country), nearly half of the respondents reported only using AIVs and a quarter of the sample used AIVs equally or more frequently than exotics.

The current survey was carried out in the two urban centres of Cotonou and Lokossa and their peri-urban environs. Both locations are known for their dynamic vegetable production sites and offered an interesting contrast. Cotonou is the administrative capital of the country and the biggest city in Benin (at the last census in 2002, the Littoral Department, which coincides with the city of Cotonou, numbered 665,100 inhabitants living in an area of 79 square kilometres; INSAE, 2004). It is the site of most government and commercial activity in the country, with good air, sea and road access. The port in Cotonou is very active, and shipments of goods destined for Lagos, in Nigeria, are often routed to Cotonou instead (Ugboma et al, 2004) and transported by road to Lagos. The roads to the Nigerian and Togolese frontiers are in good condition.

Lokossa is a much smaller town, located in the *département* of Mono, in the south-west of Benin, bordering Togo. Exact population figures are not available; but the whole commune numbers 77,065 inhabitants over 260 square kilometres (INSAE, 2004). The area is well known for the high diversity of AIVs, which are produced for local consumption, but which are also destined for the markets of other cities in the country (particularly Cotonou).

Kenya

Kenya is located in East Africa, lying across the Equator, and is bordered by Ethiopia to the north, Somalia to the east, the Indian Ocean to the south-east, Tanzania to the south-west, Lake Victoria and Uganda to the west, and Sudan to the north-west. The agricultural sector is a significant contributor to Kenya's economy, accounting for 28 per cent of gross domestic product (GDP) and generating 60 per cent of foreign exchange earnings and employing about 70 per cent of the population directly or indirectly. Poverty remains a significant challenge, although levels have been declining. In 2005/2006, the national absolute poverty was 46.1 per cent; but there were strong regional disparities – for example, poverty levels in Central Province were 30.3 per cent and in Nyanza Province were 47.9 per cent (World Bank, 2008). In 2004, the Human Poverty Index was 30.8; thus the country ranks 60th among 108 developing countries for which the index has been calculated. The 2005 human development index was 0.521, which gives the country a rank of 148th out of 177 assessed countries (UNDP, 2008b).

Kenya has a wide variety of agro-ecological zones, ranging from humid highlands to arid and semi-arid regions. About 20 per cent of the Kenyan land surface can support rain-fed agriculture. Only 7 per cent is high-potential arable land (Ekbom et al, 2001). In both the survey cities, rainfall has a bimodal pattern, with short rainy (September to November) and long rainy (March to June) seasons.

Traditionally, AIVs have played an important role in the diets of Kenyan communities. As mentioned in the introduction, 210 species are known to communities across the country (Maundu, 1996). However, with the focus of official agricultural policy being placed on crops with potential for export, the production and use of AIVs had begun to decline. This trend reversed in the 1990s with awakened interest in the commercial and nutritional potential of AIVs, resulting in the Indigenous Food Plants Programme initiated by the National Museums of Kenya together with two non-governmental organizations (NGOs) (www.unesco.org), and which has been followed up with other successful promotional and development programmes (Irungu et al, 2006).

Nairobi is the capital and largest city of Kenya and also the capital of Nairobi Province. The city has an elevation of 1661m above sea level. It is the fourth largest city in Africa and the most populous city in East Africa, with an estimated urban population of between 3 and 4 million occupying approximately 684 square kilometres.

Kisumu is the administrative headquarters of Nyanza Province and the main commercial centre in western Kenya. It is the third largest city in Kenya, with a population of 504,000 over a land area of 919 square kilometres. The main industries are subsistence agriculture and a fishery on Lake Victoria. Kisumu is, however, a net food importer and experiences high levels of unemployment (30 per cent), the highest average urban poverty levels at 48 per cent against a national average of 29 per cent, and registers 53.4 per cent incidence of food poverty. Growing poverty has been attributed to the reduction of shipping services to Uganda and Tanzania, the decline of railway services, adverse weather conditions, infestation of the lake by the water hyacinth, and closure and downscaling of operations by many major industries. Currently, over half of the working population is engaged in informal activities.

South Africa

South Africa is the southernmost country of the continent, sharing its northern borders with Namibia, Botswana, Zimbabwe and Mozambique. Approximately 57 per cent of individuals in South Africa were living below the poverty income line in 2001, a statistic that has not changed from 1996. Limpopo and the Eastern Cape had the highest proportion of poor, with 77 per cent and 72 per cent of their populations living below the poverty income line, respectively. The Western Cape had the lowest proportion in poverty (32 per cent), followed by Gauteng (42 per cent) (Schwabe, 2004). In 2004, the Human Poverty Index was 23.5, ranking South Africa as 55th among 108 developing countries for which the index has been calculated. The 2005 Human Development Index for South Africa is 0.674, which gives the country a rank of 121st out of 177 assessed countries (UNDP, 2008c).

Previous work in South Africa shows that traditional foods, including AIVs, are an integral part of people's diets (Shackleton and Shackleton, 2006); however, a limited number of wild species are cultivated other than pumpkins and sweet potatoes. Instead, there is a heavy reliance on species collected either from fields where they occur as weeds or from the rangelands and forests. Various studies have shown that there is a tendency for increased use of wild foods at times of household crisis – for example, during drought (Shackleton et al, 2000; Shackleton, 2003) or unemployment (Dovie et al, 2002).

Soshanguve is a former township that is part of the northern part of Tshwane Metropolitan Council in Gauteng Province and is situated at an altitude of 1307m above sea level. Gauteng is the most urbanized province of South Africa and is responsible for over one third of South Africa's GDP. Soshanguve was established in 1974 and is an acronym for Sotho, Shangaan, Nguni and Venda (different ethnic groups in South Africa) and has about 700,000 residents. It is estimated that about 37.3 per cent of residents in the northern part of Tshwane are poor and that the unemployment rate is 45.5 per cent (Shilowa, 2004). Soshanguve has a slightly undulating terrain and has an average annual rainfall of 672mm. Most of the rainfall is in the form of thunderstorms, with 85 per cent of the precipitation occurring between October and March. Agriculture contributes 0.7 per cent to jobs in Gauteng, with commercial to subsistence farmers found here. About 30,000 smallholdings can be found in the province, mostly in the peri-urban areas.

Durban lies within the KwaZulu-Natal Province of South Africa, under the eThekwini Metropolitan area. The modern city of Durban dates from 1824. A sugar cane industry was established during the 1860s, and due to conflict with the local Zulu population, thousands of Indian labourers were brought to Durban on five-year contracts; as a result, Durban became home to the largest population of Indians outside of India. Today, Durban is the busiest container port in Africa, the world's ninth busiest port, and a popular tourist destination. Durban is characterized by a humid subtropical climate and a mean annual rainfall of 1000 mm (CEROI, 1999). The study area has a total population estimated at 4 million, and a land area of 2292 square kilometres (CEROI, 1999). The region is culturally diverse and has a majority of black residents (68 per cent), a large Indian community (20 per cent) and a white and coloured minority (9 and 3 per cent, respectively) (South African Census, 2001). The area has a large and diversified economy, with strong manufacturing, tourism, transportation, finance and government sectors. Although Durban is the third most populous city in South Africa, the area still has relatively high unemployment, with only 37 per cent of the total population employed, and with 28 per cent unemployed and 35 per cent of citizens not economically active (South African Census, 2001). It is estimated that 16 per cent of the labour force is active in the informal sector.

The producer profile

When examining the data presented in the tables in the following sections the reader should note that, in some cases, the figures add up to slightly more than 100 per cent for certain locations (e.g. in Table 6.1a, the first column adds up to 101 per cent). This is because of rounding up.

Gender and age

In Kenya and South Africa, women constituted the majority of the sample in peri-urban locations (64 per cent in Nairobi, Kisumu and Pretoria, and 78 per cent in Durban) and around half of the sample in urban areas (44 per cent in Nairobi and 50 per cent in Kisumu). In Benin, there was a marked contrast between the two cities. Although women dominated the sample in Lokossa urban and peri-urban areas (79 and 61 per cent, respectively), in Cotonou, these proportions were only 5 and 11 per cent.

Table 6.1a shows the age distribution of the respondents. The age profile of respondents in Benin was almost the opposite of that found in South Africa. Whereas in South Africa over 85 per cent of the sample was over the age of 40, and, indeed, older respondents above the age of 60 made up 34 and 41 per cent of the Durban and Soshanguve samples, in Benin young respondents dominated the sample in all locations except peri-urban Lokossa, and overall very few respondents over the age of 60 were interviewed. It was notable that in Cotonou urban zones, over 70 per cent of the respondents were young men under the age of 40, similar to figures reported in Sodjinou and Assogba Komlan (2008). In Kenya, there was a greater spread between age categories; but there was a higher proportion of young people in Kisumu compared to Nairobi. The age profile of men and women was similar in all cities except Durban, where nearly half of the male sample was over the age of 60, compared to just under one third of women (see Table 6.1b).

Education and main occupation

In terms of education, in all three countries there was a tendency for respondents located in or near the larger cities to have attained higher levels than respondents in the smaller cities (see Table 6.2). This pattern was particularly noticeable in Benin, where 55 and 71 per cent of the Lokossa urban and peri-urban respondents had not received any form of education, compared to the figures in Cotonou of only 15 and 11 per cent. Also, in Lokossa, fewer respondents reached secondary school level. However, it should be noted that these overall figures are strongly affected by the high numbers of women farmers in Lokossa and the absence of women farmers in Cotonou. Indeed, in Lokossa, illiteracy rates amongst the female respondents reached 62 and 94 per cent in urban and peri-urban locations, compared to only 30 and 35 per cent for men. In Kenya, the differences between cities were less marked and around half of the respondents in all locations had received primary school education and one third had gone to secondary school, high school or college/university. In South Africa, the majority of respondents only had primary-level education (around 40 per cent) and a significant number of respondents had received no formal education, reflecting the comparatively older age profile. In both countries there was a tendency for men to have reached higher educational levels than women, although the gender differences were not as marked as in Benin. It should be noted that the educational systems of the countries are not directly comparable at all levels. In South Africa, the term high school encompasses what in other countries is known as secondary school, which explains the apparent lack of respondents at 'secondary' level in South Africa.

Farming was the primary occupation for the vast majority of respondents in Benin and Kenya (89, 89, 91 and 98 per cent in urban and peri-urban Cotonou and Lokossa, respectively; and 78, 85, 95 and 95 per cent of urban and peri-urban respondents in Nairobi and Kisumu, respectively). However, fewer of the respondents in South Africa considered themselves to be full-time farmers (48 and 35 per cent of male and female farmers in Durban; 37 and 50 per cent in Soshanguve). Interestingly, a high proportion of respondents considered themselves primarily as unemployed, particularly in Soshanguve (27, 25, 54 and 30 per cent). 22 and 13 per cent of women farmers in Durban and Soshanguve stated that their primary occupation was being a housewife.

Ethnicity and origin

Farmers were questioned on their language and origin to determine whether or not they were migrants to the area where they were working. Additionally, farmers were queried on where they grew up as children, as well as whether or not they consider that place to be their home.

In the case of Benin, both the urban and peri-urban areas of Cotonou showed a high linguistic/ethnic diversity (up to 12 groups); however, overall, around half of the respondents belonged to the Fon group. In Lokossa, urban areas over 80 per cent of respondents belonged to the Kotafon group, whereas in peri-urban areas the dominant groups were Adja (nearly half the sample) and Sahoue (nearly one third).

Table 6.3 shows that for Lokossa there is a general correspondence between the area where the respondent had grown up and the area where they were farming. Respondents also indicated that they considered the place where they had grown up as their village of origin (85 and 88 per cent of the sample in urban and peri-urban locations). Unsurprisingly, perhaps, the picture for Cotonou is different. Although nearly half of the farmers in urban Cotonou had grown up in the city itself, 40 per cent of the remaining ones originated from other parts of Benin. A high proportion of farmers in peri-urban locations were from peri-urban communes (46 per cent); but 25 per cent of them had migrated from other parts of Benin and 21 per cent were from other West African countries (Togo, Nigeria and Ghana). In Cotonou, farmers' sense of origin and belonging to place was clearly influenced by multiple factors. While it might be expected that respondents who had spent their childhood in a town or village outside of Cotonou might retain a sense of belonging to that place (as was the case for 56 per cent of farmers who had migrated to Cotonou), it was interesting to note that 70 per cent of farmers who had grown up in Cotonou did *not* feel this to be their place of origin. Presumably these farmers retain a strong feeling of connection to the village or town of origin of their family, even if they themselves have not spent any significant time there.

Table 6.1a *Farmer age in urban and peri-urban areas of the six survey cities (percentage of respondents)*

Age	*Urban Cotonou (n = 59)*	*Peri-urban Cotonou (n = 28)*	*Urban Lokossa (n = 47)*	*Peri-urban Lokossa (n = 50)*	*Urban Nairobi (n = 9)*	*Peri-urban Nairobi (n = 55)*	*Urban Kisumu (n = 38)*	*Peri-urban Kisumu (n = 39)*	*Peri-urban Durban (n = 164)*	*Peri-urban Soshanguve (n = 148)*
<30	34	21	23	12	0	5	21	15	1	1
30–40	37	32	38	24	22	24	37	28	12	7
40–50	14	25	19	40	56	36	18	15	21	20
50–60	14	21	15	22	11	22	8	18	33	31
>60	2	0	4	2	11	13	16	23	34	41

Table 6.1b *Farmer age in Durban and Soshanguve broken down by gender (percentage of respondents)*

Age	*Durban men (n = 36)*	*Durban women (n = 128)*	*Soshanguve men (n = 54)*	*Soshanguve women (n = 94)*
<30	0	2	0	1
30–40	14	11	11	5
40–50	6	25	15	22
50–60	33	33	31	31
>60	47	30	43	40

Source: Indigeno Veg survey data, 2006

Table 6.2 *Educational levels of farmers in urban and peri-urban zones in the six survey cities (percentage of respondents)*

Education	*Urban Cotonou (n = 59)*	*Peri-urban Cotonou (n = 28)*	*Urban Lokossa (n = 47)*	*Peri-urban Lokossa (n = 51)*	*Urban Nairobi (n = 9)*	*Peri-urban Nairobi (n = 55)*	*Urban Kisumu (n = 38)*	*Peri-urban Kisumu (n = 39)*	*Peri-urban Durban (n = 164)*	*Peri-urban Soshanguve (n = 148)*
No education	15	11	55	71	0	9	8	18	23.5	31.1
Primary	39	30	30	24	56	48	53	49	43.8	46.6
Secondary	24	44	11	6	22	37	37	23	–	–
High school	17	15	2	0	11	2	0	0	29.6	20.3
College/university	3	0	2	0	11	4	3	10	1.2	1.4
Technical	2	0	0	0	0	0	0	0	1.2	0.7
Other	0	0	0	0	0	0	0	0	0.6	0

Source: Indigeno Veg survey data, 2006

Table 6.3 *Place of origin of farmers in urban and peri-urban zones in Cotonou and Lokossa (percentage of respondents)*

Place of origin	*Urban Cotonou (n = 58)*	*Peri-urban Cotonou (n = 28)*	*Urban Lokossa (n = 46)*	*Peri-urban Lokossa (n = 51)*
Cotonou peri-urban	9	46	0	0
Cotonou urban	48	7	0	0
Lokossa peri-urban	0	0	13	69
Lokossa urban	0	0	74	10
Other parts of Benin	40	25	11	18
Outside of Benin	2	21	9	2

Source: Indigeno Veg survey data, 2006

In Kenya there were also clear differences with respect to language and ethnic origin of farmers between the two cities. Respondents in Kisumu were predominantly Luo (76.6 per cent), while those in Nairobi were predominantly Kikuyu (71.9 per cent). Other communities involved in indigenous vegetable production in Nairobi were the Swahili speakers (17.2 per cent), Kamba (6.3 per cent) and Masai (1.6 per cent), while those in Kisumu included the Luhya (15.6 per cent), Swahili speakers (3.9 per cent) and Nandi (2.6 per cent). Average length of residence in the cities was similar (26.9 years for Nairobi and 28.6 years for Kisumu). Over half of the respondents in the Nairobi sites had grown up in peri-urban locations around Nairobi, whereas the majority of others (over 35 per cent) came from other locations in Kenya (most frequently, central Kenya). Hardly any respondents were from Nairobi itself. In Kisumu peri-urban locations, the majority of respondents were also from peri-urban locations (though not necessarily the same locations where they were currently farming), whereas in urban Kisumu respondents had grown up in Kisumu, peri-urban Kisumu or were from other parts of Kenya in equal proportions. It should be noted that even those who were not within the Kisumu catchment areas most frequently originated from western Kenya. The majority of respondents in both cities (84 and 85 per cent) considered the location in which they had grown up as their home village.

In Soshanguve, the farmers consisted mainly of Pedi (Sepedi speakers – 38 per cent) and Shangaan (Tsonga speakers – 24 per cent) ethnic groups (see Table 6.4), two of the dominant groups in the northern parts of South Africa. These trends were also reflected in the main language of the spouse and the first childhood language learned (data not shown). The main regions of origin also reflected the language results as Limpopo, Gauteng, Northwest and Mpumalanga are the provinces dominated by Sotho (speakers of Sepedi, Tswana and Ndebele), Venda and Shangaan (Tsonga speakers) nations. The results suggest that Soshanguve has been the growth point for urbanization for these provinces, with few people of the more southern and eastern provinces (Xhosa and Zulu) migrating here (these groups tend to migrate to Johannesburg). Over 85 per cent of respondents reported that their parents lived in their childhood towns and 79 per cent also retained a strong association with these towns, indicating that most of the respondents are first-generation migrants to Soshanguve. Indeed, the average length of stay in Soshanguve was just over 16 years. When the establishment of Soshanguve (1974) and the age of the respondents (usually older than 50) are taken into account, the reason for this is obvious.

In Durban, the most common language (main language) spoken by the farmers and their spouses was Zulu, with more than 90 per cent speaking Zulu as their main language. This was followed by Xhosa, English, English/Hindi and Zulu/Xhosa. The average number of years spent living in the Durban Metropolitan area was 31 years, and many have spent their entire lives there. None of the respondents had moved to the city within the last two years. The majority of the respondents grew up in villages in what are now part of the

Table 6.4 *Main language and place of origin for Soshanguve farmers (percentage of respondents)*

Main Language	*Percentage farmers (n = 148)*	*Place of origin*	*Percentage farmers (n = 148)*
Sepedi	38	Limpopo	43
Tsonga	24	Gauteng	18
Tswana	9	Northwest	16
Ndebele	9	Mpumalanga	15
Zulu	7	Lesotho	2
Swazi	4	Mozambique	1.3
Xhosa	2.7	Freestate	1.3
Venda	2	Eastern Cape	1.3
Sesotho	2.7	Malawi	0.7
Afrikaans	0.6	KwaZulu-Natal	0.7
		Venda	0.7
		Western Cape	0.7

Source: IndigenoVeg survey data, 2006

Table 6.5 *Place of origin for Durban farmers (percentage of respondents)*

Place of origin	*Overall (n = 164)*	*Males (n = 36)*	*Females (n = 128)*
Within the current Durban Metro	77	82	76
Elsewhere in the same province	13	12	13
Other provinces (Eastern Cape)	10	6	11

Source: IndigenoVeg survey data, 2006

Durban Metropolitan area, or within a short radius of it (see Table 6.5). Across the female farmers, 57 childhood villages and towns were identified, whereas only 29 were mentioned by the male farmers. The majority of respondents (84.9 per cent) also reported that their parents continued to reside in the town/village where they grew up, while 85.3 per cent still consider this home. A greater proportion of male farmers (91.4 per cent men versus 84.1 per cent women) considered the place where they grew up as home, no doubt a reflection of a patrilineal society.

Experience in the production of AIVs

Table 6.6 show the number of years of experience in the production of AIVs. Strikingly, in the Lokossa urban area, over half of the respondents had been involved in the production of AIVs for less than six years, contrasting with Cotonou urban areas, where this proportion was only 17 per cent. The data were grouped together into three categories (<6 years of experience, 6 to 15 years, and over 15 years of experience) for each location and evaluated with a chi-square test for independence. The results showed that there was a significant association between location and experience ($\chi^2(_6) = 24.039$, p = 0.001). An inspection of the data suggested that the significant result was explained by the disproportionately high numbers of inexperienced farmers in urban Lokossa (<6) and the high numbers of experienced farmers in urban Cotonou (15+). In order to identify the observations which are primarily responsible for the significant result, a series of comparisons that are meaningful at a theoretical and/or practical level can be conducted (Sheskin, 2007). In this context, a simple comparison was carried out on these two groups, which gave the results $\chi^2_{(1)} = 17.133$, $p<0.001$, confirming that there is a significant difference in experience between the two groups.

This difference (which was not linked to either age or gender characteristics of the respondents) may be explained by the fact that through NGO efforts, vegetable production has been promoted in Cotonou (and Porto Novo) since the early 1970s. There were no similar interventions for other cities in Benin. However, in recent years there has been an expansion in the civil administration sector in Lokossa (leading to increased numbers of civil servants residing in the town), as well as a new Chinese-owned textile factory, probably resulting in increased demand for vegetables.

An inspection of the data in Table 6.6 suggested that peri-urban farmers in Kisumu were more experienced than farmers in urban Kisumu and peri-urban Nairobi (the sample size in urban Nairobi was too small to allow for inclusion into any statistical analysis).

The farmer data were grouped into three categories of experience, and were evaluated with a chi-square for independence test. The results were: $\chi^2_{(4)} = 11.562$, $p = 0.021$. Two simple comparisons were carried out: one to compare the low experienced and high experienced categories of farmers in peri-urban Kisumu and in urban Kisumu, and one to compare the same categories in peri-urban Kisumu and peri-urban Nairobi. Since the comparisons had not been planned in advance, in order to avoid inflating the type I error rate, each of the comparisons was evaluated using the tabled critical chi-square value at the 0.025 level, rather than the 0.05 level. The outcome of the first comparison was $\chi^2_{(1)} = 5.853$ and the second one was $\chi^2_{(1)} = 6.492$. In both cases, the χ^2 value achieves significance at the 0.025 level, confirming the observation that farmers in peri-urban Kisumu are more experienced than farmers in either urban Kisumu or peri-urban Nairobi.

Farmer experience was also grouped into the three categories for Durban and Soshanguve. The test results were: $\chi^2_{(2)} = 94.618$, $p<0.001$. Examination of the data and a subsequent χ^2 test confirmed that farmers in Durban were much more experienced than farmers in Soshanguve. However, as can be seen in Table 6.6, and assuming that the number of years in this question is consecutive, in Durban there was a decrease in the proportion of farmers growing AIVs between 10 and 20 years ago, which then increased 10 years ago. Of the male farmers, 42.3 per cent have grown AIVs for more than 20 years compared to the 61.4 per cent of their female counterparts that have done so, an interesting finding when considered against the age profiles (80 per cent of male farmers were over the age of 50, compared to only 63 per cent of female farmers).

Access to cultivable land resources

Table 6.7 shows the patterns of landownership, renting or free use/occupation in the six cities. A number of farmers had portions of land under different tenure arrangements.

Landownership was high both in South Africa and Kenya (ranging from 78 per cent in urban Nairobi to 100 per cent in peri-urban Soshanguve and Kisumu), but in Benin it was about 60 per cent in Lokossa, and much lower in urban and peri-urban Cotonou (15 and 21 per cent). In Soshanguve, however, the high rates of landownership are partly attributable to the very limited kitchen garden space (generally less than 10 square metres) that is attached to their houses. In Kenya, in both cities a greater proportion of women owned land, whereas there were no marked gender differences for either of the South African cities or Lokossa urban areas. However, in Lokossa peri-urban areas landownership rates were highest for men (70 versus 48 per cent).

The highest rates of renting were found in peri-urban Cotonou and peri-urban Kisumu. In Kenya, in both cities a greater proportion of men rented land, whereas in Durban a greater proportion of female farmers rented land. The latter case may possibly be related to women's access to land. All of the male farmers who rent also own land, while not all female farmers who rent also own land.

In all three countries, farmers used land which they neither owned nor rented, noticeably so for Cotonou, Soshanguve and Lokossa. In Cotonou urban and peri-urban areas, this was explained by the fact that farmers had been given free access to land owned by the state. However, farmers' access to land remains insecure for several reasons. For example, one main production location is in the airport area. One site actually belongs to the airport and a second site is located just behind the airport. This second site was sold in 2007 to a Lebanese developer so that a hotel and supermarket could be built; as a result, about 100 farmers were evicted. The airport site is also under threat over concerns about the safety of food produced here because of landing aircraft passing directly overhead. For the moment, no action has been taken; but given the evident development value of vacant plots of land, food safety may well become a convenient and easy excuse to get farmers to vacate the site. In other locations, farmers may use municipal grounds (e.g. roadsides and under power lines), vacant land which has not yet been developed, or land belonging to the church, schools or clinics.

Use of land resources

Farmers were also asked how much cultivable land they had access to, and how much of this was under vegetable production. Figures 6.1 and 6.2 show the median and inter-quartile range for cultivable areas and areas under vegetable production, respectively, in urban and peri-urban locations for the six cities.

Table 6.6 *Number of years of experience in the production of AIVs in urban and peri-urban areas of Cotonou, Lokossa, Nairobi and Kisumu, and for male and female farmers in Durban and Soshanguve (percentage of respondents)*

Years of experience	*Urban Cotonou (n = 59)*	*Peri-urban Cotonou (n = 28)*	*Urban Lokossa (n = 47)*	*Peri-urban Lokossa (n = 49)*	*Urban Nairobi (n = 9)*	*Peri-urban Nairobi (n = 55)*	*Urban Kisumu (n = 38)*	*Peri-urban Kisumu (n = 39)*	*Peri-urban Durban men (n = 36)*	*Peri-urban Durban women (n = 128)*	*Peri-urban Soshanguve men (n = 54)*	*Peri-urban Soshanguve women (n = 94)*
0–2	3	14	6	0	22	16	21	8	11	11	20	19
3–5	14	21	51	22	33	27	24	21	17	13	52	38
6–10	24	18	9	22	11	25	32	8	19	9	17	24
11–15	19	21	19	24	11	9	3	13	8	3	2	12
16–20	15	11	0	12	0	5	8	10	3	3	4	6

Source: IndigenoVeg survey data, 2006

Table 6.7 *Farmers' access to land in urban and peri-urban areas of Cotonou, Lokossa, Nairobi, Kisumu, Durban and Soshanguve (percentage of respondents)*

Land tenure system	*Urban Cotonou (n = 59)*	*Peri-urban Cotonou (n = 28)*	*Urban Lokossa (n = 47)*	*Peri-urban Lokossa (n = 51)*	*Urban Nairobi (n = 9)*	*Peri-urban Nairobi (n = 55)*	*Urban Kisumu (n = 38)*	*Peri-urban Kisumu (n = 39)*	*Peri-urban Durban (n = 151)*	*Peri-urban Soshanguve (n = 166)*
Land owned	15	21	60	59	78	82	82	100	93	100
Land rented	1.7	25	9	12	44	13	18	26	9	0
Land neither rented nor owned	100	86	51	49	–	20	29	10	17	100

Source: IndigenoVeg survey data, 2006

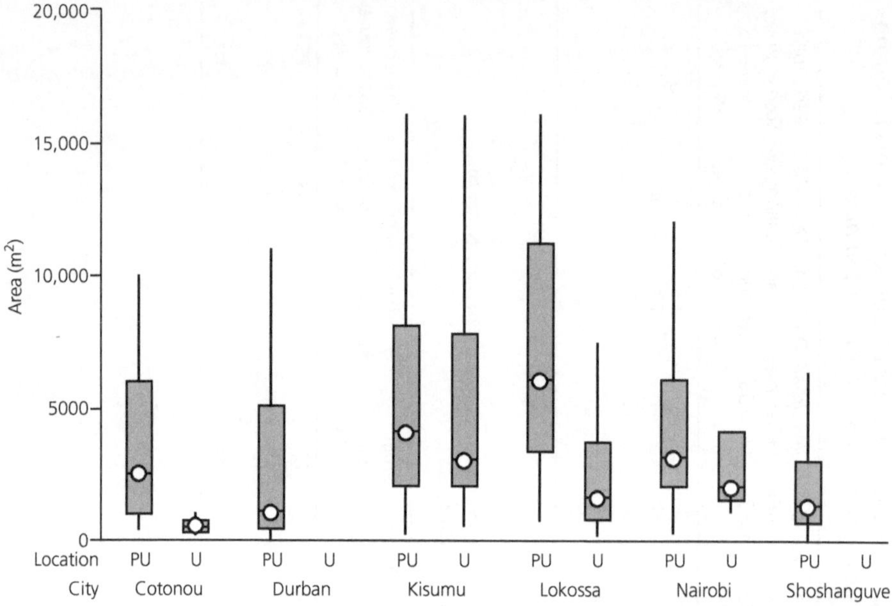

Source: IndigenoVeg survey data, 2006

Figure 6.1 *Cultivated areas in urban and peri-urban locations of the six survey cities (median values and inter-quartile ranges are displayed)*

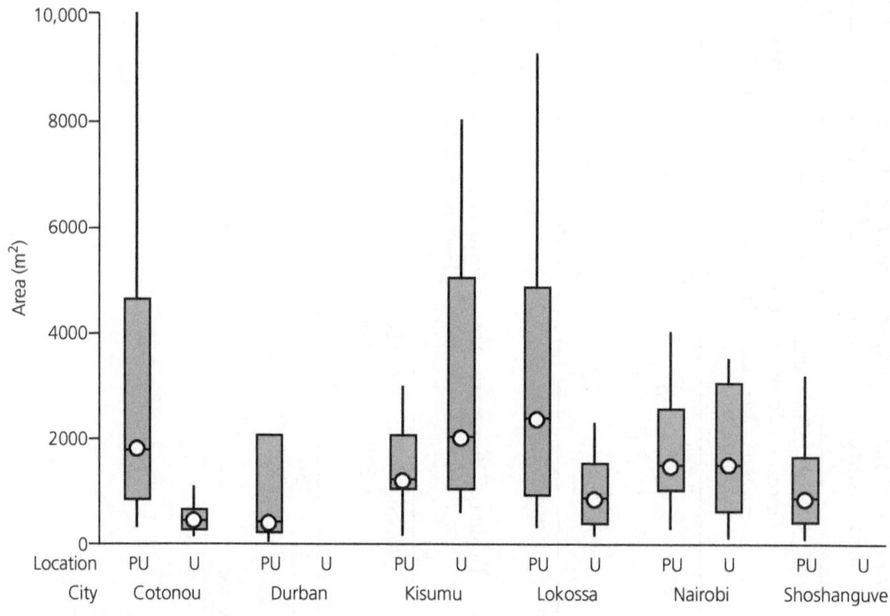

Source: IndigenoVeg survey data, 2006

Figure 6.2 *Vegetable production areas in urban and peri-urban locations of the six survey cities (median values and inter-quartile ranges are displayed)*

Table 6.8 shows the number of farmers using different proportions of land for vegetable production (100, 50–99 and <50 per cent) in the two cities for Benin, Kenya and South Africa.

In the case of Benin, a chi-square test of independence showed that there was a significant association between location and proportion of land under vegetable production ($\chi^2(_6) = 83.124$, $p<0.001$). An analysis of the standardized residuals (Haberman, 1973, as cited in Sheskin, 2007) indicated that the significant result was given by the higher than expected numbers of farmers exclusively producing vegetables in urban Cotonou, and lower than expected numbers in peri-urban Lokossa; higher than expected numbers in peri-urban Lokossa using less than half their land for vegetable production, and lower than expected numbers in all other locations; and lower than expected numbers using between 50 and 99 per cent of their land in urban Cotonou.

Three complex comparisons were performed:

1 a comparison of land use in Cotonou overall to that in Lokossa overall;
2 a comparison between farmers exclusively producing vegetables and farmers producing a mixture of vegetables and other crops in urban and peri-urban Cotonou;
3 a comparison between farmers exclusively producing vegetables and farmers producing a mixture of vegetables and other crops in urban and peri-urban Lokossa.

In order to avoid the overall type I error rate being greater than 0.05, the alpha level used for each comparison was 0.05/3 = 0.0167. For the first comparison, $\chi^2(_2) = 67.274$, $p< 0.001$. In Cotonou, a higher proportion of farmers exclusively produced vegetables, whereas in Lokossa a higher proportion produced vegetables on less than 50 per cent of their land.

However, the results of the second and third comparisons were $\chi^2(_1) = 11.774$, $p = 0.001$, and $\chi^2(_1) = 3.930$, $p = 0.047$. Thus, in Cotonou the urban or peri-urban location is associated with the proportion of land under vegetable production (more specifically, farmers located in urban areas are more likely to be producing vegetables exclusively). In Lokossa, the result was not significant at the determined alpha level.

In the case of Kenya (urban Nairobi was excluded from the analysis because of the low number of respondents), a chi-square test of independence did not show a significant association between location and proportion of land under vegetable production.

For the South African data, a chi-square test of independence showed significant association between location and proportion of land under vegetable production ($\chi^2(_2) = 24.264$, $p< 0.001$). Analysis of the standardized residuals (Haberman, 1973, as cited in Sheskin, 2007) indicates that the number of Durban farmers using 100 per cent and between 50 and 99 per cent of their land for vegetable production was, respectively, significantly higher and lower than expected ($p<0.01$). The number of Soshanguve farmers using

Table 6.8 *Proportion of land used for vegetable production in urban and peri-urban areas of the six survey cities (count data)*

	100% land used for vegetable production	*50–99% land used for vegetable production*	*<50% of land used for vegetable production*
Urban Cotonou	55**	1**	3**
Peri-urban Cotonou	18	8*	2
Urban Lokossa	15	6	23*
Peri-urban Lokossa	8**	13	28**
Urban Nairobi	2	3	2
Peri-urban Nairobi	14	20	19
Urban Kisumu	17	8	10
Peri-urban Kisumu	11	11	16
Durban	31*	20**	26
Soshanguve	30	89*	28

Notes: * = standardized residual significant at 0.05 level; ** = standardized residual significant at 0.01 level.
Source: IndigenoVeg survey data, 2006

between 50 and 99 per cent of their land for vegetable production was significantly higher than expected.

There were no gender differences with respect to proportion of land used for vegetable production in Kenya or Benin; however, there was in Soshanguve (South Africa) ($\chi^2(_2) = 0.3271$, $p< 0.01$), where a lower than expected proportion of men used less than half their land for vegetable production. This is explained by the fact that under traditional customs, men are responsible for producing maize for household consumption, whereas women produce vegetable crops.

Diversity of African indigenous vegetables

Tables 6.9a to 6.9c show the percentage of farmers growing specific AIVs in the survey areas. In Benin *Solanum macrocarpon* and *Amaranthus* spp. (see Plates 6.1 and 6.2) are the most widely grown species for all four areas (a finding which corroborates work by Assogba-Komlan et al, 2007). The patterns emerging between cities and between urban and peri-urban areas suggest that there is a degree of specialization in the types of crops grown in a particular area. For example, *Vernonia amygdalina* was widespread in Cotonou urban and peri-urban areas (over 70 per cent of respondents grew this species), and hardly found in Lokossa. In contrast, *Corchorus olitorius* (see Plates 6.3 and 6.5) was a key species for Lokossa urban and peri-urban farmers (over 85 per cent of the sample), but much less so for Cotonou. In the case of *Abelmoschus esculentus* (see Plate 6.4) there was a marked difference between Lokossa urban and peri-urban areas (13 versus 67 per cent). The pattern was reversed for *Solanum scabrum/villosum*, which was much more important in Lokossa urban areas compared to its peri-urban environs (45 versus 10 per cent).

In Kenya, widely grown species for both cities comprised *Amaranthus* spp., *Solanum scabrum/villosum*, *Cleome gynandra* (see Plate 6.6) and *Vigna unguiculata* (see Plate 6.7). As in the case of Benin, there were differences between locations. *Crotalaria brevidens* (see Plate 6.8) was a widespread species in Kisumu, but not in Nairobi. A higher proportion of respondents in Nairobi reported growing *Cucurbita* spp. than in Kisumu. *Urtica massaica* was only found in Nairobi.

The two South African cities differed quite markedly. In Durban, 69 per cent or more of respondents reported growing *Colocasia esculenta* (see Plates 6.9 and 6.10), *Ipomoea batatas* and *Cucurbita* spp.; but in Soshanguve the most widespread crop, *Cucurbita* spp., was reported only by around half of the respondents. Dahl (which is a broad term for various leguminous crops) was only found in Durban, whereas *Cleome gynandra* was only found in Soshanguve. *Cleome gynandra* and *Corchorus olitorious* are generally wild and only grow in the northern parts of South Africa. *Colocasia esculenta* does not grow very well in Soshanguve as it requires a lot of water and no frost. Dahl is grown specifically for the Indian markets found in and around Durban. The data suggest that Durban farmers grow the same main crops, whereas in Soshanguve there is more variation between farmers, which is perhaps a reflection of the greater ethnic diversity in the area.

Although there were slight differences in the proportions of male and female farmers growing specific crops in each country (see, for example, Table 6.9c for a breakdown of the South African data), there was no indication that specific crops were exclusively grown by either men or women. However, in Soshanguve, there was a tendency for older women to hold to the traditional concept that cash crops (i.e. *Abelmoschus esculentus* and *Ipomoea batatas*) are grown by men and that household food (*Cleome gynandnra*, *Corchorus* spp., *Vigna unguiculata*, *Amaranthus* spp. and *Cucurbita* spp.) is grown mainly by women. This is, however, changing in the younger generations. Similarly, in Lokossa, Benin, where *Corchorus olitorius* is considered in local traditions to be a woman's crop, it was evident that though this was a popular crop for women (at least 95 per cent of women respondents in urban and peri-urban areas grew it), it was also produced extensively by men (95 per cent of men in peri-urban areas and 60 per cent in urban areas) and in similar areas (median plot size for both men and women was 400 square metres).

The number of crops mentioned by each farmer is plotted against each location in Figures 6.3a, 6.3b and 6.3c (in addition, the South African data are broken down by gender). Overall, Kenyan farmers appeared to grow a higher number of AIVs than their Beninoise or South African counterparts.

Production practices

Farmers were asked to identify their top three vegetables grown the previous season and to describe the specific production practices. Overall, the most commonly cited vegetables in Benin were *Solanum macrocarpon*, *Amaranthus*

Table 6.9a *Percentage of farmers growing specific AIVs in urban and peri-urban areas of Cotonou and Lokossa*

Species	*Urban Cotonou (n = 59)*	*Peri-urban Cotonou (n = 28)*	*Urban Lokossa (n = 47)*	*Peri-urban Lokossa (n = 51)*
Adansonia digitata	0	0	4	0
Abelmoschus esculentus	20	30	13	67
Amaranthus spp.	100	89	94	73
Corchorus olitorius	5	22	87	96
Hibiscus sabdariffa	24	67	53	24
Occimum gratissimum	29	56	0	0
Solanum aetiopicum	0	0	17	10
Solanum macrocarpon	98	96	98	96
Solanum scabrum/villosum	0	0	45	10
Vernonia amygdalina	76	70	13	4
Vigna unguiculata	0	4	51	24

Table 6.9b *Percentage of farmers growing specific AIVs in urban and peri-urban areas of Nairobi and Kisumu*

Species	*Urban Nairobi (n = 9)*	*Peri-urban Nairobi (n = 55)*	*Urban Kisumu (n = 38)*	*Peri-urban Kisumu (n = 39)*
Amaranthus spp.	78	80	74	51
Brassica carinata	22	36	29	26
Cleome gynandra	56	64	87	72
Corchorus olitorius	44	24	55	69
Crotalaria brevidens	11	9	76	77
Cucurbita spp.	67	73	26	38
Ipomoea batatas	0	0	8	0
Manihot esculenta	0	0	5	0
Solanum scabrum/villosum	67	93	74	56
Urtica massaica	11	20	0	0
Vigna unguiculata	89	58	89	97
Other	22	2	8	0

Table 6.9c *Percentage of farmers growing specific AIVs in peri-urban areas of Durban and Soshanguve*

Species	*Durban men (n = 36)*	*Durban women (n = 128)*	*Soshanguve men (n = 54)*	*Soshanguve women (n = 94)*
Abelmoschus esculentus	3	0	11	6
Amaranthus spp.	11	13	16	28
Cleome gynandra	0	0	18	30
Colocasia esculenta	69	80	4	1.2
Corchorus olitorius	0	0	7	11
Cucurbita spp.	91	93	33	53
Ipomoea batatas	86	84	12	6
Other gourds	0	0	1.2	0
Dahl (unidentified)	40	33	0	0
Solanum scabrum/villosum	3	2	0	0
Solanum aethiopicum	0	2	0	0
Vigna unguiculata	51	44	22	39

Source: IndigenoVeg survey data, 2006

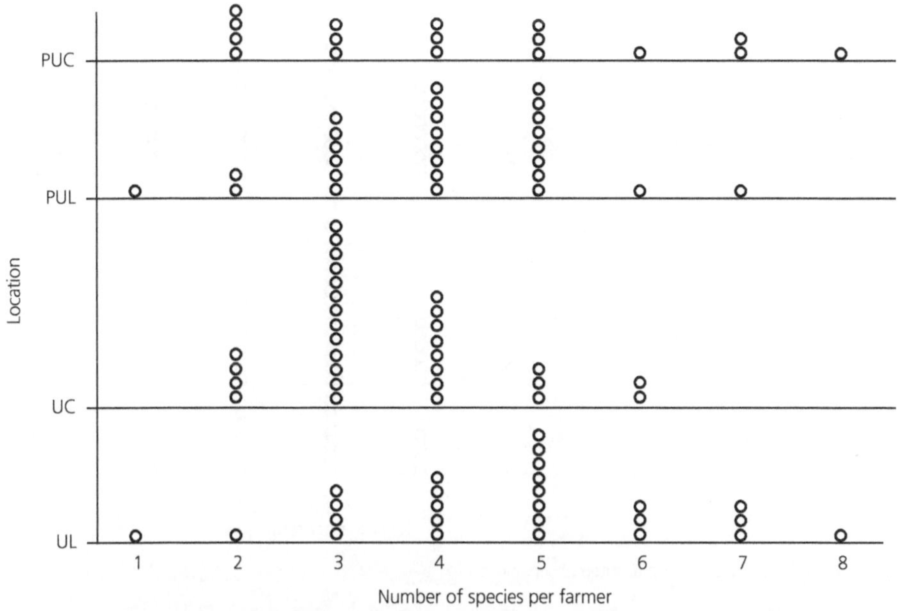

Note: Each symbol represents up to two observations. PUC = peri-urban Cotonou; PUL = peri-urban Lokossa; UC = urban Cotonou; UL = urban Lokossa.

Figure 6.3a *Number of species grown by respondents in urban and peri-urban locations in Cotonou and Lokossa*

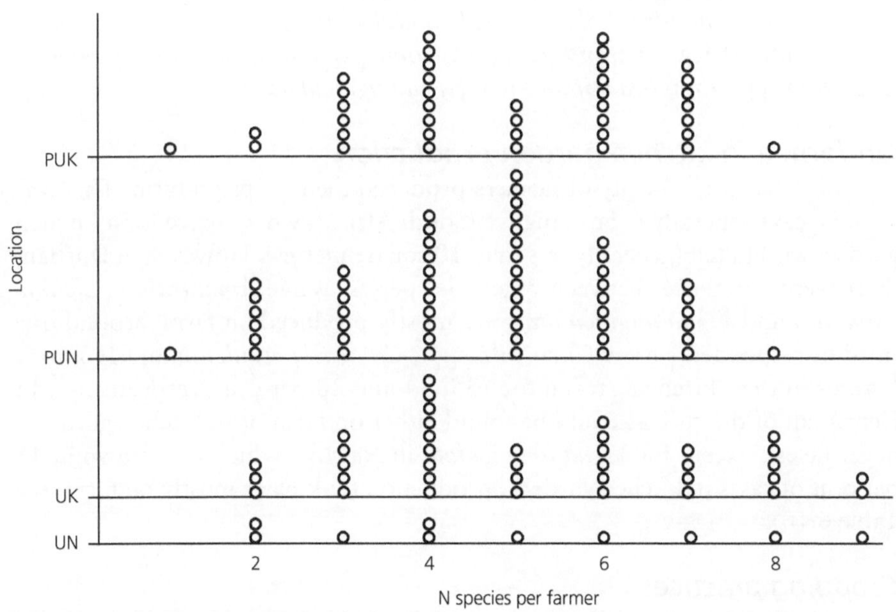

Note: Each symbol represents up to two observations. PUK = peri-urban Kisumu; PUN = peri-urban Nairobi; UK = urban Kisumu; UN = urban Nairobi.

Figure 6.3b *Number of species grown by respondents in urban and peri-urban locations in Nairobi and Kisumu*

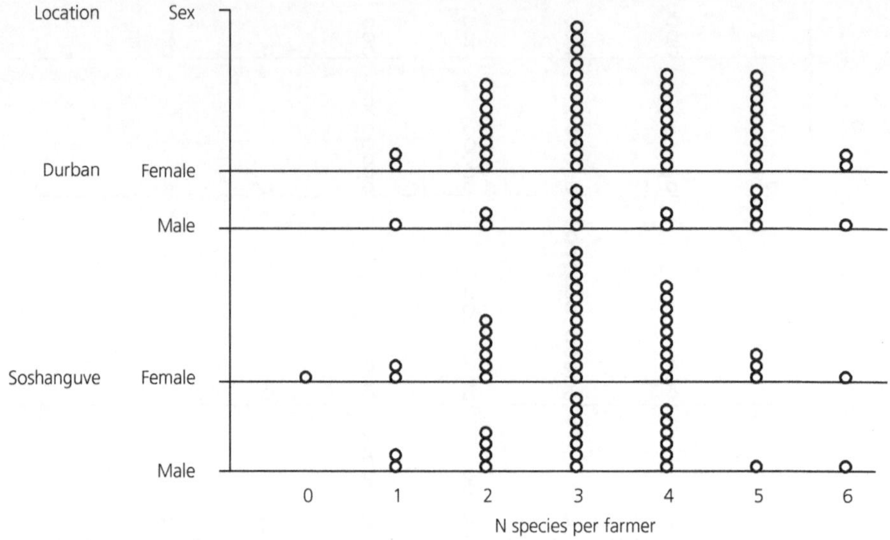

Note: Each symbol represents up to three observations.
Source: IndigenoVeg survey data, 2006

Figure 6.3c *Number of species grown by female and male respondents in Durban and Soshanguve*

spp., *Corchorus olitorius* and *Vernonia amygdalina*; in Kenya *Amaranthus* spp., *Cleome gynandra*, *Solanum scabrum/villosum* and *Vigna unguiculata*; and in South Africa *Amaranthus* spp., *Cleome gynandra*, *Colocasia esculenta*, *Cucurbita* spp., *Ipomoea batatas* and *Vigna unguiculata*.

On-farm and kitchen garden production

In Benin, the vast majority of farmers produced their crops on farm. This was also the case generally in Soshanguve (South Africa), where space for a kitchen garden was limited (generally less than 10 square metres). However, in Durban, there were differences between vegetable species. While *Amaranthus*, *Cleome gynandra* and *Vigna unguiculata* were mostly produced on farm, around one third of respondents grew *Cucurbita* spp., *Colocasia esculenta* and *Ipomoea batatas* in their kitchen garden too (35, 34 and 28 per cent, respectively). In Kenya, all of the species could be found either on farm or in kitchen gardens; nevertheless, except for *Vigna unguiculata* in Nairobi, which was grown in 45 per cent of cases in kitchen gardens, production took place mostly on farm (see Table 6.10).

Cropping practices

Cropping practices differed between countries. In Benin, farmers tended to grow the crops in pure stands. This was the case for 76 per cent of farmers growing *Solanum macrocarpon* (n = 156), 71 per cent growing *Amaranthus* spp. (n = 121), 95 per cent growing *Corchorus olitorius* (n = 87) and 77 per cent growing

Table 6.10 *Percentage of farmers producing a specific vegetable on their farms or in the kitchen garden in Nairobi and Kisumu*

Species	*Nairobi farm*	*Nairobi kitchen garden*	*Kisumu farm*	*Kisumu kitchen garden*
Amaranthus spp.	65% (n = 26)	35% (n = 14)	50% (n = 4)	50% (n = 4)
Cleome gynandra	70% (n = 19)	30% (n = 8)	73% (n = 37)	27% (n = 14)
Solanum scabrum/villosum	72% (n = 36)	28% (n = 14)	67% (n = 29)	33% (n = 14)
Vigna unguiculata	55% (n = 17)	45% (n = 14)	63% (n = 27)	37% (n = 22)

Source: IndigenoVeg survey data, 2006

Vernonia amygdalina (n = 53). In the case of *Solanum macrocarpon*, there was, however, a noticeable difference by location. In urban and peri-urban areas of Lokossa and the peri-urban areas of Cotonou, the vegetable was grown in pure stands; but in urban Cotonou, 60 per cent of farmers preferred to intercrop it, usually with *Amaranthus* spp., claiming that it improved yield through decreased pest damage. In addition, because the growing cycle of *Solanum macrocarpon* is quite long, by intercropping it with the shorter-cycle *Amaranthus* spp., farmers can obtain a crop fairly quickly, which they sell and then use the revenue to continue tending the *S. macrocarpon*.

In Kenya, intercropping was a widespread practice in both cities (see Table 6.11a), although it appeared to be slightly more common in Kisumu. There were no remarkable differences by urban and peri-urban location; however, women tended to intercrop more frequently than men for all species except *Cleome gynandra*.

In South Africa, the majority of farmers intercropped AIVs; however, in Durban, *Colocasia esculenta* and *Ipomoea batatas* were grown mainly in pure stands (see Table 6.11b). There were no obvious gender differences in the patterns of intercropping.

Plot size

This section presents available data for land area under key AIVs for the three countries. The results should be interpreted with caution for two reasons. First, farmers may not necessarily be able to estimate plot size reliably using a metric system (particularly if they have not had any formal schooling). Second, the data can be strongly affected by the practice of intercropping. To illustrate this point, in Kenya, in the case of nearly 30 per cent of respondents, the sum of the areas under their three main AIVs exceeded the total land area under vegetable production. This was because a farmer who had, for example, 1200 square metres of land for vegetable production would report that each of the three main vegetables occupied 1200 square metres. This could indicate that all three vegetables (and possibly more) were intercropped in the same area or grown in close succession in the same space.

As farmers in Benin tended to grow their crops in pure stands, it has been possible to examine the data in more depth, evaluating the effect of location type and vegetable species on mean plot sizes in the two cities.

Table 6.11a *Patterns of intercropping of four AIVs by location and gender in Kenya*

Species	*Nairobi*	*Kisumu*	*Men*	*Women*
Amaranthus spp.	46% (n = 19)	88% (n = 7)	39% (n = 7)	61% (n = 19)
Cleome gynandra	63% (n = 17)	63% (n = 32)	67% (n = 20)	60% (n = 29)
Solanum scabrum/villosum	46% (n = 23)	63% (n = 27)	47% (n = 21)	60% (n = 29)
Vigna unguiculata	50% (n = 16)	71% (n = 42)	56% (n = 22)	69% (n = 36)

Table 6.11b *Patterns of intercropping of six AIVs by location in Durban and Soshanguve*

Species	*Durban*	*Soshanguve*
Amaranthus spp.	68% (n = 22)	73% (n = 60)
Cleome gynandra	0% (n = 1)	88% (n = 82)
Colocasia esculenta	8% (n = 101)	0% (n = 8)
Cucurbita spp.	78% (n = 108)	69% (n = 108)
Ipomoea batatas	11% (n = 108)	58% (n = 12)
Vigna unguiculata	76% (n = 17)	66% (n = 77)

Source: IndigenoVeg survey data, 2006

Table 6.12 shows the mean, median, minimum and maximum plot sizes under each of the four crops for urban and peri-urban locations in Cotonou and Lokossa. Between subjects, factorial analysis of variance was used to evaluate the effect of location type (urban or peri-urban) and type of vegetable grown on mean plot sizes in Lokossa and Cotonou (see Table 6.13). Prior to the analysis, the data were normalized through a logarithmic transformation.

In both cities, the main effects were significant; but in Cotonou, the interaction effect was also significant. In Lokossa, the average plot size was significantly smaller in urban locations compared to peri-urban locations. Multiple comparisons (carried out using Tukey's procedure) showed that *Corchorus olitorius* plots were significantly larger than both *Solanum macrocarpon* and *Amaranthus* spp. plots.

The significant interaction effect in Cotonou indicates that the effect of one of the factors was not consistent across all levels of the other factor. For example, although the overall trend in Cotonou is also for peri-urban plots to be larger than urban plots, an examination of the means suggests that the difference appears to be more pronounced for *Vernonia amygdalina* and *Solanum macrocarpon* than for *Amaranthus* spp.

To investigate this further, the procedure of testing for simple effects (i.e. carrying out an analysis of variance evaluating all the levels of one factor across only one level of the other factor) as described by Sheskin (2007) was followed. This confirmed that plot size was significantly larger in peri-urban areas compared to urban areas for *Solanum macrocarpon* and *Vernonia amygdalina*, but not *Amaranthus* spp. (see Table 6.14). It also revealed that whereas in peri-urban areas the plot sizes for the three species are more or less

Table 6.12 *Areas under selected AIV production in urban and peri-urban locations in Cotonou and Lokossa (m²)*

African indigenous vegetable (AIV)	*Measure*	*Urban Cotonou (n = 52)*	*Peri-urban Cotonou (n = 19)*	*Urban Lokossa (n = 33)*	*Peri-urban Lokossa (n = 16)*
Amaranthus spp.	Mean	185	232	162	269
	Median	141	240	100	200
	Minimum	30	50	40	50
	Maximum	1080	500	800	1024
Solanum macrocarpon	Mean	215	551	189	750
	Median	120	454	103	400
	Minimum	30	57.6	22	50
	Maximum	1500	2880	1260	5184
Vernonia amygdalina	Mean	100	426		
	Median	78	264		
	Minimum	12	55		
	Maximum	360	1536		
Corchorus olitorius	Mean			362	981
	Median			225	661
	Minimum			16	64
	Maximum			1536	5760

Source: IndigenoVeg survey data, 2006

the same, in urban areas, plot size for *Vernonia amygdalina* was significantly smaller than for the other two species (see Table 6.14). The fact that farmers do not increase the portion of land under *Amaranthus* spp. production even when more land is available, as in peri-urban areas, may be explained by the fact that it has a lower market value than either of the other two species (it fetches a good price only in the dry season). The significantly smaller areas under *V. amygdalina* production in urban locations compared to peri-urban locations might be explained by the fact that *V. amygdalina* is a perennial shrub and, if planted extensively, could create a problem of shading for the other vegetables. Furthermore, where land constraints are very severe or when tenure is uncertain, farmers will be less willing to invest in long-term crops and are, instead, more likely to want to keep the majority of their land under production with short- or medium-cycle crops so that they can have flexibility in the crops they are growing over the course of the year and produce the highest-value vegetables for each season.

In Kenya, because the sum of the areas under their three main AIVs exceeded the total land area under vegetable production for one third of respondents, these data were analysed separately from data given by farmers whose total land area under vegetable production was equal to, or higher than, the sum of the land area under their three main crops. In Table 6.15, it has been assumed that in the latter cases, the farmers were giving areas for each species grown as pure stands; however, this assumption may not hold true in all cases. Because of the small sample sizes, no attempt has been made to differentiate between urban and peri-urban areas, or to do any statistical analyses.

Table 6.13 *Results for between subjects analysis of variance, evaluating the effect of location type (urban or peri-urban) and vegetable species grown on plot size (transformed data) in Benin*

City	*Source of variation*	*Degrees of freedom*	*Sum of Squares*	*Mean of squares*	*F ratio*	*Probability value*
Lokossa	Vegetable species	2	4.9348	2.4674	15.04	0.000
	Location	1	7.1882	7.1882	43.82	0.000
	Interaction	2	0.7196	0.3598	2.19	0.114
	Within groups	194	34.2842	0.1640		
Cotonou	Vegetable species	2	1.1344	0.5672	4.87	0.009
	Location	1	4.6513	4.6513	39.93	0.000
	Interaction	2	1.0671	0.5335	4.58	0.011
	Within groups	194	22.5977	0.1165		

Source: IndigenoVeg survey data, 2006

Table 6.14 *Outcome of simple effects tests in Cotonou*

	F ratio	*probability value*	
Cotonou urban	9.55 (2, 139)	0.000	* *Solanum macrocarpon*> *Vernonia amygdalina* * *Amaranthus* spp.> *Vernonia amygdalina*
Cotonou peri-urban	2.72 (2, 55)	ns	
Amaranthus spp.	2.32 (1, 65)	ns	
Solanum macrocarpon	18.09 (1, 75)	0.000	PUC>UC
Vernonia amygdalina	23.24 (1, 50)	0.000	PUC>UC

Note: * = outcome of Tukey's multiple comparisons; ns = no significance; PUC – peri-urban Cotonou; UC = urban Cotonou

Source: IndigenoVeg survey data, 2006

In South Africa, the sum of the areas under the three main AIVs was usually equal to or greater than the total area used for vegetable production (the exceptions were discarded from the analysis). Mean, median, minimum and maximum plot sizes under the six crops are shown in Table 6.16a and 6.16b. However, because of the high rates of intercropping for all species except *Colocasia esculenta* and *Ipomoea batatas*, and a problem of small sample sizes in Durban, there has been no attempt at statistical analysis.

Irrigation methods

In South Africa, all respondents who answered the question about source of water produced AIVs under rain-fed conditions. In addition, bucket irrigation was used by 52 and 23 per cent of respondents in Soshanguve and Durban, respectively. Other methods in Durban included sprinkler (4 per cent) and water conservation methods (8 per cent), and hosepipes used in kitchen gardens in Soshanguve (4 per cent). In Kisumu, 90 per cent of farmers were producing AIVs primarily under rain-fed conditions. A smaller number of farmers irrigated during the dry season with buckets (18 per cent). In Nairobi,

Table 6.15 *Pure stand and intercropped areas under selected AIV production in Nairobi and Kisumu (m²)*

AIV	*Measure*	*Pure stand Nairobi*	*Intercropped Nairobi*	*Pure stand Kisumu*	*Intercropped Kisumu*
Amaranthus spp.	Mean	384	2504	287	
	Median	250	2000	160	
	Minimum	8	50	10	
	Maximum	1000	11,500	1000	
Cleome gynandra	Mean	398	3392	515	1918
	Median	500	1500	290	500
	Minimum	50	200	25	100
	Maximum	1000	12,000	4000	8000
Solanum scabrum/ villosum	Mean	566	1192	493	1893
	Median	500	500	225	1000
	Minimum	71	50	106	400
	Maximum	2000	4000	2000	8000
Vigna unguiculata	Mean	406	3004	658	2623
	Median	400	1000	300	1000
	Minimum	16	50	25	150
	Maximum	1000	12,000	4000	16,000

Source: IndigenoVeg survey data, 2006

only 70 per cent of respondents reported rain-fed farming. Overall, sprinkler irrigation and bucket irrigation was practised by 31 and 22 per cent of farmers (these figures group farmers who were producing under rain-fed conditions together with farmers who presumably were only farming in the dry season).

Very few farmers in urban and peri-urban Cotonou stated that they practised rain-fed vegetable production, whereas in Lokossa urban and peri-urban areas, 32 and 60 per cent did so. In Cotonou, urban and peri-urban

Table 6.16a *Areas under different crops in Soshanguve (m²)*

	Cucurbita *spp.* *(n = 109)*	Amaranthus *spp.* *(n = 61)*	Vigna unguiculata *(n = 77)*	Cleome gynandra *(n = 82)*
Mean	573	311.5	221.8	307.0
Median	100	266	120	220
Minimum	17	12	17	16
Maximum	1260	1000	1170	1000

Table 6.16b *Areas under different crops in Durban (m²)*

	Cucurbita *spp.* *(n = 37)*	Colocasia esculenta *(n = 28)*	Ipomoea batatas *(n = 24)*
Mean	573	527	768
Median	100	110	100
Minimum	6	4	10
Maximum	10,000	5000	10,000

Source: IndigenoVeg survey data, 2006

farmers predominantly used watering cans (93 and 81 per cent), whereas in urban and peri-urban areas of Lokossa, they used buckets and bowls (80 and 63 per cent).

The use of inputs

Table 6.17 shows the use of pesticides, inorganic and organic fertilizers in the three countries. Benin contrasts sharply with both Kenya and South Africa for its widespread use of chemical inputs on AIVs. Furthermore, the use of organic inputs was also widespread, particularly in Cotonou. In Kenya, the use of inputs varied between locations. More farmers in Nairobi reported using inputs than farmers in Kisumu. In the latter city, quite a high proportion of farmers did not use any inorganic or organic fertilizers (37 per cent in urban Kisumu and 21 per cent in peri-urban Kisumu), compared to only 13 per cent in peri-urban Nairobi. In South Africa, inputs are used by more farmers in Durban than in Soshanguve. Indeed, in Soshanguve, around half of the respondents did not use any fertilizers at all (39 per cent of men and 52 per cent of women).

The frequency of application of inputs in Cotonou greatly surpassed that of the other cities (see Tables 6.18a and 6.18b). For example, whereas a strong majority of respondents (at least 85 per cent) in Lokossa, Kisumu and Nairobi applied pesticides no more than five times, in Cotonou the majority (58 per cent) applied pesticides six or more times. Furthermore, in the over ten times category, which made up one third of the sample, ten farmers sprayed 10 to 15 times; six farmers sprayed 16 to 20 times; five farmers sprayed 21 to 30 times; four farmers sprayed 31 to 40 times; one farmer sprayed 41 to 50 times; and three farmers sprayed 51 to 66 times. These figures may be inflated if farmers were counting applications for all the different crops that they grow and with different types of pesticides (rather than provide an average use across crops); but they are, nevertheless, cause for concern when it is considered that the recommendation for *Amaranthus* spp. is three applications and *Solanum macrocarpon* three to four times depending upon the type of pest or disease. In contrast, the highest recorded pesticide application rate was 14 in peri-urban Nairobi and 12 in urban Kisumu. The patterns of inorganic and organic fertilizer use are similar. In South Africa, hardly any respondents in Soshanguve provided information on the frequency of application of chemical inputs (therefore, Table 6.18c only shows the frequency of application of organic fertilizer). In Durban, a total of 7 men and 27 women gave information on pesticide use: overall, 32 per cent applied them once a month; 18 per cent twice a month; and 15 per cent once a week. Five male respondents and 14 female respondents provided information on the frequency of application of inorganic fertilizers, and the majority applied them once only (89 per cent). The highest recorded application rates for fertilizer were six times and once a month.

Table 6.19 shows the type of pesticides used by farmers in Benin. It should be noted that while pesticides such as Maneb, Deltamethrin and Thiophanate-methyl used by farmers in Cotonou are recommended for vegetable

Table 6.17 *Use of inputs in urban and peri-urban areas in Cotonou, Lokossa, Nairobi and Kisumu and for male and female farmers in Durban and Soshanguve (percentage of respondents)*

Inputs	*Urban Cotonou (n = 59)*	*Peri-urban Cotonou (n = 28)*	*Urban Lokossa (n = 47)*	*Peri-urban Lokossa (n = 51)*	*Urban Nairobi (n = 9)*	*Peri-urban Nairobi (n = 55)*	*Urban Kisumu (n = 38)*	*Peri-urban Kisumu (n = 39)*	*Peri-urban Durban men (n = 36)*	*Peri-urban Durban women (n = 128)*	*Peri-urban Soshanguve men (n = 54)*	*Peri-urban Soshanguve women (n = 94)*
Pesticides	97	100	87	98	33	36	39	13	28	27	2	1
Inorganic fertilizer	97	100	81	86	33	53	24	5	19	15	7	6
Organic fertilizer	97	89	60	57	89	82	55	77	89	91	52	43

Source: IndigenoVeg survey data, 2006

Table 6.18a *Frequency of application of various inputs in Cotonou and Lokossa (percentage of respondents)*

Frequency of application		*<3*	*3–5*	*6–9*	*>10*
Pesticides	Cotonou (n = 86)	9	33	24	34
	Lokossa (n = 83)	48	38	8	6
Inorganic fertilizer	Cotonou (n = 86)	8	26	24	42
	Lokossa (n = 63)	60	32	5	3
Organic fertilizer	Cotonou (n = 83)	17	43	13	27
	Lokossa (n = 16)	75	12.5	12.5	–

Table 6.18b *Frequency of application of various inputs in Nairobi and Kisumu (percentage of respondents)*

Frequency of application		*1–2*	*3–4*	*>5*
Pesticides	Nairobi (n = 21)	70	20	10
	Kisumu (n = 20)	55	30	15
Inorganic fertilizer	Nairobi (n = 29)	97	3	–
	Kisumu (n = 11)	100	–	–
Organic fertilizer	Nairobi (n = 43)	100	–	–
	Kisumu (n = 50)	98	2	–

Table 6.18c *Frequency of application of organic inputs in Durban and Soshanguve (percentage of respondents)*

Frequency of application		*1*	*2*	*3*	*>4*
Organic fertilizer	Durban (n = 135)	97	0	1	2
	Soshanguve (n = 68)	76	10	7	4

Source: IndigenoVeg survey data, 2006

production, Methamidophos and Lambdacyhalothrin + dimethoate or Lambdacyhalothrin + profenofos are not. The main insecticide used in Lokossa – Endosulfan – should not be used on vegetables either as it is actually formulated for cotton.

Reasons for production

The main reasons for growing AIVs across all four areas in Benin were economic as they related to the existence of marketing outlets, the possibility of generating extra income and the good price fetched by the vegetables (see Table 6.20). A high proportion of farmers in urban and peri-urban Lokossa (68 and 73 per cent) stated that they grew AIVs for cultural reasons; however, rather surprisingly, they also declared that they did not use them for home consumption. The situation in Cotonou was quite different as here one quarter of urban farmers and nearly 40 per cent of peri-urban farmers did produce for home consumption. There were no particularly marked gender differences in terms of reasons for production.

Table 6.19 *Types of pesticide used by farmers in Cotonou and Lokossa*

Category of pesticide	*Cotonou (n = 87)*	*Lokossa (n = 98)*
Fungicides	Maneb (18%) Mancozeb (13%) Copper oxide (2%)	Copper oxide (2%) Unknown (1%)
Nematicide/fungicide	Thiophanate-methyl (22%)	–
Insecticide	Deltamethrin (25%), Methamidophos (12%) Lambdacyhalothrin + dimethoate or Lambdacyhalothrin + profenofos (9%) Nine others (22%)	Endosulfan (39%), Malathion + Cyfluthrin (8%) Six others (11%)
Herbicide	Monate (26%)	–
Rodenticide	–	Alpha-chloralose (14%)

Source: IndigenoVeg survey data, 2006

In Kenya, the same economic reasons were important; but in contrast to Benin, auto-consumption was the reason mentioned most frequently in all locations (see Table 6.20). In terms of gender differences, a higher proportion of women compared to men stated that auto-consumption was a reason (92 versus 76 per cent), whereas a higher proportion of men cited production experience (17 versus 7 per cent).

The dominant reason given in South Africa was auto-consumption and, to a lesser extent, cultural reasons, as well as the opportunity to gain extra income (see Table 6.20). There were no marked gender differences, apart from women mentioning cultural reasons more frequently than men; but there were interesting patterns related to age category. The proportion of respondents mentioning auto-consumption increased with the age categories (71, 80, 87, 87 and 100 per cent). No young people (<30 years) mentioned cultural reasons or experience (however, 30 per cent of the oldest category did mention experience). Extra income was mentioned by half of the respondents in younger categories, but more often in the two oldest categories (61 and 70 per cent). The greater importance of extra income in Durban when compared to Soshanguve is possibly because of the ready markets in and around Durban for dahl and *Colocasia esculenta*. Markets in Soshanguve are mainly visited by the same local groups as the farmers and no attempts to address the needs of other African emigrant groups were observed.

Table 6.20 *Main reasons for production of AIVs in urban and peri-urban areas in Cotonou, Lokossa, Nairobi and Kisumu, and for male and female farmers in Durban and Soshanguve (percentage of respondents)*

Reasons	*Urban Cotonou (n = 59)*	*Peri-urban Cotonou (n = 28)*	*Urban Lokossa (n = 47)*	*Peri-urban Lokossa (n = 51)*	*Urban Nairobi (n = 9)*	*Peri-urban Nairobi (n = 55)*	*Urban Kisumu (n = 38)*	*Peri-urban Kisumu (n = 39)*	*Peri-urban Durban men (n = 36)*	*Peri-urban Durban women (n = 128)*	*Peri-urban Soshanguve men (n = 54)*	*Peri-urban Soshanguve women (n = 94)*
Good price	63	71	81	65	44	53	55	67	14	5	0	1
Contract with grower	3	11	0	0	0	0	3	3	0	2	0	0
Production experience	39	50	23	24	11	7	16	13	0	5	4	5
Available market	76	71	55	51	78	71	74	69	3	0	0	0
Opportunity to earn extra income	46	54	57	63	44	64	71	33	36	53	28	23
Cultural reasons	24	18	68	73	11	11	8	5	39	45	22	29
Auto-consumption	25	39	0	0	89	82	76	97	94	95	89	94
Other	10	11	4	2	0	22	18	5	3	2	0	0

Source: IndigenoVeg survey data, 2006

Principal features of indigenous vegetable farming systems in Benin, Kenya and South Africa

The main findings of the six city surveys have been condensed into an overview table (see Table 6.21), bringing into relief the differences between countries and cities. Although there is still a need to triangulate the survey results with other methods, the observed patterns strongly indicate that AIVs play very different roles in the urban and peri-urban farming systems of the three countries, and, indeed, in some aspects also in the different cities within the same country. Such a wide array poses both opportunities and constraints: opportunities because it shows that AIVs can fill a variety of niches, in terms of species available and used, role players involved and production systems. Consequently, there should be an appropriate model of AIV production for almost any city in SSA. However, this also poses constraints because at the moment it is impossible to make too many generalizations since the systems are so widely different and site specific. Consequently, more studies are required to facilitate development of general models and understandings. The developmental challenge now is how to harness this amazing diversity and complexity to stimulate AIV production and marketing without decreasing diversity and promoting only a few species or models, which has been the case with exotic vegetables and conventional agriculture.

In Benin, respondents produce AIVs primarily for commercial reasons, particularly in Cotonou, where a majority of respondents were full-time specialist vegetable producers. Many of the respondents were motivated by an available market and the fact that AIVs also fetched a good price (although this is the case mainly during the dry season). The predominantly commercial rationale of the system can be also seen in the fact that a high proportion of farmers produce in pure stands and with very heavy use of inputs.

It is likely that Cotonou farmers choose to specialize in the production of vegetables partly because being the biggest city in Benin, with good connections to neighbouring countries, it offers excellent market opportunities and easier access to farming inputs (fertilizers and pesticides developed for vegetables), and partly because there are severe land constraints (particularly in urban areas). Under these conditions, it is unsurprising that farmers prioritize high-value crops. In this respect, it was significant that the total amount of land under the three main AIVs (produced the previous year) was a large proportion of the overall land area used for vegetable production. Indeed, for a quarter of farmers in Cotonou, the entire vegetable production area the previous year had been occupied only by AIVs.

Interestingly, in Lokossa, this proportion was much higher as half of the farmers appeared to have used the portion of their vegetable production land exclusively for the production of AIVs. This finding, when viewed in the light of the high numbers of respondents indicating that they farmed AIVs for cultural reasons (but, at the same time, did not produce for home consumption), suggests that urban consumers in Lokossa have a greater demand for

Table 6.21 *Principal features of indigenous vegetable farming systems in six cities in Benin, Kenya and South Africa*

	Benin		*Kenya*		*South Africa*	
	Cotonou	*Lokossa*	*Nairobi*	*Kisumu*	*Durban*	*Soshanguve*
Farmer gender	Male majority (~90%)	Female majority (60–80%)	Female majority (~65%) in peri-urban areas; equal numbers in urban areas	Female majority (~65%) in peri-urban areas; equal numbers in urban areas	Female majority (~80%)	Female majority (~65%)
Farmer age	Young respondents – 70% under 40	Young respondents in urban areas Middle-aged in peri-urban areas	Middle aged	Young respondents in urban areas Middle aged in peri-urban areas	Older respondents – 34% over 60	Older respondents – 41% over 60
Farmer main occupation	Majority full-time farmers; specialized vegetable producers	Majority full-time farmers; in peri-urban areas, high numbers using <50% land for vegetables	Majority full-time farmers	Majority full-time farmers	<50% full-time farmers;large unemployed category	<50% full-time farmers; large unemployed category
Farmer socio-linguistic groups	~50% Fon Remainder ethnically diverse; ~20% foreign nationals	~80% Kotafon in urban areas Half Adja and one third Sahoue in peri-urban areas	~70% Kikuyu	~75% Luo	38% Pedi and 24% Shangaan; other groups from neighbouring provinces	~90% Zulu
Farmer experience in the production of AIVs	More experienced in urban areas (40% versus 25% sample >15 years' experience)	~50% inexperienced in urban areas (<6 years)	Majority sample <11 years' experience	Peri-urban more experienced than urban	60% female versus 40% male farmers >20 years' experience	Majority sample <11 years' experience
Main reasons for production	Commercial	Commercial; cultural	Auto-consumption; commercial	Auto-consumption; commercial	Auto-consumption; extra income; cultural Cultural reason more important for women	Auto-consumption; extra income; cultural Cultural reason more important for women

Production systems	On farm; majority pure stands	On farm; majority pure stands	On farm and kitchen garden; majority intercropping	On farm and kitchen garden; majority intercropping	On farm and kitchen garden depending upon species; intercropping depends upon species	On farm; majority intercropping
Pesticide use	Widespread and frequent use of pesticides; wide range of pesticides	Widespread use of pesticides Endosulfan, which is formulated for cotton, is common	Some pesticide use; majority no more than four applications	Some pesticide use in urban areas; majority no more than four applications	Around one third use pesticides monthly or bi-monthly	Virtually no pesticides used
Fertilizer use	Widespread and frequent use of inorganic and organic inputs	Widespread use of inorganic and organic inputs	Majority use organic inputs, once or twice	Majority use organic inputs once or twice; but 37% and 21% in urban and peri-urban areas use nothing	Majority use organic inputs once	Around half use no fertilizer; the remainder use organic inputs once
Diversity of species	Three to four species per farmer	Four to five species per farmer	Four to six species per farmer	Four to six species per farmer	Three to five species per farmer; preference for multipurpose plants	Three to four species per farmer; preference for multipurpose plants
City specialization for specific vegetables	*Vernonia amygdalina*	*Corchorus olitorius*, *Abelmoschus esculentus* in peri-urban	*Crotalaria brevidens*	*Cucurbita* spp.	*Colocasia esculenta*, *Ipomoea batatas*	*Cleome gynandra*
Widely cultivated vegetables common to both cities in each country	*Solanum macrocarpon* *Amaranthus* spp.		*Amaranthus* spp., *Solanum villosum*, *Vigna unguiculata* and *Cleome gynandra*		*Cucurbita* spp.	

AIVs than consumers in Cotonou. This demand may be motivated by cultural eating habits, but also by economic factors. A recent consumer survey (from an ongoing project involving two of the authors of this chapter) in the cities of Porto Novo, Dassa-Zoumé and Parakou revealed a widespread perception that traditional vegetables were less expensive than exotics (unpublished data). Because of its larger population and greater employment opportunities, Cotonou may have a larger middle class able to afford exotic vegetables. It should be noted that as land resources are more readily available in Lokossa, not only in peri-urban but also in urban areas, farmers were using significant proportions of their land for the production of other agricultural products (staples such as maize and potatoes, poultry and, to a lesser extent, cattle). It would appear, therefore, that for Lokossa farmers, AIVs are part of a wider livelihood strategy. The vegetables fetch a good price and are therefore a good source of additional income; but being a rather small urban centre, Lokossa offers much more limited marketing opportunities – thus, farmers diversify their use of land. Vegetables with a longer shelf life or that can be conserved, such as okra (*Abelmoschus esculentus*), which was widely cultivated by peri-urban farmers, would be an exception because from the smaller urban centre of Lokossa they can be transported to other locations in Benin.

In Kenya, the position of AIVs within the farming systems of the two cities was similar. A majority of respondents produced them for home consumption; but there was a significant sale component. A high proportion of respondents considered that there was a market for the vegetables and that the price obtained was good. The home consumption aspects appeared to be more important for Kisumu peri-urban locations, whereas the commercialization aspects appeared to be more important in urban areas. This could explain the younger profile of urban respondents and the fact that while women constitute the majority in peri-urban areas, in urban areas numbers are more or less balanced.

It is not possible to compare the patterns for Nairobi urban and peri-urban areas because the sample size in Nairobi urban areas was only nine respondents. The fact that few respondents were identified might be an indication that AIVs are not a priority for urban farmers; however, land is very scarce in urban Nairobi, plots tend to be small and scattered, and, furthermore, urban respondents may not be full-time farmers (and may not have been on locations when the enumerators visited the sites). This question cannot be resolved conclusively without further investigation into the overall numbers of farmers in urban locations and the relative numbers producing AIVs.

For the majority of respondents, farming was a full-time occupation; but there were no differences between locations in the numbers of farmers choosing to specialize in vegetables, using 50 per cent or more of their land for vegetables, or using less than half their land for vegetables. Important agricultural products for Nairobi and Kisumu included maize (around 80 per cent in both cities), potatoes (about 50 per cent in both cities), fruit (about 60 per cent in both cities), poultry (50 and 68 per cent) and cattle (53 and 74 per cent). The production systems for AIVs were consistent with the mixed home consumption/commercial rationales. The AIVs were produced both on farm and in kitchen gardens, and intercropping was a widespread practice. Pesticides were used in Nairobi peri-urban areas and Kisumu urban areas. A higher proportion of Nairobi farmers used inputs, presumably because they had better access to them and were possibly better able to afford them. A higher proportion of farmers in urban Kisumu did not apply any type of fertilizer compared to peri-urban farmers. It is probable that these farmers could not afford inorganic fertilizers and did not have enough land to keep livestock (unlike farmers in peri-urban areas).

The picture in South Africa differs sharply from the one in Benin and Kenya. In the first place, the profile of the farmers is very different. Indigenous vegetable producers were much older (there was a significant over-60 category in both cities), and less than half considered themselves to be full-time farmers. Of the others, a large proportion considered themselves to be primarily unemployed or housewives, signifying that they considered farming to be a side activity.

In both cities, farmers cultivated a variety of crops, AIVs being only one component. In Soshanguve, all respondents produced exotic vegetables and, in addition, farmers grew maize (74 per cent) and potatoes (19 per cent). Poor rainfall and land insecurity mean that farmers are unlikely to invest in long-term crops or crops with high water requirements. The municipal by-laws in Soshanguve also prohibit keeping cattle and limit the numbers of poultry. In Durban, agricultural products comprised exotic vegetables (64 per cent), maize (86 per cent), potatoes (77 per cent), fruit (32 per cent) poultry (40 per cent) and cattle (22 per cent). The AIVs were produced for home consumption and to generate additional income (although at the same time respondents did not perceive that there was an available market for the vegetables). It is notable that four of the key species for the two cities are multipurpose. All species can be used for their leaves; but *Colocasia esculenta* is primarily used for the corms, *Cucurbita* spp. for the fruit and *Ipomoea batatas* for its tuber. *Vigna unguiculata* can also be used for its seed.

There were some noteworthy differences between the two cities in terms of cultivation practices of the top-ranking crops. *Colocasia esculenta* and *Ipomoea batatas* (primarily produced in Durban) were mostly cultivated in pure stands, whereas all of the other vegetables were intercropped. Soshanguve farmers used no pesticides, and around half of the respondents would use no fertilizers at all. In contrast, the majority of Durban farmers used organic fertilizers, and pesticides were used by nearly 30 per cent of the sample. The greater use of organic inputs in Durban compared to Soshanguve can be explained by the fact that around the former, city livestock keeping is fairly common. The more significant use of pesticides in Durban might be explained by greater accessibility, the provision of agricultural extension services (which does not happen in Soshanguve) or perhaps by the types of crops grown.

Overall, AIVs do not appear to occupy an important place in the peri-urban production systems of the two South African cities. However, it should be noted that in both cities, the collection of wild AIVs was widely reported by respondents. In Durban, the majority reported collecting and using weedy species such as *Bidens* (96 per cent) (see Plate 6.11), *Chenopodium* (70 per cent) and *Amaranthus* (91 per cent) (see Plate 6.12), and in Soshanguve, the fact that the volumes traded by farmers were much higher than volumes produced (data not shown) is also an indicator of significant collection.

Looking at patterns across all the cities surveyed, relatively few differences were apparent between urban and peri-urban zones. Typically, but with exceptions, the urban production systems were dominated by younger farmers, working smaller areas of land with higher inputs of inorganic fertilizers and pesticides than their peri-urban counterparts. Security of tenure was somewhat more precarious in the urban centre, with fewer farmers owning the land they work (e.g. in Cotonou, a higher proportion of urban farmers mentioned lack of land or land tenure insecurity as a constraint: 29 versus 21 per cent). The underlying reasons for these differences currently remain speculative, but can be explained by a model of peri-urban regions being inhabited by more settled long-term residents, in contrast to the high in-migration into urban centres of younger people seeking opportunities. However, the general residency data from the IndigenoVeg surveys do not support this hypothesis. Consequently, further work is required before explanatory models can be developed. In the interim, these differences demand different extension messages and interventions within the two locations.

Research and development for the promotion of African indigenous vegetables: Opportunities and constraints

This analysis of the state of production of AIVs in urban and peri-urban locations in Benin, Kenya and South Africa clearly shows that research and development interventions to support the promotion of AIVs in urban and peri-urban production systems may have to differ quite considerably from country to country.

In countries such as Benin where AIVs are already widely integrated within the urban and peri-urban production systems, and are produced with a strong commercial orientation, there could be interesting opportunities for crop diversification. Interventions could focus on the domestication of wild species with existing important commercial importance, supporting their production in urban and peri-urban areas. In Benin, for example, *Vitex doniana* is widely found in markets across the country; but currently it is only collected in the wild, where it is reported as being increasingly rare.

There is also a need for breeding programmes to develop improved varieties of existing cultivated species and for studies to assess and, where relevant, to improve existing management practices. The reported frequent and heavy use of pesticides in Cotonou and the use of inappropriate pesticides in both cities, but particularly Lokossa, is cause for concern. The situation is undoubtedly caused by a real problem of pest infestation, coupled with the low availability of pesticides formulated specifically for vegetables, and, at the same time, the widespread availability of pesticides formulated for cotton but with little accompanying awareness of the consequences of misusing these inputs. There is undoubtedly a need for further research work on the pest complexes of AIVs, the factors influencing the degree of pest infestation (species, varieties, intercropping, crop rotation, climate, etc.), pest resistance, and options for pest control. Recent examples include work on mite infestation on *Amaranthus cruentus* and *Solanum macrocarpon* in Benin (Adango et al, 2006) and control of thrip infestation on snap bean by intercropping with *Cleome gynandra* in Kenya (Waiganjo et al, 2006).

One common argument put forward in favour of promoting AIVs is their ease of cultivation with low inputs (e.g. Opole et al, 1991; Modi, 2003). While the survey results in Benin clearly show that AIVs are not always produced under low-input conditions, in countries such as Kenya, where the production rationale is home consumption and surplus sale, and therefore adequate but not large yields are required, there could be an opportunity to promote low-input production for urban marketing and consumption under an 'organic' label.

A study in Tanzania found that consumers widely believed that AIVs are healthier than exotics because of lower pesticide and fertilizer use (Weinberger and Msuya, 2004). If consumer perceptions are similar in Kenya, given that existing farming practices in highland areas are based on the application of organic fertilizer and low pesticide use, then interventions could focus, on the one hand, on supporting farmers to switch to organic production and get certification, and, on the other, to stimulate consumer interest in 'healthy' AIVs.

The fact that marketing and consumption patterns can be changed quite significantly is shown by the rapidly growing interest in AIVs in the country. In 2005, a survey of nearly 800 households in Nairobi found that the use of AIVs was low, involving only 34 per cent of households (Kimiywe et al, 2007). Consumption patterns were influenced by ethnicity and by respondent occupation, with casual labourers/unemployed people more likely to use AIVs than respondents in full-time occupation/business. Significantly, over 20 per cent of respondents reported that they did not have the time to go and shop for AIVs and prepare them, underscoring the importance of accessibility issues and easy-to-prepare recipes.

This situation is, no doubt, changing. As a result of various interventions which started during the 1990s, there has been a considerable demand-driven market development in Nairobi (Irungu et al, 2006). One intervention initiated in 2004 by AVRDC – World Vegetable Center and Family Concern, which implemented various approaches to improve poor and vulnerable farmers' market access, helped small-scale growers from Kiambu district to successfully penetrate various supermarket outlets in Nairobi, such as Uchumi and Nakumatt (Farm Africa, 2006).

A cautionary note must be sounded with regards to the potential impacts of encouraging the commercialization of indigenous vegetable production systems. While the survey results from peri-urban locations in South Africa and Kenya support the view that it is primarily women who are involved in the production of AIVs, in and around Cotonou women constituted a very small proportion of the sample. This finding suggests that where agricultural production becomes primarily a commercial venture, land resources are in short supply (characteristically in heavily urbanized areas, or even in peri-urban areas near large, fast-growing cities), women's access to land is limited under traditional customs, and/or they may have limited access to financial resources to hire labour for irrigation or ploughing, women may end up being excluded from an activity which was in origin female dominated.

Furthermore, interventions to improve and expand the production and marketing of AIVs could have unexpected effects at other stages of the chain. This happened, for example, in the palm oil industry in Benin (Bélières et al, 2002). During the mid 1970s, the government ceased investing in the palm oil sector, which, however, remained a fairly significant local enterprise, and 80 per cent of the palm oil processors were women. Women's access to land in the south of the country is very limited, and the processors sourced the palm oil primarily from small family farms, where the palms were integrated within other cropping systems. In the early 1990s, the government intervened, privatizing the industrial-scale processing facilities (which had been very inefficient under public management) and encouraging private enterprises. This encouraged newcomers to the sector, mostly men, who used their plots entirely for the growing of palm oil. While the development initially benefited the women processors, the introduction of processing equipment, which has been adopted by the new planters, is leading to the exclusion of women from the sector. Moreover, there has been an impact of land tenure strategies, with new planters adopting more aggressive strategies to accessing land, and an increasing polarization between the new form of commercial farming and more traditional enterprises. Although this study was conducted in rural areas, it still serves to illustrate the point.

In countries such as South Africa, at present, the prospects for increasing indigenous vegetable production in urban and peri-urban areas are poor. Home consumption is an important contribution to household food security and a strategy to save cash resources which may then be used to purchase other household requirements, such as school fees (Webb, 1998; Shackleton and Shackleton, 2004). This can be of significant value. For example, Shackleton (2003) showed that the direct-use value of home consumption of AIVs across several rural sites was just over 1000 South African rand (approximately US$130 per year valued at the turn of the millennium). The value in urban and peri-urban settings would be greater because of the higher local prices. However, in the current survey, relatively small numbers of farmers earned reasonable cash incomes from their farming activities in the urban and peri-urban areas (the mode was 30 rand, or US$4 per month), indicating that most of the sample could be classified as 'home subsistence farmers', rather than entrepreneurial as per the typologies of May and Rogerson (1995) and van Veenhuizen (2006). This suggests that scaling up to more cash-orientated production systems is unlikely or impossible for most due to lack of land, labour and capital.

The lack of involvement of young farmers in the production of AIVs is of concern, and other studies in South Africa have highlighted the shift in food use towards exotic vegetables by young people and the widespread perception of AIVs as poverty foods (Vorster et al, 2007a). This suggests that interventions in South Africa will need, in the first instance, to focus on improving the image of AIVs, especially targeting young urban consumers. This will be a considerable challenge, though, because few consumers under the age of 40 in urban and peri-urban areas are likely to have been exposed to AIVs.

Conclusions

The evidence presented in this chapter shows that AIVs are integrated to very different degrees in the UPA production systems of the three countries. Undoubtedly, the patterns that were detected in terms of the production systems for AIVs are influenced by a complex combination of the dynamics of household or commercial decision-making, and within-city and between-city differences in terms of resource endowments, intensity of development, land tenure systems, etc. (refer to Chapter 1 for a broader discussion around the factors which may influence and define the spaces and places for AIV production in urban and peri-urban locations). Any interventions to promote and intensify the production of AIVs in any country will therefore need to be underpinned by city-specific investigations of the current status of AIV production, an in-depth understanding of the rationale behind and the forces shaping the systems of production, and a considered analysis of the potential impacts ensuing from any intervention.

Acknowledgements

The survey data presented in this chapter were collected through the IndigenoVeg project, a Coordination Action funded under the European Union Sixth Framework Programme priority FP6-2003-INCO-DEV-2 A.3.2 Bio-Diverse, Bio-Safe and Value-Added Crops. The survey was carried out with the help and participation of numerous people and institutions. The authors would like to thank, in particular, the teams in Benin, Kenya and South Africa who administered the survey, compiled the data and developed country reports; Dr G. Pichop and Dr K. Weinberger for developing the survey protocol, standardized questionnaire and data input sheets; Mr F. Maro and Ms R. Marealle for their specific support on data analysis for some of the country reports; and last, but not least, the farmers who gave valuable time and shared their knowledge on the production of AIVs. Additional information can be found in the country reports by Akplogan et al (2007), Abukutsa-Onyango et al (2007), Vorster et al (2007b) and Shackleton et al (2007).

Note

1 The project is entitled Conservation of Biodiversity in Traditional West African Vegetables Species in West Africa and is funded by the Darwin Initiative. It is carried out in collaboration between Bangor University (UK), the Institut d'Economie Rurale (Mali) and the Institut National de Recherches Agricoles du Benin (Benin). Further references in this chapter to unpublished results of an ongoing study in Benin refer to the above-mentioned project.

References

Abasse, T. A., Gouzayé, A., Woltering, L. and Pasternak, D. (2007) 'The role of indigenous leafy vegetables on daily diet and rural and urban economy of Niger: I – International conference on indigenous vegetables and legumes. Prospectus for fighting poverty, hunger and malnutrition', *Acta Horticulturae*, vol 752, pp35–40

Abukutsa-Onyango, M. O., Mwai, G. N., Otiato, D. A. and Onyango, J. C. (2007) 'IndigenoVeg Kenya country report: Draft 2', unpublished IndigenoVeg Survey Report

Adango, E., Onzo, A., Hanna, R., Atachi, P. and James, B. (2006) 'Mite pests of major importance on indigenous leafy vegetables in Benin: The search for appropriate control strategies. I – International conference on indigenous vegetables and legumes. Prospectus for fighting poverty, hunger and malnutrition ISHS', *Acta Horticulturae*, vol 752, pp311–318

Akpavi, S., Batawila, K., Wala, K., Kossi-Titrikou, K., Amouzou, K., Kokou, K., Chango, A., Butare, I. and Akpagana, K. (2008) 'Sociocultural factors influencing the availability and the consumption of food plants in Togo', Paper presented to Theme 1: Just Famine Food? What Contributions can Underutilised Plants Make to Food Security? at the International Symposium on Underutilised Plants for Food, Nutrition, Income and Sustainable Development, Arusha, Tanzania, 3–7 March 2008

Akplogan, F., Adegbola, P. and Assogba-Komlan, F. (2007) 'Production et commercialisation des legumes locaux dans les zones urbaines et peri-urbaines au Benin', unpublished IndigenoVeg Survey Report

Assogba-Komlan, F., Anihouvi, P., Achigan, E., Sikirou, R., Boko, A., Adjè, C., Ahlé, V., Vodouhe, R. and Assa, A. (2007) 'Pratiques culturales et teneur en éléments anti nutritionnels (nitrates et pesticides) du *Solanum macrocarpum* au sud du Bénin', *African Journal of Food, Agriculture, Nutrition and Development*, vol 7, no 4, www.ajfand.net/Vol7No4.html

Bélières, J. F., Bosc, P. M., Faure, G., Fournier, S. and Losch, B. (2002) *What Future for West Africa's Family Farms in a World Market Economy?*, International Institute for Environment and Development, Issue Paper 113

CEROI (1999) *Durban Metro State of the Environment and Development*, Cities Environmental Reports on the Internet, www.ceroi.net/reports/durban/, accessed 5 September 2007

Colin, B. and Heyd, F. (1991) 'La situation des légumes feuilles dans la production maraîchère au sud-Bénin', *Tropicultura*, vol 9, pp129–133

Dovie, D. B. K., Shackleton, C. M. and Witkowski, E. T. F. (2002) 'Direct-use values of woodland resources consumed and traded in a South African village', *International Journal of Sustainable Development and World Ecology*, vol 9, pp269–283

Ekbom, A., Knutsson, P. and Ovuka, M. (2001) 'Is sustainable development based on agriculture attainable in Kenya? A multidisciplinary case study of Murang'a district', *Land Degradation and Development*, vol 12, pp435–447

FAO (United Nations Food and Agriculture Organization) (1988) *Traditional Food Plants: A Resource Handbook for Promoting the Exploitation and Consumption of Food Plants in Arid, Semi-Arid and Sub-Humid Lands of Eastern Africa*, FAO, Rome

Farm Africa (2006) 'Improved health and livelihoods with African indigenous vegetables', *Appropriate Technology*, vol 33, pp10–12

Gockowski, J., Mbazo'o, J., Mbah, G. and Fouda Moulende, T. (2003) 'African traditional leafy vegetables and the urban and peri-urban poor', *Food Policy*, vol 28, pp221–235

Grubben, G. J. H. and Denton, O. A. (eds) (2004) *Plant Resources of Tropical Africa 2: Vegetables, PROTA Foundation*, Wageningen, The Netherlands

Guidi, S. M. G. (2007) 'Analyse socio-économique de la consommation des légumes à Cotonou: contraintes, attitudes et opportunités', Thèse Diplôme d'Ingénieur Agronome, Université d'Abomey-Calavi, Benin

Haberman, S. J. (1973) 'The analysis of residuals in cross-classified tables', *Biometrics*, vol 29, pp205–220

Iaquinta, D. L. and Drescher A. W. (2000) 'Defining periurban: Understanding rural–urban linkages and their connection to institutional contexts', Paper presented at the Tenth World Congress, IRSA, Rio de Janeiro, 1 August 2000

INSAE (2004) *Cahiers des Villages et Quartiers de Ville*, Département du Littoral, Direction des Études Démographiques, Cotonou

Irungu, C., Mburu, J., Maundu, P., Grum, M. and Hoeschle-Zeledon, I. (2006) 'Marketing of African leafy vegetables in Nairobi and its implications for on-farm conservation of biodiversity. I – International conference on indigenous vegetables and legumes. Prospectus for fighting poverty, hunger and malnutrition', *Acta Horticulturae*, vol 752, pp197–202

Kimiywe, J., Waudo, J., Mbithe, D. and Maundu, P. (2007) 'Utilisation and medicinal value of indigenous leafy vegetables consumed in urban and peri-urban Nairobi', *African Journal of Food Agriculture Nutrition and Development*, vol 7, pp1–15

Maundu, P. (1996) 'Utilisation and conservation status of wild food plants in Kenya', in Maesen, L. J. G., van der Burgt, X. M. and van Medenbach de Rooy, J. M. (eds) *The Biodiversity of African Plants*, Proceedings of the XIVth AETFAT Congress, 22–27 August 1994, Wageningen, Kluwer Academic Publishers, Dordrecht, The Netherlands, pp678–683

Maundu, P., Ngugi, G. W. and Kabuye, C. H. S. (1999) *Traditional Food Plants of Kenya*, National Museum of Kenya, Nairobi, Kenya

May, J. and Rogerson, C. M. (1995) 'Poverty and sustainable cities in South Africa: The role of urban cultivation', *Habitat International*, vol 19, pp165–181

Modi, A. T. (2003) 'What do subsistence farmers know about indigenous crops and organic farming? Preliminary experience in KwaZulu-Natal', *Development Southern Africa*, vol 20, pp675–684

Opole, M., Chweya, J. A. and Imungi, J. K. (1991) *Indigenous Vegetables in Kenya: Field and Laboratory Experience Report*, KENGO, Nairobi

Pasquini, M. W. and Young, E. M. (2006) 'Networking to promote the sustainable production and marketing of indigenous vegetables through urban and peri-urban agriculture in sub-Saharan Africa (IndigenoVeg): I – International conference on indigenous vegetables and legumes. Prospectus for fighting poverty, hunger and malnutrition', *Acta Horticulturae*, vol 752, pp41–48

Schwabe, C. (2004) *Fact Sheet: Poverty in South Africa*, Human Sciences Research Council, www.sarpn.org.za/documents/d0000990/index.php

Shackleton, C. M. (2003) 'The prevalence of use and value of wild edible herbs in South Africa', *South African Journal of Science*, vol 99, pp23–25

Shackleton, C. M. and Shackleton, S. E. (2004) 'The importance of non-timber forest products in rural livelihood security and as safety-nets: Evidence from South Africa', *South African Journal of Science*, vol 100, pp658–664

Shackleton, C. M. and Shackleton, S. E. (2006) 'Household wealth status and natural resource use in the Kat River Valley, South Africa', *Ecological Economics*, vol 57, pp306–317

Shackleton S. E., Shackleton, C. M. and Cousins, B. (2000) 'Re-valuing the communal lands of southern Africa: A new understanding of rural livelihoods', *ODI Natural Resource Perspectives*, vol 62, pp1–4

Shackleton, C. M., Ernst, L., Mthembu, T., Paumagarten, F., Dlamini, P., Gabela, M., Mchunu, P., Mtolo, B., Zuma, Z. and Auerbach, R. (2007) 'Marketing of African indigenous vegetables in the Durban Metropolitan Area, South Africa', unpublished IndigenoVeg Survey Report

Sheskin, C. (2007) *Handbook of Parametric and Nonparametric Statistical Procedures*, Chapman and Hall/CRC, Taylor and Francis Group, Boca Raton, FL

Shilowa, M. (2004) 'Welcoming address for the opening of the community health centre in Soshanguve', www.uclgafrica.org/Documents/OfficialDocuments/Welcome%20Address%20by%20Mbazima%20Shilowa%20ENG.doc, accessed June 2007

Sodjinou, E. and Assogba Komlan, F. (2008) 'Besoins en financement et définition de stratégies de mise en place de crédits au profit des agriculteurs de Cotonou', unpublished technical report, IAGU, Cotonou, Benin

South African Census (2001) *South African Census: Municipal Level*, Statistics South Africa, Pretoria. www.statssa.gov.za, accessed 5 September 2007

Ugboma, C., Ibe, C. and Ogwude, I. C. (2004) 'Service quality measurements in ports of a developing economy: Nigerian ports survey', *Managing Service Quality*, vol 14, pp487–495

UNDP (United Nations Development Programme) (2008a) *2007/2008 Human Development Report. Benin. The Human Development Index: Going Beyond Income*, http://hdrstats.undp.org/countries/country_fact_sheets/cty_fs_BEN.html, accessed October 2008

UNDP (2008b) *2007/2008 Human Development Report. Kenya. The Human Development Index: Going Beyond Income*, http://hdrstats.undp.org/en/2008/countries/country_fact_sheets/cty_fs_KEN.html

UNDP (2008c) *2007/2008 Human Development Report. South Africa. The Human Development Index: Going Beyond Income*, http://hdrstats.undp.org/en/2008/countries/country_fact_sheets/cty_fs_ZAF.html

van Veenhuizen, R. (2006) *Cities Farming for the Future: Urban Agriculture for Sustainable Cities*, RUAF Foundation, IDRC and IIPR, the Philippines

Vorster, H. J., Jansen van Rensburg, W. S., Van Zijl, J. J. B. and Venter, S. L. (2007a) 'The importance of traditional leafy vegetables in South Africa', *African Journal of Food, Agriculture, Nutrition and Development*, vol 7, no 4, www.ajfand.net/Vol7No4.html

Vorster, H. J., Pichop, G. N., Maro, F. and Marealle, R. (2007b) 'Indigenous vegetable supply chain in South Africa: Significance, constraints and opportunities – case of Soshanguve (City of Tswane), unpublished IndigenoVeg Survey Report

Waiganjo, M. M., Muriuki, J. and Mbugua, G. W. (2006) 'Potential of indigenous leafy vegetables as companion crops for pest management of high-value legumes: a case study of *Gynandropsis gynandra* in Kenya. I – International conference on indigenous vegetables and legumes. Prospectus for fighting poverty, hunger and malnutrition', *Acta Horticulturae*, vol 752, pp319–322

Webb, N. (1998) 'Urban cultivation: Food crops and their importance', *Development Southern Africa*, vol 15, pp201–213

Weinberger, K. and Msuya, J. (2004) *Indigenous Vegetables in Tanzania: Significance and Prospects*, Technical Bulletin 31, AVRDC, Taiwan

World Bank (2008) 'Data profile report', World Bank Key Development Data and Statistics, http://web.worldbank.org/WBSITE/EXTERNAL/DATASTATISTICS/0,,contentMDK:20535285~menuPK:1192694~pagePK:64133150~piPK:64133175~theSitePK:239419,00.html accessed October 2008

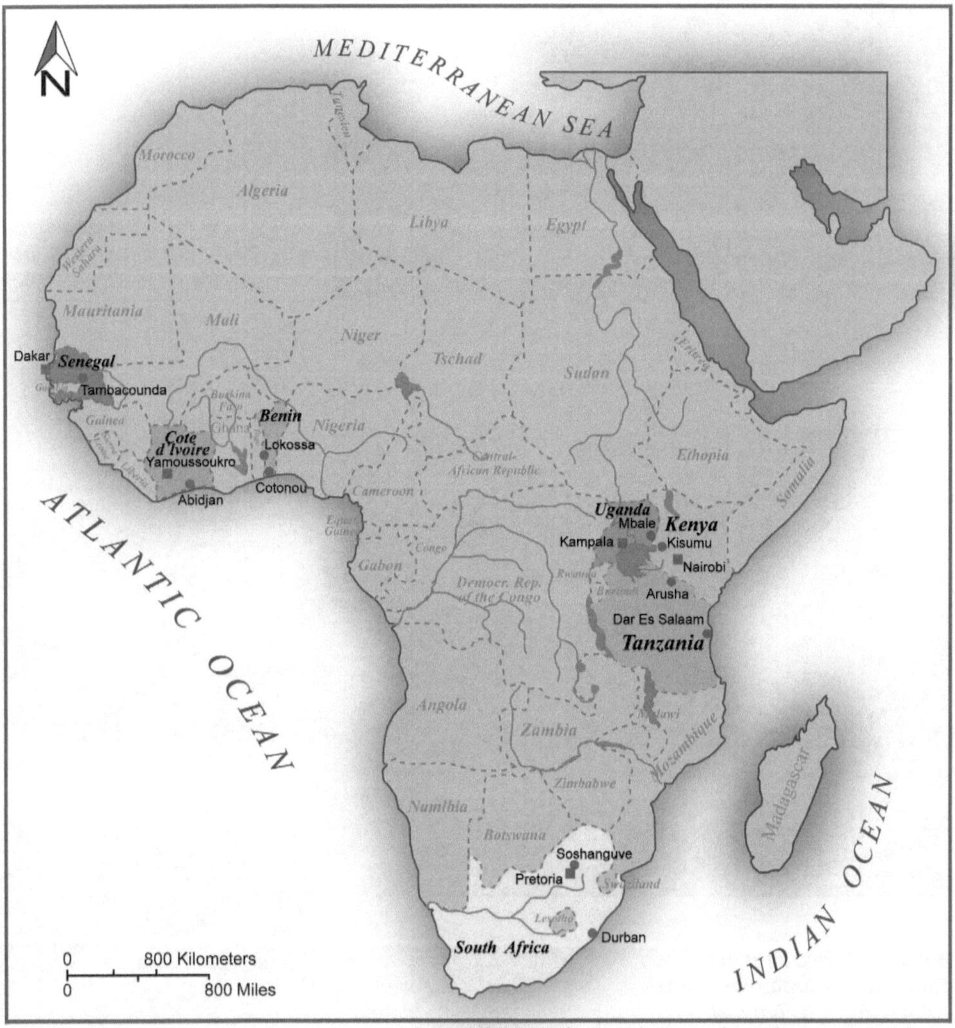

Cartography: Institut für Physische Geographie, Freiburg; K.- D. Lickert

◂**Plate 1.1** *Chagga home garden, Tanzania*
Source: Photo courtesy of A. Drescher

▾**Plate 1.2** *Peri-urban farm in Soshanguve, South Africa*
Source: Photo courtesy of ARC-Roodeplaat

▾**Plate 2.1** *Space-saving solutions for urban vegetable production – a compost mound in a home garden in Kampala, Uganda*
Source: Photo courtesy of ARC-Roodeplaat

▸**Plate 2.2** *Space-saving solutions for urban vegetable production –* Amaranthus dubius, *Kitui district, Kenya*
Source: Photo courtesy of Y. Morimoto

◂**Plate 3.1** Cucurbita pepo *and* Cucumis sativus *sold at a market in Ouagadougou, Burkina Faso*
Source: Photo courtesy of E. Achigan-Dako

▸**Plate 3.2** Cucurbita maxima *at a market in Niamey, Niger*
Source: Photo courtesy of E. Achigan-Dako

◂**Plate 3.3** Cucurbita moschata
Source: Photo courtesy of E. Achigan-Dako

◂**Plate 3.4** *Seeds (*egusi*) of* Citrullus lanatus *are widely consumed in West Africa*
Source: Photo courtesy of E. Achigan-Dako

▾**Plate 3.5** Telfairia occidentalis
Source: Photo courtesy of E. Achigan-Dako

◂**Plate 3.6** Trichosanthes cucumerina
Source: Photo courtesy of E. Achigan-Dako

▸**Plate 4.1** *In various African countries African indigenous vegetables are served as a sauce to accompany a starchy staple, which is often a stiff porridge made from cereals*
Source: Photo courtesy of ARC-Roodeplaat

◄**Plate 4.2** Moringa oleifera *leaves deserve greater promotion for household consumption as they are relatively high in β-carotene, iron, ascorbic acid and minerals, and low in oxalate*
Source: Photo courtesy of P. Maundu

►**Plate 6.1** Amaranthus graecizans
Source: Photo courtesy of P. Maundu

◄**Plate 6.2** Amaranthus dubius
Source: Photo courtesy of P. Maundu

◄**Plate 6.3** Corchorus olitorius
Source: Photo courtesy of P. Maundu

▼**Plate 6.4** Abelmoschus esculentus
Source: Photo courtesy of ARC-Roodeplaat

◄**Plate 6.5** Corchorus olitorius *Ajae type*
Source: Photo courtesy of ARC-Roodeplaat

◄**Plate 6.6** Cleome gynandra
Source: Photo courtesy of P. Maundu

▼**Plate 6.7** Vigna unguiculata
Source: Photo courtesy of ARC Roodeplaat

◄Plate 6.8 Crotalaria brevidens
Source: Photo courtesy of ARC-Roodeplaat)

▲Plate 6.9 Colocasia esculenta *leaves*
Source: (Photo courtesy of ARC-Roodeplaat)

◄Plate 6.10 Colocasia esculenta *corms*
Source: Photo courtesy of P. Maundu

►Plate 6.11 *Bidens pilosa*
Source: Photo courtesy of P. Maundu

►**Plate 6.12** Amaranthus *spinosus*
Source: Photo courtesy of P. Maundu

◄**Plate 7.1** *Dried vegetables, market in Limbe, Malawi*
Source: Photo courtesy of P. Maundu

►**Plate 7.2** *Shredding* Corchorus *spp. leaves, Mali*
Source: Photo courtesy of IER, Mali, Darwin Initiative project

7

Marketing of African Indigenous Vegetables along Urban and Peri-Urban Supply Chains in Sub-Saharan Africa[1]

Katinka Weinberger and Germain N. Pichop

Introduction

Indigenous (or traditional) vegetables have historically played an important role in farming and consumption systems across Africa. Many of these species – for example, amaranth (*Amaranthus* spp.), African nightshade (*Solanum scabrum*), African eggplant (*S. aethiopicum*), jute mallow (*Corchorus olitorius*) and bitter leaf (*Vernonia amygdalina*) – are particularly suitable for resource-poor farmers as they are easy to grow and require minimal external inputs, unlike most exotic vegetables (Opole et al, 1991). They are also an integral component of many traditional dishes. However, modernization and the progression of the market economy in Africa have meant that scientific agronomic research and development has shifted over to exotic crops suitable for export (Weinberger and Lumpkin, 2007). This may have resulted in the notion that African indigenous vegetables (AIVs) are grown by subsistence farmers only – 'hunger food' that people consume in times of need and drought (Humphry et al, 1993), or as a safety net during social unrest and war (Smith et al, 1996a, 1996b), or as important contributors to dietary requirements of isolated communities (Grivetti and Ogle, 2000).

Few authors have highlighted the economic importance of AIVs for income generation and livelihoods (High and Shackleton, 2000; ; Gockowski et al,

2003; Weinberger and Msuya, 2004; Ngugi et al, 2007). Among these, indigenous wild vegetables were found to be a vital component of the rural economy in South Africa (High and Shackleton, 2000). In a comparative value analysis of wild plant resources harvested from home gardens and arable plots against other domesticated crops, it was found that traditional vegetables (both domesticated and wild) contributed 31 per cent of the value of all plants grown on residential plots. This study also found that the value of all plants (both domestic and wild) grown in home gardens was comparable to, or even better than, the mean wage paid to agricultural labourers in the vicinity. In addition, although most of the trade was within the village or region, a growing supply and trade to larger regional centres was observed. In Tanzania, in a study examining production and commercialization aspects of different indigenous vegetables grown as field crops (Weinberger and Msuya, 2004), it was found that AIVs, on average, contributed 13 per cent of all household income of farmers, and that 88 per cent of farmers marketed a share of their crops; also in addition, of all produce, approximately half (46 per cent) was being sold. In the Kiambu District (Kenya) Ngugi et al (2007) showed how farmer groups successfully penetrated the high-value segment of markets for leafy indigenous vegetables through collective action and collaboration with a support system. Market restructuring and support to supply chain development through Farm Concern International (formerly Family Concern), a development agency, assisted small farmers in linking to high-value markets (e.g. supermarkets), thereby eliminating brokers, guaranteeing markets for farmer produce all-year round and maximizing income. Farmers organized in groups were able to realize profits higher by 35 to 72 per cent compared to farmers not organized in groups.

The studies by High and Shackleton (2000), Weinberger and Msuya (2004) and Ngugi et al (2007) all indicate the economic importance of AIVs to rural economies and highlight that these are crops that are being marketed, rather than serving as subsistence crops only. Turning towards the importance further along the supply chain in wholesaling and marketing, Gockowski et al (2003) investigated the importance of leafy AIVs for urban and peri-urban livelihoods in Yaoundé (Cameroon). They found a high level of participation by women in AIV marketing since little capital is required for entry, allowing even the poorest households to participate. In this study, while earnings per product were found to be significantly higher for exotic vegetables than for traditional leafy vegetables, it was also found that there was no significant difference in gross margins between retailers of exotic and traditional leafy vegetables. In this study, the low elasticities for this group of food items in higher expenditure classes indicated that demand for these vegetables may decline as incomes increase. However, other examples from South-East Asia show that indigenous vegetables have the potential for commercial exploitation (Weinberger, 2007). Urban supermarkets in this region increasingly stock a wide variety of indigenous vegetables for affluent consumers.

Box 7.1 Criteria for selection of locations

Physical criteria

- street patterns;
- housing density.

Functional criteria

- communication systems;
- employment levels;
- transportation networks.

Social and socio-psychological criteria

- urban life quality;
- general social life of the people;
- perceived influence of the city on this peripheral location.

Administrative criterion

- local government authority boundaries.

For small farmers and other actors to benefit from participation in markets for indigenous vegetables, whether in domestic markets or for high-value market segments, depends to a large extent upon how the market is organized. It is thus useful to look at the production and marketing of AIVs in a supply chain context. A supply chain can be defined as 'the network of organizations that are involved, through upstream and downstream linkages, in the different processes and activities that produce value in the form of products and services in the eyes of the ultimate consumer' (Christopher, 1998). To understand the contribution of market access to poor farmers, intermediaries and retailers to livelihoods, one must understand what stakeholders are involved in the supply chains, what their roles and interests are, and what barriers to entry there are. Hence, the objective of this research was to provide an overview and status assessment of the marketing of AIVs in several representative regions of sub-Saharan Africa (SSA). It is the first such continent-wide survey of AIVs in urban and peri-urban markets, using a standardized approach so that regional results are comparable. We sought to assess the scope and value of economic activities related to marketing of AIVs by documenting different supply chains, the range of different actors involved, the volumes moved along the chain, as well as the value of produce in several countries representing different geographic regions. The survey was conducted in seven countries drawn from the western (Benin, Côte d'Ivoire and Senegal), eastern (Uganda, Tanzania and Kenya) and southern region (South Africa) in sub-Saharan Africa. Two cities were selected in each country. Respondents were sampled from urban and peri-urban areas based on predetermined criteria (see Box 7.1) and using a

Table 7.1 *Sample size by country, site and actor*

Country	*City*	*Farmers*	*Total*	*Intermediaries*	*Total*	*Retailers*	*Total*	*Sample size*
Benin	Cotonou	87	185	26	55	68	145	385
	Lokossa	98		29		77		
Côte d'Ivoire	Abidjan	70	140	25	32	69	140	312
	Yamoussoukro	70		7		71		
Senegal	Dakar	67	131	67	95	44	143	369
	Tambacounda	64		28		99		
Kenya	Nairobi	64	141	75	138	80	158	437
	Kisumu	77		63		78		
Tanzania	Arusha	69	150	27	29	89	179	358
	Dar es Salaam	81		2		90		
Uganda	Kampala	90	162	50	63	71	153	378
	Mbale	72		13		82		
South Africa	Durban	165	331	–	–	55	88	419
	Soshanguve	166		–		33		
Total			1240		412		861	2658

Source: IndigenoVeg survey data, 2006

snowballing approach, moving upstream from producers to retailers. Three standardized questionnaires were used to interview producers, intermediaries and retailers. Table 7.1 details the sample size ranging between 240 (in Benin) to 437 (in Kenya).

The content of this chapter is based on the results of this multi-country survey. The chapter is divided into four sections. The following section describes production systems for AIVs, supply chains and major actors involved. In the third section, the market volume, value and value-adding activities are determined. The final section discusses major findings in the context of interventions required. The chapter provides the marketing and market information to complement the farming information presented in Chapter 6.

African indigenous vegetables supply chains in sub-Saharan Africa

Important African indigenous vegetables

All actors were asked to rank their top three AIVs in terms of importance to their turnover and their household livelihood (see Table 7.2 for the detailed list). The following AIVs were found to be important in the majority of the surveyed countries' urban and peri-urban areas: amaranth (*Amaranthus* spp.), African nightshade (*Solamun scabrum*, *S. villosum*, *S. nigrum* and *S. americanum*), African eggplant (*Solanum macrocarpon*, *S. aethiopicus* and *S. anguivi*), vegetable cowpea (*Vigna enguiculata*), Ethiopian mustard (*Brassica carinata*), jute mallow (*Corchorus olitorius*), okra (*Abelmoschus esculentus*), spider plant (*Cleome gynandra*) and pumpkin (*Cucurbita moschata*). Amaranth was ranked as important in all of the countries surveyed.

Table 7.2 *Most important AIVs produced and traded in sub-Saharan Africa urban and peri-urban areas*

Crops	*Scientific name*	*Country*						
		Benin	*Côte d'Ivoire*	*Senegal*	*Kenya*	*Tanzania*	*Uganda*	*South Africa*
Amaranth	*Amaranthus* spp.	✗	✗	✗	✗	✗	✗	✗
African nightshade	*Solanum scabrum, S. villosum, S. nigrum, S. americanum*		✗	✗	✗	✗		
African eggplant	*Solanum macrocarpon, S. aethiopicum, S. anguivi*	✗	✗	✗		✗	✗	
Black jack	*Bidens pilosa*							✗
Bitter leaf	*Vernonia amygdalina*	✗						
Cassava leaves	*Manihot esculentum*			✗				
Vegetable cowpea	*Vigna unguiculata*		✗		✗	✗	✗	✗
Ethiopian mustard	*Brassica carinata*	✗				✗	✗	
Jute mallow	*Corchorus olitorius*	✗	✗		✗		✗	✗
Hibiscus	*Hibiscus sabdarifa*			✗			✗	
Okra	*Abelmoschus esculentus, A. caillei*		✗			✗	✗	✗
Pumpkin	*Cucurbita maxima, C. pepo, C. moschata, C. mixta*				✗	✗	✗	✗
Spider plant	*Cleome gynandra*		✗				✗	✗
Sweet potato leaves	*Ipomea batatas*		✗			✗		
Crotalaria	*Crotalaria ochroleuca, C. brevidens*				✗			
Moringa	*Moringa oleifira*			✗				

Source: IndigenoVeg survey data, 2006

Major production systems

AIVs can either be collected or cultivated. Wild collection activities are mainly encountered in rural areas; but AIVs may also grow as volunteer plants in urban and peri-urban areas where soil conditions are suitable. In South Africa, most of the AIVs in the market are collected, not cultivated (see Table 7.3), other than pumpkin leaves, okra, sweet potato, cassava and cowpea. Such collection activities occur from people's own fields, disturbed sites or on commercial farms where landowners permit harvesting of what they deem to be 'weeds' (also see Chapter 6).

Table 7.3 *Importance of collection activities among retailers in Soshanguve, South Africa*

*Quantity collected**	*Percentage share of respondents*	
	Frequency	*%*
None	8	24.2
Less than half	3	9.1
Half	3	3.0
All	19	57.6
Total	33	100.0

Note: * = represents proportion of collected AIVs in relation to the total quantity consumed traded during the surveyed period.
Source: IndigenoVeg survey data, 2006

In all of the other countries surveyed, AIV production is commercial in nature in both urban and peri-urban areas. Three main production systems were encountered. The intensive urban (IU) production, which lies within the inner city limits; the semi-intensive peri-urban (SIPU) production, which lies just outside the city, extending approximately 30km from the city limits (both use a mono-cropping system); and the extensive peri-urban (EPU) production, which lies beyond 30km of the city limits and produces AIVs in a mix-cropping mode in association with staple crops (Gockowski et al, 2003). The area used for production is usually very small, about 0.25ha or less.

Characterization of actors

Producers

The origin of producers is much more diverse in urban than in peri-urban and rural areas. Whereas farmers in rural and, to some extent, in peri-urban areas consist exclusively of locals, they are of various ethnic and national origins in urban areas. One example for this is Abidjan, where 77 per cent of producers are foreign nationals. Often, these are people who migrated from rural or resource-poor areas to urban centres in search of jobs or improved livelihoods.

The education level of producers is generally low as most producers have only attained up to primary school education and lack modern agricultural production skills. Most producers still rely on indigenous technical knowledge for AIV production.

Age-wise, urban and peri-urban farmers are younger than their rural counterparts. They are also younger in West and East Africa, compared to South Africa. This is interpreted as the peri-urban and urban workforce being composed of young migrants to cities in search of job opportunities. They participate in the agricultural sector while searching (often unsuccessfully) for employment in the formal sector because it is something they know from their rural homes.

Intermediaries

AIV intermediaries perform many different activities and, thus, include different categories: collectors, brokers, wholesalers and transporters. Each category performs a specific task along the chain. Collectors essentially perform a pooling activity, gathering vegetables from several small producers into quantities large enough to satisfy wholesalers' demand. Collectors are often, but not always, commissioned by wholesalers. Brokers' activity is similar to collectors; however, they are not generally commissioned by other chain actors. Brokers aim to identify price differential in the market that can enable them to earn a profit by buying where prices are low and selling where prices are high. Wholesalers source produce from production areas or markets in order to supply urban and peri-urban markets. Intermediaries are generally young and, in common with farmers, most of them have no or limited formal education.

Retailers

Most actors, regardless of their position in the supply chain, undertake retail activities. Farmers sell their produce to final consumers at farm gates, on the street, or at the traditional market place in villages or in towns. Intermediaries, transporters and wholesalers, especially, often do not hesitate to sell to final consumers. However, the bulk of the retailers are made of wet market vendors, hawkers and street vendors. The wet market is the typical African public market setting, where the produce is displayed either on a counter, a stall or on the floor. Retailers are generally younger than other actors along the supply chain. Their level of education is also low.

Gender involvement along the supply chain

Gender distribution differs according to the position in the chain and also between rural and urban areas. In rural areas, both production and marketing of AIVs are undertaken mainly by women. Women usually engage in production activities, while men only perform tasks that require a great amount of physical strength. The case is different in the urban and peri-urban areas where men are more involved in production activities. This is especially evident in the urban areas of West Africa where production is very intensive. Results of the IndigenoVeg survey showed that 86 per cent of AIV producers in the urban and peri-urban areas of Dakar and Tambacounda (Senegal) were men. In Abidjan and Yamoussoukro (Côte d'Ivoire), men comprised 84 per cent of producers, while in Cotonou (Benin) 61 per cent of the farmers were men. The percentage of men engaging in AIV production in Lokassa, the other survey site in Benin, was, however, much lower due to the high number of women producing jute mallow in the urban and peri-urban zones. The intensive and delicate harvesting methods used for this crop discourage men from engaging in production. Men also comprised 66 per cent of the producers in East Africa, particularly in the urban and peri-urban areas of Dar es Salaam and Arusha. The same applies to Kampala and Mbale (Uganda). Even in Kenya, where most of the producers are women, the proportion of male farmers is higher (41 per cent) than in the

Table 7.4 *Share of women along the supply chain across countries (percentage share of respondents)*

Category	*Country*						
	Benin	*Côte d'Ivoire*	*Senegal*	*Kenya*	*Tanzania*	*Uganda*	*South Africa*
Farmers	40	16	14	59	34	37	65
Intermediaries	78	100	26	95	75	69	na
Retailers	100	100	58	86	58	68	86

Note: na = not applicable.
Source: IndigenoVeg survey data, 2006

intermediary and retailer categories (5 and 15 per cent, respectively). However, the situation is different in South Africa, where most actors are women (see Chapter 6).

Urbanization increases competition for land use and, thus, reduces the amount of land available for agriculture. In order to maximize yields and returns, urban farmers generally engage in very intensive commercial production. Such commercial production is very labour intensive and often necessitates hired labour, which is mainly offered by young men.

Another explanation for the comparatively high share of men in urban AIV production is that in the countries involved in the study, the production of cash crops has traditionally been reserved for men, while women have been confined to the production and marketing of subsistence crops. As AIV production becomes a cash-generating activity, more men become involved.

However, the IndigenoVeg survey also shows that marketing (both as middle(wo)men and retailers) is still a domain of women, with the exception of Senegal, where women only make up 25 per cent of intermediaries, and Kampala, where women make up 37 per cent of them (see Table 7.4). In Côte d'Ivoire, all (100 per cent) surveyed intermediaries were women, both in Abidjan and Yamoussoukro.

Communication and cooperation

Cooperation and communication among and between actors is very important in a supply chain because it can help increase market coordination and efficiency by reducing transaction costs. Clear differences exist across countries in the way and extent that supply chain actors cooperate with others (see Table 7.5). Very little cooperation takes place in South Africa. Only 20 per cent of retailers shared information there, or transportation and storage. On the other hand, 43 per cent of the retailers in Tanzania and 50 per cent in Uganda collectively purchased AIV in bulk. In Senegal, 60 per cent of all retailers supplied vegetables to other retailers when there was need of this.

Marked differences also exist in the degree of market penetration with contractual arrangements (see Table 7.6). In most countries, more intermediaries and retailers enter contractual arrangements than farmers; for instance, in

Table 7.5 *Cooperation among AIV retailers*

Type of cooperation	*Benin*	*Côte d'Ivoire*	*Senegal*	*Country* *Kenya*	*Tanzania*	*Uganda*	*South Africa*
Lending/borrowing money	0.0	1.4	2.1	23.4	43.6	27.5	2.3
Supplying vegetables to others	6.2	4.3	59.4	17.7	39.7	32.7	2.3
Sharing storage	0.0	0.7	9.1	6.9	39.7	14.4	15.9
Buying vegetable together	20.7	27.1	8.4	23.4	43.0	52.9	7.9
Sharing information	31.7	2.1	25.9	25.9	17.3	50.3	19.3
Sharing transport	4.1	5.0	7.0	8.8	19.7	19.0	18.2

Note: Data generated from a multiple-response question.
Source: IndigenoVeg survey data, 2006

Kenya, nearly half of all retailers interviewed purchased AIVs from their contracted suppliers. In some countries (Kenya, Côte d'Ivoire and Benin), approximately 20 per cent of farmers produce AIVs under contract. Most of these contract arrangements are oral in nature.

Produce flow along the chain

The actors in the AIV supply chain are of several categories: the farmers who produce the crops, the intermediaries, the retailers and the consumers. Because of disorganization and weak market integration in the sector, many actors do not perform specific roles, but undertake many or even all of the tasks that exist along the chain. Wholesalers sell their products not only to other traders, but also, in large part, to households (consumers). Similarly, it is common to find farmers who produce, wholesale and retail AIVs. The potential for arbitrage is high and creates situations where trade between chain actors of the same category and the same level occurs. The supply chain varies from very simple to

Table 7.6 *Share of actors with contract arrangements (percentage of total respondents)*

Contract arrangements	*Benin*	*Côte d'Ivoire*	*Senegal*	*Country* *Kenya*	*Tanzania*	*Uganda*	*South Africa*
Verbal contracts							
Farmers	23.8	19.3	9.2	19.1	0.0	2.5	1.8
Intermediaries	36.4	46.9	40.0	52.9	17.2	12.7	–
Retailers	25.5	17.9	24.5	48.1	3.9	16.3	5.7
Written contracts							
Farmers	2.2	0.7	0.0	2.8	0.0	1.2	0.6
Intermediaries	0.0	0.0	0.0	5.8	0.0	1.6	–
Retailers	2.1	0.0	0.0	5.1	2.8	0.0	1.1

Source: IndigenoVeg survey data, 2006

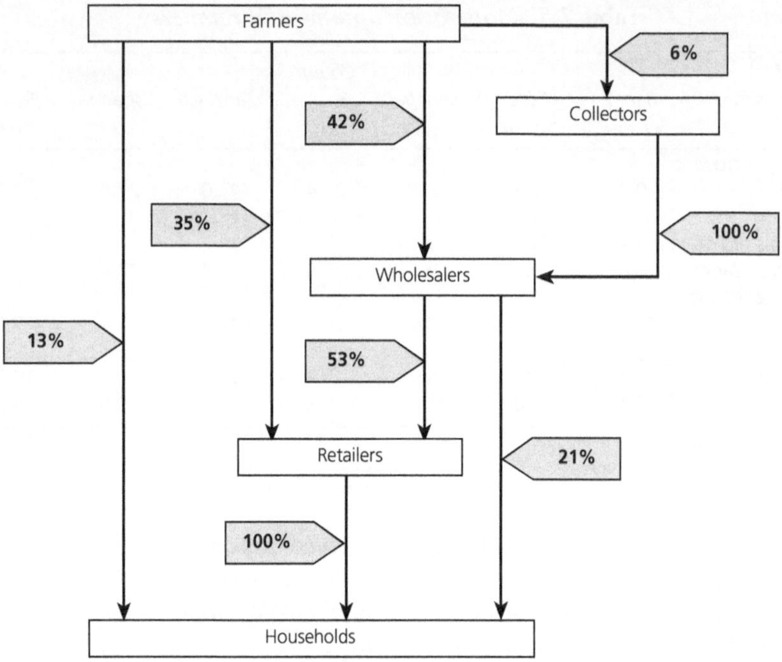

Source: IndigenoVeg survey data, 2006

Figure 7.1 *AIV supply chain in the urban and peri-urban areas of Abidjan and Yamoussoukro (Côte d'Ivoire); the percentages indicate the share of actors who consider their counterparts as main partners*

very complex, depending upon the country. Actors therefore use direct and indirect distribution channels to dispose of their products. In some West African countries, some wholesalers dictate what should be produced, how much should be produced and how it should be produced by supplying producers with inputs and harvesting the produce themselves on farmers' fields.

In South Africa, there is a direct link between farmers/collectors and consumers. Intermediaries are usually non-existent and farmers sell directly to consumers at home, at farm gates, on the roadside or at local markets. In the other African countries, the supply chains tend to be more complex, with actors sometimes performing all activities along the chain. Farmers harvest and sell their produce to intermediaries, retailers and households. Figure 7.1 illustrates the normal flow of AIVs in the urban and peri-urban environments of Abidjan and Yamoussoukro in Côte d'Ivoire.

Market size

Value of the AIV market

Table 7.7 provides an overview on the total volume and value of the three most important AIVs traded at retail level in each country. The average volume sold per retailer was largest in Kenya at 27,115 kg per annum, very similar in

Table 7.7 *Volume and value of annual sales for three most important AIVs to consumers by retailers*

Sub-Saharan Africa	*n*	*Total volume (kg)*	*Average volume (kg) per retailer*	*Average price (US$/kg)*	*Total turnover (US$)*	*Average annual turnover per retailer (US$)*
Benin	145	656,602	4528	0.61	401,578	2769
Côte d'Ivoire	140	99,877	713	0.54	53,544	382
Uganda	153	582,338	3806	0.31	179,884	1176
Tanzania	179	1,986,760	11,099	0.23	451,789	2,524
Senegal	143	1,654,474	11,570	1.47	2,437,867	17,048
South Africa	88	27,324	311	3.31	90,486	1028
Kenya	158	4,284,120	27,115	0.44	1,900,007	12,025
Total		9,291,495	13,631		5,515,155	

Note: Exchange rates used were for 1 January 2007; n = toal sample size.
Source: IndigenoVeg survey data, 2006.

Senegal, Tanzania and Benin at around 11,000kg to 13,000kg per annum, and lowest in Côte d'Ivoire (713kg) and South Africa (311kg). Individual retailers' turnover was highest in Senegal (US$17,048 per annum) and Kenya (US$12,025 per annum), and lowest in Côte d'Ivoire (US$382) and South Africa (US$1028). Clearly, this is a significant market with nearly 10,000 tonnes traded over the span of one year amongst our sample of retailers. The combined turnover of our sample at this level in the supply chain was US$5.5 million. Considering that our sample represents only a portion of retailers of the main AIVs (not all) in two cities in 7 of the 53 African countries, it is safe to deduct that the AIV market is worth billions of US dollars across sub-Saharan Africa.

As Table 7.8 shows for various AIVs in the sampled countries, the profit margin at the retailer level is high over the year. For many of these crops, the average profit margin is well above 30 per cent and often above 40 per cent. Only in two countries, and in both instances for amaranth, was the average profit margin negative. This indicates that retailing AIVs is a profitable enterprise, although price ranges can be very high over the year. For instance, in Kampala, they were found to range from US$0.45 during the off-season to US$0.06 during peak season. Average price across all species was related to the country's gross domestic product (GDP).

Since profit margins are high for retailers, it is worth asking about the relative share in profits for other actors along the supply chain. Table 7.9 shows the mean and median minimum selling price in US dollars that various actors along the supply chain received for their produce over the period of one year. Figure 7.2 depicts the share for different actors in the final selling price to consumers that is available to compensate for labour and other input cost. We compare this data to results from Koenig et al (2008) who assessed the value chains of two significant exotic vegetable crops, onion and tomato, in Tanzania and Kenya.

Table 7.8 *Retail prices and profit margins for selected crops*

	Benin		*Côte d'Ivoire*		*Kenya*		*South Africa*		*Senegal*		*Tanzania*		*Uganda*	
	P	*M*	*P*	*M*	*P*	*M*	*P*	*M*	*P*	*M*	*P*	*M*	*P*	*M*
African eggplant	0.59	34.0	0.66	44.0					1.13	39.9	0.31	33.8	0.29	43.4
Amaranth	0.46	39.6	0.39	49.5	0.43	45.5	3.32	−121.1	1.38	−32.6	0.19	31.5	0.29	34.0
Cowpea	0.30	29.8			0.43	45.9	3.20	77.5	3.64	24.3	0.22	15.9	0.30	45.2
Ethiopian mustard			0.19	50.0	0.66	41.7					0.23	37.1	0.34	41.0
Jute mallow	0.49	38.6	0.65	36.5	0.50	40.5	1.74	86.9					0.28	42.5
Nightshade	0.32	26.5	0.28		0.45	44.6							0.30	33.1
Okra	1.36	33.8	1.25	70.1			1.77	26.2	1.80	38.6	0.34	38.4	0.17	33.3
Pumpkin leaves					0.45	54.6	2.84	33.3			0.23	36.8		
Spider flower					0.41	39.6							0.30	44.3
Sweet potato									1.47	41.2	0.19	40.7		

Note: P = selling price (US$/kg); M = profit margin in percentage (difference between purchase and selling price as a share of the selling price).
Source: IndigenoVeg survey data, 2006

Table 7.9 *Minimum selling price for selected crops for different actors along the supply chain (US$)*

Country	*Crop*	*Farmer*			*Trader*			*Retailer*		
		Mean	*Median*	*Standard deviation*	*Mean*	*Median*	*Standard deviation*	*Mean*	*Median*	*Standard deviation*
Kenya	Onion		0.07			0.17			0.42	
	Tomato		0.11			0.31			0.32	
	AIVs	0.24	0.14	0.21	0.32	0.27	0.29	0.38	0.29	0.30
Tanzania	Onion		0.08			0.17			0.42	
	Tomato		0.13			0.18			0.29	
	AIVs	0.09	0.08	0.05	0.16	0.15	0.08	0.20	0.19	0.11
Uganda	AIVs	0.09	0.08	0.04	0.17	0.17	0.07	0.25	0.28	0.10
Benin	AIVs	0.19	0.19	0.08	0.27	0.23	0.19	0.31	0.25	0.20
Senegal	AIVs	0.90	0.58	1.13	1.06	0.58	1.60	1.33	1.16	1.25
Côte d'Ivoire	AIVs	0.19	0.13	0.16	0.35	0.23	0.32	0.51	0.34	0.43
South Africa	AIVs	2.35	1.24	2.16	–	–	–	2.86	1.77	4.02

Note: Exchange rates used were for 1 January 2007.
Source: IndigenoVeg survey data, 2006; data for onion and tomato in Kenya and Tanzania are derived from Koenig et al (2008) for the high season when prices are low

For AIVs, the farmer share in consumer selling price is lowest in Uganda, at 29 per cent and in Côte d'Ivoire at 38 per cent. It is highest in Benin (76 per cent) and South Africa (70 per cent). In Kenya, one of the two countries with comparable data between exotic vegetables and AIVs, the share of farmers in the final consumer price is higher for AIVs than for onions and tomatoes. In Tanzania, it is approximately the same as for tomatoes and much higher than for onions. Especially in countries where the farmer share in final consumer price is very high, the data suggest that relatively few post-harvest activities, processing and

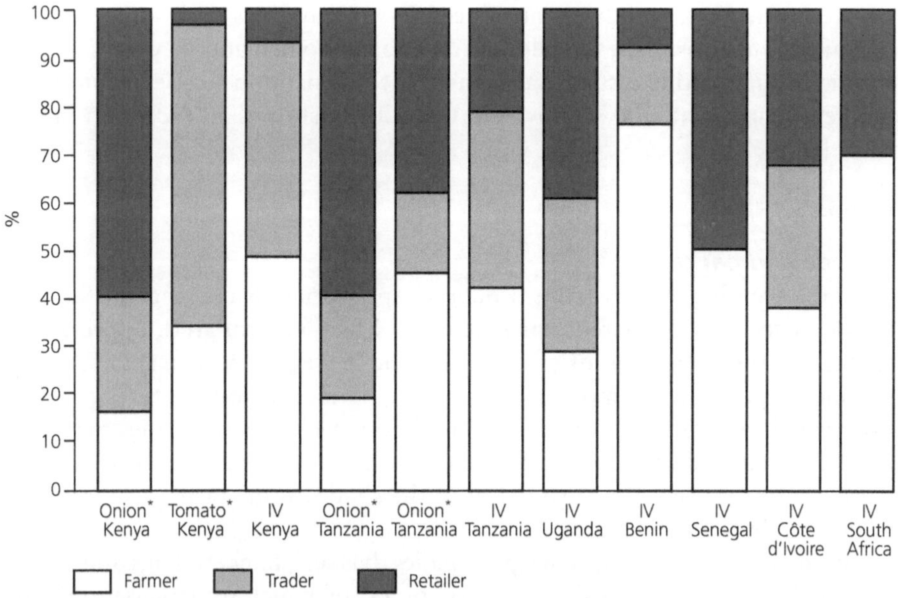

Note: Data are based on minimum price achieved and median value within actors.
Source: IndigenoVeg survey data, 2006; data for onion and tomato in Kenya and Tanzania are derived from Koenig et al (2008)

Figure 7.2 *Share of different supply chain actors in final selling price*

value addition take place. It also shows that the market power of the retail level is relatively limited, leaving farmers in a position of comparative advantage.

Place and form of sale to consumers

AIVs are generally sold in traditional public markets or on the streets, also known as wet markets, where infrastructure is usually lacking. Markets are often muddy and dirty, which raises questions about the safety of the produce sold there. It is common to encounter situations where the produce is displayed on the ground on sheets of paper or fabric. Most public markets lack appropriate storage facilities and cooling is generally not available. Intermediaries (wholesalers, collectors and brokers) purchase the produce in loose form and/or packed in sacks or bags. The produce sold to the final consumers is generally bundled up. Some intermediaries may go all the way to farm gates to source their products. Some other common places where AIVs are sold are farm gate, along the streets or at public gathering places.

In rural areas, the produce is sold at farm gates, on the road side, door to door or at the village market. Village markets meet a few times a week on set days. The majority of traders in these markets are producers who retail their produce to final consumers. Some intermediaries visit the village markets in order to purchase produce that they will eventually sell on urban and peri-urban markets.

Recently, AIVs have been introduced to supermarkets with some success. In Nairobi supermarkets, the sales of AIVs have boomed and they are sold out by midday, indicating a strong and unsatisfied demand in this segment of the market (Ngugi et al, 2007). However, this still constitutes an exception rather than the rule.

Value addition

Storage and packaging

Vegetables are highly perishable products. Appropriate packaging and storage is often necessary to increase product shelf life. Unfortunately, appropriate storage infrastructure is often non-existent in vegetable markets. Unsold produce is generally dumped. This severely limits the volume traded as the traders reduce their risk of losses by purchasing very low quantities that they can sell by the end of the market day. Immediately after harvest, most farmers store their produce on the ground or in the shade. Some farmers place their produce in baskets and crates.

Products are sold loose in heaps, bundles, baskets, buckets, bags and sacks. Packaging is minimal, consisting mostly of a rope which traders use to tie up the produce in bundles when sold to final consumers. Fruit vegetables such as African eggplant, pumpkin and okra are sold either in heaps, bowls or buckets. Heaps and bundles can be displayed on the ground over a sheet of paper/fabric.

Transportation

The average distance travelled, and time used for travelling to sell produce is not long, as the example from Uganda (see Table 7.10) shows. Produce rarely travels for more than one hour, and an average distance of between 11km to 14km to intermediaries, and between 1km and 7km to retailers. An exception to this is sales between intermediaries. Overall, however, Table 7.10 indicates that produce sold at urban markets is produced close to cities.

Many chain actors carry their produce themselves to the point of sale. Usual means of transportation are by motorcycle, bicycle, pickup truck, public transportation and walking. Walking is one of the most popular means of transportation, especially in rural and peri-urban settings. Rural farmers have to carry their produce to the village market because of poor road conditions and limited availability of other affordable means of transportation. They will also carry their produce on their heads during their hawking activities. In urban and peri-urban areas, the proximity of the urban markets encourages producers and other traders to transport their produce to markets on foot, especially when the quantity transported is small. However, most traders in urban and peri-urban areas will use public transportation (taxis or motorbikes) when the distance increases and the quantity brought to the market becomes larger. Wholesalers who generally source very large quantities of AIVs mainly use pickup trucks. Other means of transportation commonly used are human or animal power carts, bicycles and private cars.

Table 7.10 *Average distance and time between the product source and the point of sale (between trading partners) for Ugandan urban and peri-urban AIV traders*

Distance and time	*Intermediaries*				*Retailers*			
	Kampala (n = 50)		*Mbale (n = 13)*		*Kampala (n = 71)*		*Mbale (n = 82)*	
Average distance (km) from	*Mean*	*Standard deviation*	*Mean*	*Standard deviation*	*Mean*	*Standard deviation*	*Mean*	*Standard deviation*
Farmers	11.1	11.3	5.6	8.2	6.7	7.1	7.3	8.5
Collectors	12.1	4.8	–	–	1.0	–	8.3	8.1
Intermediaries	40.0	–	12.0	–	3.9	5.9	8.0	8.3
Wholesalers	14.1	13.0	3.5	2.1	1.2	1.8	2.9	1.3
Average time (hours) from								
Farmers	1.1	0.6	0.8	0.5	0.9	0.6	1.0	0.7
Collectors	1.0	–	–	–	1.3	1.1	0.8	0.4
Intermediaries	1.0	–	1.0	–	0.7	0.4	0.5	0.4
Wholesalers	1.2	0.7	1.3	1.0	0.1	0.1	0.6	0.3

Source: IndigenoVeg survey data, 2006

During transportation, the produce is packed in plastic, polyethylene or jute bags. When carried on the head, AIVs are placed in baskets, plastic or aluminium containers.

Processing

Processing adds value to the produce, but also extends its life and allows for use during off-seasons. Most of the processing occurs at retail level (see Table 7.11) and is done by approximately one third of retailers in Uganda and South Africa and, to a lesser degree, in other countries. The major processing techniques applied to AIVs are sun-drying (see Plate 7.1) slicing (see Plate 7.2) and, in South Africa, also blanching. Sun-drying can include simple sun-drying; sun-drying and grinding into powder; and cutting/slicing and sun-drying. Those who engage in processing do not add a significant mark-up to their prices, but instead use it for product differentiation.

In South Africa, fresh AIVs are rarely available in the markets during the off season (winter) because most of the AIVs consumed are not cultivated. However, drying for use in winter is commonplace in rural areas (Shackleton et al, 1998; Vorster et al, 2007), and some finds it way onto city markets, as

Table 7.11 *Share of actors doing processing (percentage)*

	Benin	*Côte d'Ivoire*	*Senegal*	*Kenya*	*Tanzania*	*Uganda*	*South Africa*
Farmers	0	9.3	0.8	2.1	0	1.2	10.3
Intermediaries	0	3.1	4.2	14.5	0	3.2	–
Retailers	1.4	22.1	17.5	21.5	2.8	35.3	32.9

Source: IndigenoVeg survey data, 2006

prices are high during winter. Most farmers and traders indicate that they process AIVs in order to have them available during the winter.

Discussion and policy implications

The IndigenoVeg survey has shown that the production and marketing of AIVs in SSA is a complex activity that involves many actors and contributes to the earnings and livelihoods of many millions of farmers, traders and retailers all across SSA. Much of the cultivation of AIVs sold in cities takes place close to these cities, probably because road infrastructure is so poor in many of the countries in SSA, and fresh produce cannot be easily transported over large distances (see Chapters 6 and 8). AIVs have a value in providing employment and livelihoods for those with a low asset base – often uneducated young women or migrants in rapidly growing urban conglomerations. Yet, to reap the full positive impacts that AIVs can have upon the poor requires policy interventions at scale to improve the functioning of market chains. Interventions required include:

- improvement of market infrastructure;
- timely access to affordable financial and business services, especially market information systems;
- promotion and implementation of efficient business organization practices;
- better access to appropriate processing technologies; and
- greater inclusion and empowerment of the disadvantaged, including women.

Public markets in SSA are poorly adapted for fresh vegetable sales. These markets are usually compartmentalized hangars, sometimes equipped with counters and simple cabinets for storage. Moreover, the markets may not be or are only partially covered and inconsistently cleaned, which causes them to become muddy during rainy days. In addition, markets do not have proper storage facilities that may help to keep the produce fresh for longer periods. Public markets are not equipped with cold rooms and chain actors cannot afford refrigerated trucks. This forces the chain actors to seek strategies to limit transaction costs in order to increase the efficiency of their operations. Consequently, vendors avoid having to store their products. They limit the quantities purchased to the strict minimum that they anticipate selling for the day, which results in the volume traded being low and the unsold produce dumped at the end of the market day. Enhancing the market infrastructure by building paved market spaces, by providing parking and unloading spaces, by building roofs, by enlarging market spaces, by electrifying markets, and by providing cold rooms and areas for storage would contribute to public health and food safety and allow more actors to enter markets, thus contributing to enhanced competitiveness.

Market information systems (MIS) enhance competition in the market by increasing transparency for all participants, particularly the weakest, who are smallholder farmers. Having a functioning MIS in place empowers farmers by strengthening their bargaining power in order to increase their share of the retail proceeds of their produce (Schubert, 1988; Robbins, 1998; Giovannucci, 2003; Tollens, 2006). Despite these benefits, donors and government agencies remain sceptical as to the need for long-term MIS support. This lack of support is based on several factors, including poor past performance of government-managed MIS, the long-term nature of the funding requirement, and lack of evidence to show the value of market information as a tool that can assist farmers and farmer groups in making better marketing decisions and increasing their incomes. There are no examples of market information services operating in sub-Saharan Africa that have the efficiency and participation levels of those operating in all developed countries. However, in Uganda, the introduction of a low-cost market information network (FOODNET) for grain crops, for instance, resulted in the participation of the majority of farmers, stabilizing their incomes and enhancing their decision-making (Ferris and Robbins, 2004; Ferris et al, 2008).

Limited market transparency results in weak vertical coordination of the supply chain and inefficient markets, which explains the huge price differences in between points of sale. Retailers in Nairobi who purchase AIVs in supermarkets and then resell them at local public markets is one example of this. The analysis of the Kampala AIV supply chain showed backward flows of produce between retailers and wholesalers. Such abnormalities show the existence of substantial price differentials that may encourage arbitrage and contribute to overall inefficiency of markets. By supporting and promoting farmer organizations, and helping them to link directly to retail outlets, the bargaining position of farmers is strengthened, resulting in substantial income increases for farmers and also benefiting consumers through increased supply. For instance, AIV production in Nairobi, through the support of farm groups, increased the traded volume from 31 tonnes (t) per month for three to four months with an estimated farm gate value of US$6000 per month to 600t per month all year round, generating an estimated US$142,860 per month. Today, a network of 3000 farmers supplies supermarkets, markets, informal markets and kiosks on a daily basis (Ngugi et al, 2007). Strengthening direct linkages between farmers and retailers, reduction of taxes en route to the market, and simplification of the registration process for small businesses can all contribute to providing an enabling business environment.

Overall, processing and other value-addition activities were found to be poorly developed in the supply chain for AIVs. This is true not only for AIVs, but for horticulture crops, in general, as there is a lack of investment in horticultural processing in Africa (Lenné and Ward, 2008). There are also examples of processing factories being established and then going out of business for different reasons. But, as in the case of market information systems, it appears that there is the potential to look at opportunities for supporting access to processing technologies since providing access to small-scale actors for appropriate processing technologies would contribute to increased small-producer incomes and the subsequent development of small-scale business and, thus, employment opportunities. This should also include training on simple processing techniques and easier access to credit.

It can be expected that if markets for AIVs in SSA grow, they will become more formalized. Weaker actors along the chain could, however, be negatively affected and crowded out of the chain as a consequence. Our data show that many actors along the AIV supply chain are young and uneducated, often women. Our data also indicate that in urban centres, where land is scarce and expensive, the share of men in production and trading is already higher than would be expected. Thus, all activities to strengthen the AIV supply chain in urban and peri-urban areas of SSA should aim for the inclusion of disadvantaged groups, including women. This is essential since disadvantaged groups usually:

- lack access to information (e.g. standards), which will become more important as markets become more formal;
- lack access to credit, which is essential for investments to ensure food safety and certification; and
- are disadvantaged in the enforcement of legal rights, which are important, for instance, to sign contracts.

Thus, in order to further ensure that the supply chain for AIVs remains easily accessible, donors and policy-makers need to ascertain that these are all in place.

Overall, there is evidence that demand for AIVs is growing (Maundu, 2004; Weinberger, 2007), and this increasing demand needs to be encouraged and exploited for the benefit of small-scale producers, traders and retailers in and around the cities. Thus, there is an increased need for policy-makers at municipal and national levels to address the major constraints so that AIVs can continue to play a significant role in providing jobs, food security and income for the resource poor.

Future research needs

The outcome of this chapter shows that the production, handling and marketing of AIVs are activities of considerable size and value in the countries surveyed, and that AIVs comprise a lucrative market worth billions of US dollars across sub-Saharan Africa. AIVs also provide employment and livelihoods to millions of farmers and other actors. Our data also show that the farm share in the retail price is comparatively high for indigenous vegetables, sometimes higher than for exotic vegetables. This suggests that relatively few post-harvest activities, processing and value addition take place. More adaptive research on processing techniques for leafy and fruit AIVs would contribute to providing small-scale actors with access to such technologies. Some indigenous technologies may also be available, but may need to be adapted to enhance the food safety of processed food items.

Increasing commercialization of AIVs may negatively affect weaker actors along the chain, such as the young, the uneducated and women. Our data indicate that in urban centres, where land is scarce and expensive, the share of men in production and trading is higher than would be expected. Urbanization appears to lead to a displacement of small-scale farmers by wealthy urban residents and (usually male) immigrants to cities. The latter often enter commercial AIV production since this requires little start-up capital. More research is thus required to understand how commercialization of indigenous vegetables will affect the livelihoods of chain actors, especially within a spatial context, considering the role of increasing urbanization. Such research could provide the basis for the development of improved market coordination mechanisms, including for improved coordination and information.

Note

1 Data presented in this chapter were generated as a collaborative effort between a consortium of African and European institutions: IndigenoVeg. Data collection and entry were conducted by the National Agricultural Research Organization (KARI) in Uganda, Maseno University (BPS) in Kenya, the Agriculture Research Council (ARC) at Rhodes University (DES) and Rainman Landcare Foundation (RLF) in South Africa, the Institut National des Recherches Agricoles in Benin (INRAB), the Institut Senegalais des Recherches Agricoles (CDH) in Senegal, the Centre National de Recherche Agronomique (CNRA) in Côte d'Ivoire, and Horti-Tengeru in Tanzania. Support in establishing the research design is gratefully acknowledged, particularly from the University of Wales Bangor (CAZS), the Albert-Ludwigs-Universität-Freiburg (APT) and the Royal Veterinary and Agricultural University (DAS). Excellent research assistance was provided by Festo Maro and Christian Genova.

References

Christopher, M. (1998) *Logistics and Supply Chain Management: Strategies for Reducing Cost and Improving Service*, Prentice Hall and Financial Times, London

Ferris, S. and Robbins, P. (2004) *Developing Marketing Information Services in Eastern Africa*, Local, National and Regional Market Information Services, ASARECA Monograph 9, IITA, Ibadan, Nigeria

Ferris, S., Engoru, P. and Kaganzi, E. (2008) *Making Market Information Services Work Better for the Poor in Uganda*, CAPRI Working Paper no 77, IFPRI, Washington, DC

Giovannucci, D. (2003) 'Market information services', in *The Guide to Developing Agricultural Markets and Agro-Enterprises*, World Bank Group, Washington, DC

Gockowski, J., Mbazo'o, J. Mbah, G. and Moulende, T. M. (2003) 'African traditional leafy vegetables and the urban and peri-urban poor', *Food Policy*, vol 28, pp221–235

Grivetti, L. E. and Ogle, B. M. (2000) 'Value of traditional foods in meeting macro- and micronutrient needs: The wild plant connection', *Nutrition Research Reviews*, vol 13, pp31–46

High, C. and Shackleton, C. M. (2000) 'The comparative value of wild and domestic plants in home gardens of a South African rural village', *Agroforestry Systems*, vol 48, pp141–156

Humphry, C., Clegg, M. S., Keen, C. and Grivetti, L. E. (1993) 'Food diversity and drought survival: The Hausa example', *International Journal of Food Sciences and Nutrition*, vol 44, pp1–16

Koenig, T., Blatt, J. Brakel, K., Kloss, K., Nilges, T. and Woellert, F. (2008) *Market-Driven Development and Poverty Reduction: A Value Chain Analysis of Fresh Vegetables in Kenya and Tanzania*, Humboldt-Universitaet zu Berlin, Berlin

Lenné, J. and Ward, A. (2008) *Exotic and Indigenous Vegetable Value Chain Analysis for Selected Countries in East and Southern Africa*, Report submitted to the Bill and Melinda Gates Foundation

Maundu, P. (2004) *Traditional African Leafy Vegetables: From Despised to Prized*, IPGRI, Geneflow, http://news.bioversityinternational.org/Geneflow.php?itemid=674

Ngugi, I. K., Gitau, R. and Nyoro, J. K. (2007) *Access to High Value Markets by Smallholder Farmers of African Indigenous Vegetables in Kenya*, IIED, London

Opole, M., Chweya J. A. and Imungi, J. K. (1991) *Indigenous Vegetables in Kenya: Field and Laboratory Experience Report*, KENGO, Nairobi, Kenya

Robbins, P. (1998) *Review of Market Information Systems in Botswana, Ethiopia, Ghana, and Zimbabwe*, Technical Centre for Agriculture and Rural Cooperation, Wageningen, The Netherlands

Schubert, B., Juntermanns, G. and Zehrfeld, E. H. (eds) (1988) *Agricultural Market Information Services: Guidelines for the Design and Management of Market Information Projects*, GTZ Rural Development Series, BMZ, Eschborn, Germany

Shackleton, S. E., Dzerefos, C. M., Shackleton, C. M. and Mathabela, F. R. (1998) 'Use and trading of wild edible herbs in the central lowveld savanna region, South Africa', *Economic Botany*, vol 52, pp251–259

Smith, G. C., Clegg, M. S., Keen, C. L. and Grivetti, L. E. (1996a) 'Mineral values of selected plant foods common to Southern Burkina Faso and to Niamey, Niger, West Africa', *International Journal of Food Sciences and Nutrition*, vol 47, pp41–53

Smith, G. C., Dueker, S. R., Clifford, A. J. and Grivetti, L. E. (1996b) 'Carotenoid values of selected plant foods common to Southern Burkina Faso, West Africa', *Ecology of Food and Nutrition*, vol 35, pp43–58

Tollens, E. F. (2006) 'Market information systems in sub-Saharan Africa: Challenges and opportunities', Poster paper prepared for presentation at the International Association of Agricultural Economists Conference, Gold Coast, Australia, 12–18 August

Vorster, H. J., Jansen van Rensburg, W. S., Van Zijl, J. J. B. and Venter S. L. (2007) 'The importance of traditional leafy vegetables in South Africa', *African Journal of Food, Agriculture, Nutrition and Development*, vol 7, no 4, www.ajfand.net/Vol7No4.html

Weinberger, K. (2007) 'Are indigenous vegetables underutilised crops? Some evidence from Eastern Africa and South East Asia', *Acta Horticulturae*, vol 752, pp29–34

Weinberger, K. and Lumpkin, T. A. (2007) 'Diversification into horticulture and poverty reduction: A research agenda', *World Development*, vol 35, pp1464–1480

Weinberger, K. and Msuya, J. (2004) *Indigenous Vegetables in Tanzania: Significance and Prospects*, Technical Bulletin no 31, AVRDC – World Vegetable Center, Shanhua

8

Integration of Urban Agriculture into Spatial Planning

Wakuru Magigi and Axel W. Drescher

Introduction

The benefits of urban and peri-urban agriculture in the cities and towns of sub-Saharan Africa (SSA) are well recognized across a number of development challenges, including poverty alleviation, environmental conservation and redressing the negative impacts of urbanization. The benefits are appreciated at the local level by authorities and farmers, such that involvement in urban agriculture is increasing. For example, it is estimated that in Tanzania at least 30 per cent of the total urban residents practice urban agriculture as one of their primary livelihood strategies (URT, 2003). In Dar es Salaam, the United Nations Development Programme (UNDP, 1998) reports that one out of five (i.e. 20 per cent) residents in the city are engaged in urban agriculture. The local government reform report of 1998 noted that more people were engaged in urban agriculture than in the formal employment sector. This has not come about by chance, but rather by the intent and activities of urban residents who, despite at times an unsupportive policy environment, have turned to urban agriculture as a means of livelihood. This demonstrates that it is a viable and valuable occupation for urban dwellers, and that, consequently, land-use planning authorities need to make provision for urban agriculture, especially during the spatial planning process to ensure that there is sufficient land and in suitable locations.

Spatial planning in this context refers to the processes of defining land use (including residential, commercial, urban agriculture, industrial, open space, recreational facilities, parks, future urban land, infrastructural service facilities

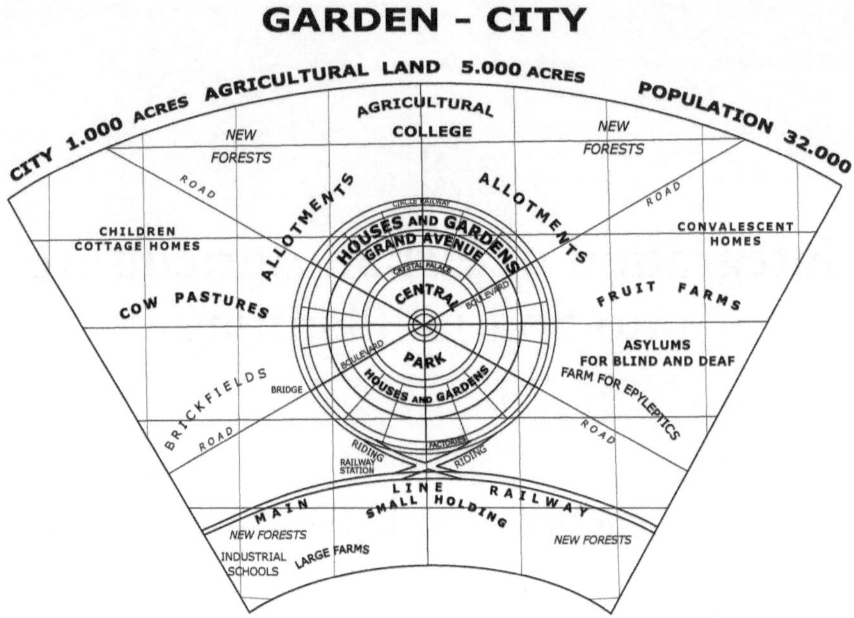

Source: redrawn by W. Magigi

Figure 8.1 *The Garden City concept and plan of Ebenezer Howard (1878)*

and utilities, areas for conservation and hazardous land) that promote equal access of residents to urban services. This includes services related to economic, health and environmental benefits for all sectors and wealth classes. The implementation of spatial planning therefore helps to create a liveable city with a healthy economy and environment. In Tanzania, this falls under the zoning regulations provided in the 2006 Urban Planning Act.

Spatial urban planning emerged during the industrialization and urbanization surge in Europe during the 19th century (Faludi, 1970, 1984; Hall, 1988). It was during this period that the idea of integration of urban agriculture evolved through the concept of the Garden City developed by Ebenezer Howard (Friedmann, 1987; Lynch, 1990). The underlying goal of the Garden City concept was to combat urban overcrowding, diseases and food insecurity in London through the establishment of garden cities, each supporting about 30,000 inhabitants by being self-contained in terms of employment, food production and recreation. Ebenezer Howard theorized a concentric plan for the Garden City, in many ways similar to the satellite towns and green belts associated with modern cities (see Figure 8.1). A central feature was that the plan included residences which were surrounded by woodland and had horticultural spaces and family gardens to meet the needs of the whole urban community.

The Garden City as a spatial planning concept was implemented in several European cities, including Letchworth and Welwyn. This was further exported

Source: Magigi (2008)

Figure 8.2 *The emergence of unorganized and informal activities in a city setting*

to various continents of the world, and in sub-Saharan Africa (SSA) found expression in Dodoma (Tanzania), Nairobi (Kenya) and Blantyre (Malawi) (Halla, 2002; Lupala, 2002). The challenge to urban agriculture today is how to become integrated within cities with mixed land uses and increased informal settlements. This is even more difficult in countries with weak economies, incapacity of local governments, poor policy enforcement strategies and different tenure modalities, as is the case in many parts of SSA (Kombe and Kreibich, 2000; Magigi and Majani, 2006).

According to UN-Habitat (2006), by the year 2025, 61 per cent of the world's population will be living in urban areas. Many African cities have urban growth rates doubling their population in less than 20 years: Kampala, Addis Ababa, Nairobi, Conakry, and Dar es Salaam are some examples. Many people will be living close to or even below the poverty line – benchmarked by the World Bank (2006) as living on less than US$1 per day. City authorities and administrations are challenged by high growth rates and influxes of people seeking new livelihoods and opportunities, many of whom have limited capital and skills. In too many instances this has resulted in unorganized and unregulated land-use activities, with authorities doing little more than harass inhabitants (see Figure 8.2). These rapid changes and unregulated environments pose challenges to spatial planning, where the different needs of urban society and land uses must be integrated in order to deliver services and complementary land-use zonation. In terms of the focus of this chapter, spatial planning can either hinder or promote urban agriculture. Thus, it is instructive to examine the nature of spatial planning approaches, and the institutional and legal basis for spatial planning to facilitate the integration of urban agriculture within urban activities and spaces.

Spatial planning approaches

Dowall and Clark (1996) and Drescher and Iaquinta (1999) show that urban planning approaches and tools include master planning, structure planning and land-use zoning. These planning approaches have been claimed to be inflexible, rigid and not adaptive to the challenges that urban land development presents (Armstrong, 1987; Iaquinta and Drescher, 2000; Drescher, 2001). Principally, none of the land-use planning approaches and tools included urban and peri-urban agriculture. Additionally, the legal basis did not allow for the inclusion of urban and peri-urban agriculture within land-use plans because it was always regarded as a temporary activity in SSA cities (Kyessi, 1997; Madaleno, 2000; Mougeot, 2005; van Veenhuizen, 2006). However, the stark reality is contrary to this and calls for more participatory planning approaches to meet urban dwellers' needs and aspirations. This would require revisions to land-use plans, policies and the legal regulations around spatial planning and around urban agriculture (Healey, 1997; Sawio, 1998; Rakodi and Lloyd-Jones, 2002).

The policy and legal frameworks of different countries with respect to urban agriculture

The legal regulatory framework around land-use planning refers to the set of laws, regulations and decrees endorsed and approved by the required authority that contributes to preparing, coordinating and implementing spatial land-use plans. It is within these land-use plans that the development of urban and peri-urban agriculture may be recognized and provided with a designated space. These plans are then used to support, or constrain certain land-use activities – in this instance, urban agriculture. Where there is an array of supportive policies or regulations, across and between the different administrative and regulatory dimensions of city management (i.e. social, economic and environmental), this can be taken to mean that urban and peri-urban agriculture is welcomed within the specific city. In instances were regulations are absent, this indicates that the city authorities are neglectful or uninformed about the benefits of urban agriculture, or might indeed be antagonistic towards it. Where supportive policies are in place, then it is incumbent upon urban planners to make the necessary spaces and places available through land-use zonation and planning.

Table 8.1 provides a brief comparative analysis of policies that are supportive of urban agriculture in several SSA cities compared with two from South America and two from Europe. It can be seen that different cities emphasize different aspects, and that not all are at the same stage of development with respect to urban agriculture:

- In Rosario (Argentina), the Urban Agriculture Programme (UAP) was established to support smallholder farmers in the city, as well as to provide support to farmers in the production of aromatic and medicinal plants, as well as traditional produce. Key components included product improvement via agro-industries and their sale through farmers' markets and fairs. Specific attention was also given to promoting farmers' social organization through a network, with special attention to women. Rosario has a very complete and effective legal and regulatory framework for the promotion of urban and peri-urban agriculture. The UAP was incorporated within the strategic plan of the city's development framework, helping to provide access for smallholder farmers to funds from financial institutions, and encouraging participatory budgets for the creation of agro-industries.

- In Brasilia (Brazil), between 1995 and 1998, the government of the federal district developed a wide array of legal and regulatory instruments in order to facilitate access to credit and inputs, to support family agro-industries, and to establish adequate spaces for the sale of urban and peri-urban agriculture products.
- In Nairobi (Kenya) there are no provincial or municipal promotion programmes or a policy basis to support urban and peri-urban agriculture. However, the activity is ongoing in road reserves, along river banks and within other open spaces. Laws from colonial days tend to restrict urban agriculture; but agricultural activities are tolerated in the city.
- Dar es Salaam (Tanzania) is involved in environmental planning and management through the Sustainable Cities Programme that was established in 1992. Urban agriculture was regarded as an important livelihood strategy and was therefore included in the city's strategic plan. Around the same time, urban and peri-urban agriculture was further enshrined in the national land development and human settlement development polices. But it has rarely been included in land-use plans and promotional practices since then.
- Kampala (Uganda) has accepted and included urban agriculture in the city setting; but poor harmonization of land development and health legislation has resulted in it being restricted largely to slum areas. The mayor has allocated a budget for urban agriculture, having realized its important social dimension. But poor urban farmers gain access to land for urban agriculture only as customary tenants on private land, and are only allowed to cultivate annual crops because their landlords and the city authorities do not allow perennial crops. Farmers are always at risk of having their land taken away for other purposes. 60 per cent of farmers are actively searching for land.
- Zimbabwe has included urban agriculture in its legal setting. However, implementation modalities in city land development plans have only begun to be effective, with both accommodative and restrictive laws in place. In Harare, an audit of the policy and legislative framework was undertaken to identify policies and legislations that affect urban agriculture. Confusing and contradictory laws and a prohibitive environment have been reported, although there has not been a general prohibition of agricultural activities. In August 2003, a meeting was held in cooperation with the UNDP, the

Table 8.1 *Selected cities with experiences of different levels of integrating and promoting urban and peri-urban agriculture (UPA) in spatial planning*

	Argentina Rosario	*Brazil Brasilia*	*Hungary Central Region*	*Germany Freiburg*	*Tanzania Dar es Salaam*	*Kenya Nairobi*	*Uganda Kampala*	*Zimbabwe Harare*	*Zambia Lusaka*
Land use for UPA	Yes	Yes	No	Yes	No	No	Yes	Yes	Yes
Support for UPA processing	Yes	No	No	No	No	No	No	No	No
Support for UPA commercialization/marketing	Yes	No	No	No	No	No	Yes	No	No
Access to credit for UPA	No	Yes	No	Yes	No	No	No	Yes	No
Access to water and other inputs for UPA	No	No	No	Yes	No	No	No	No	No
Sanitary and environmental control for UPA	No	Yes	No	Yes	No	No	No	Yes	No
Land laws allowed smallholder farmers to hold farms within a stipulated tenure or period of time	Yes	No	No	Yes	Yes	No	No	No	No
Land demarcation for UPA	Yes	No	No	Yes	Yes	No	Yes	Yes	Yes
Municipal/city UPA programmes	Yes	Yes	No	Yes	Yes	No	Yes	Yes	Yes
Physical/strategic plans for UPA	Yes	No	No	Yes	Yes	No	Yes	Yes	No

Source: Atukunda (1998); Drescher and Iaquinta (1999); Bakker et al (2000); Santandreu and Castro (2006); Van Veenhuizen (2006)

United Nations Children's Fund (UNICEF), the United Nations Food and Agriculture Organization Regional Office for Southern Africa (FAO-SAFR), the Food, Agriculture and Natural Resources Policy Analysis Network (FANRPAN), the Resource Centres on Urban Agriculture and Food Security (RUAF) Foundation and the International Development Research Centre (IDRC), where the existence and importance of urban agriculture was recognized and calls were made for its incorporation within urban development plans. Authorities from Kenya, Malawi, Swaziland, Tanzania and Zimbabwe took part.

- In Ghana, the Accra Metropolitan Assembly seeks to promote urban agriculture and to review by-laws to make them producer friendly. There are strong restrictions on sale, irrigation and livestock keeping. The Land Administration Project was launched to create land banks for agricultural development and investments.
- Swaziland shows the negative effects of continued reduction of residential plot sizes on urban agricultural activities. The food producing sector was expanding until the World Bank pressed for smaller plots.
- In Botswana, the Ministry of Agriculture argues that proper zoning encourages urban agricultural activities. Implementation of poultry zones on land considered of low potential for the development of other land uses has been successful in Gabarone. The National Nutrition Plan of Action intends to improve the nutritional situation of HIV/AIDS-affected households by supporting urban agriculture in cooperation with non-governmental organizations (NGOs), the private sector, local communities and families.

Land-use (spatial) planning processes

Land-use planning involves different stages and processes. These include initiation (i.e. setting the agenda), decision-making, preparation, implementation, evaluation and monitoring (see Figure 8.3). In SSA, spatial land-use planning decision-making follows either a top–down or bottom–up approach. The top-down decision-making occurs where communities, households and farmers are largely passive recipients of instructions and orders regarding land zonation and permissible uses. Communication is largely in one direction and passive, where the interests of the higher-level institutions are emphasized (see Figure 8.3). In Tanzania, the current policy and legal basis gives mandate and power to the local and central government planning authorities to initiate, decide and implement land-use planning processes. In most SSA countries, urban planning power and decisions are centralized even if the policy guidelines advocate a more decentralized approach and model. Decentralizing power to community level seems difficult if we take the examples of Kenya and Uganda, where town and country ordinances inherited from the colonial period are still in use. The inadequate decentralization of power to local level affects the land-use planning process, which internationally currently calls for community involve-

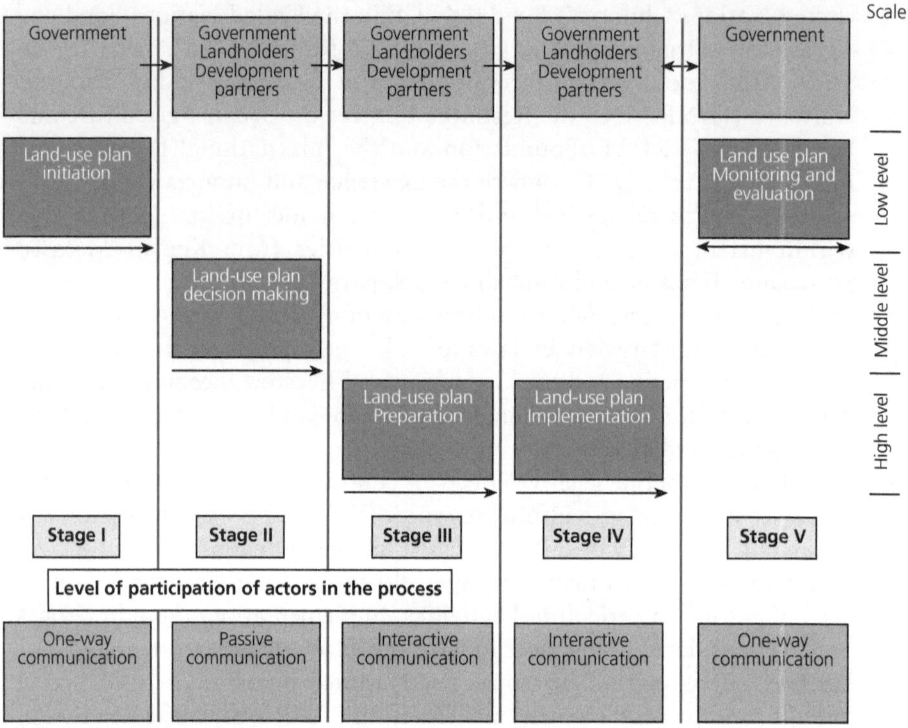

Source: Magigi (2008)

Figure 8.3 *Land-use planning processes*

ment. This lack of local community involvement in land-use planning, results in difficulties in terms of implementation and monitoring, as is discussed later.

Bottom–up land-use decision-making occurs when the local communities at the grassroots level initiate their own projects, using their own resources and those from supportive stakeholders. Consultation and interactive participation or communication are all achieved from the beginning to the end. However, long bureaucratic processes in initiation and planning approval are common, which may discourage local communities. Adopting participatory land-use planning as the favoured and entrenched planning approach helps to achieve this and therefore contributes to land-use planning effectiveness.

Proper and timely communication to different stakeholders interested in spatial land-use planning processes is important. This fosters the commitment of actors, resource mobilization and negotiations of actors during the process. Figure 8.3 shows the level of involvement and communication modes of different stakeholders. This is grouped into three parts according to a scale that includes high, middle and low. High level is when there is interactive communication and information flows among actors (Pretty, 1995). This is achieved particularly in the third and fourth stage where the preparation and implementation of land-use plans is needed. The first stage and the fifth typically involve low levels of communication. In these stages, the government appears to follow the Town and Country Planning Act of 1956. Under this act, landowners are considered passive participants rather than active stakeholders in the process. This shows a top–down approach in decision-making and implementation. Under such a model, urban and peri-urban agriculture can often be neglected. In contrast, where needs are identified and communicated by communities and solutions are initiated by them (i.e. bottom–up approaches), space is found for urban agriculture.

The second stage is typified by passive participation where landholders are not active in land-use planning processes. This particularly occurs when the government initiates the project with a top–down approach. It is also at this stage that landholders and other land developers, including private investors, engage in preparing their plan (URT, 2006, Section 16). Accurate information needs to be communicated to different stakeholders in this stage in order to facilitate its preparation and subsequent implementation as the case study settlements below demonstrate.

The case of Dar es Salaam: A look at special land-use planning processes in Goba, Ubungo Darajani and Chang'ombe 'A'

The Ministry for Lands and Human Settlements Development declared three settlements in the city of Dar es Salaam as planning areas on 13 August 1993 through Government Notice (GN) no 231. Before the declaration, these areas were zoned for different land uses according to the 1978 Dar es Salaam Master Plan – namely, industrial areas, hazardous land for future industrial use and agricultural land for future settlement development.

The zoning of the settlements through the master plan approach in 1978 in Dar es Salaam occurred despite extensive mushrooming of private housing developments that already existed since the 1960s in the area (see Table 8.4). It was also prepared without taking into account landholders' views, preferences and inputs. Likewise, the proposed plan was not implemented at that time and the consequence was non-adherence. This scenario is common in many SSA cities, including Lagos and Nairobi (Arimah, 1999; Olima and Kreibich, 2002).

Landholders in the settlements were acting as passive participants as their interests were not considered during the initiation, decision-making and subsequent preparation stages. This resulted in increased land-use conflicts during implementation where more than 75 per cent of landholders resisted the implementation of the spatial land-use plan in 2005. Urban agriculture was not included and therefore exclusion still exists in practice (see Figure 8.4).

In the case of Ubungo Darajani, the initiation was conducted in 1997 by a community-based organization (CBO) called Ubungo Darajani Settlement Development Association (UDASEDA). The organization members who were landholders (65 per cent) were motivated after the government declaration in 1990 to acquire their land. The government intention here was to implement the 1978 Dar es Salaam Master Plan, which zoned these areas as industrial,

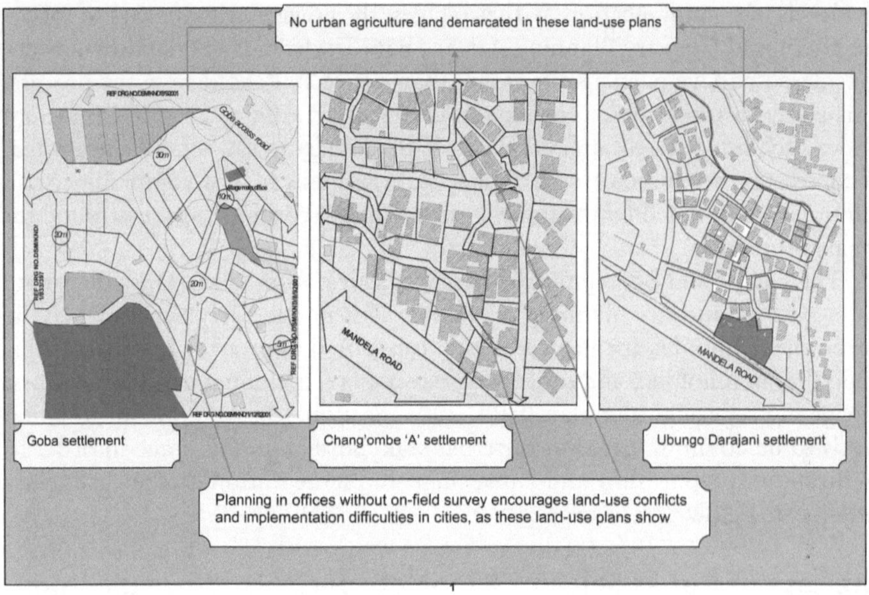

Source: Magigi (2008)

Figure 8.4 *Extract from land-use plans*

hazardous land and agricultural land, despite the presence of surveyed plots within the areas. 30 per cent of the landholders were motivated due to the fact that they had already begun surveying their plots before the new zoning initiatives, but had lapsed due to prohibitive costs. Some also complained that they had been swindled and had lost money to unregistered surveyors (5 per cent). These three factors created solidarity among the settlement because landholders faced common problems. Interviews with the sub-ward leader revealed that most of the landholders were not even aware of the existence of the 1978 Master Plan. Many therefore continued to subdivide their land for sale without regard to the 1978 plan proposal. Subdivisions affect urban agriculture implementation as land is getting smaller and smaller, which is common to other settlements studied.

The initiation in Ubungo Darajani took two years while the other cases took only six months. The latter entailed writing orders and directives to planning authorities and local leaders. The case of Ubungo Darajani involved establishing contacts with the local authority and consultations with various institutions for support. The preparation and subsequent land-use planning in Ubungo Darajani therefore included cadastral surveying, infrastructure provision, including drainage and water provision, and land registration. Community social capital mobilization in terms of funds, consultative meetings with Ardhi University (by then called University College of Land and Architecture Studies, UCLAS), and interactive communication with local and central government were the key strategies for success. Funds were mobilized within the community and used for paying consultants. For infrastructure provision, residents were able to provide stones and sand for drainage improvement, while those who had money contributed funds to buy cement and bricks.

Monitoring and evaluation are the processes of tracking and documenting the implementation (or the progress toward implementation) of land-use plan decisions. This needs to be done at least at intervals of two to five years, according to the Urban Land Use Planning Act of 2006 in Tanzania, and should be documented in the form of a tracking record or reports and upgrading. Reports must be available for public review and should describe management actions proposed or undertaken to implement land-use plan decisions, forming the basis for annual budget documents. In all three cases, there has been no such process. For example, the land-use plans for Ubungo, Goba and Chang'ombe 'A' were prepared and approved in 2004, 2001 and 1997, respectively. There has been no assessment by local authorities on the results of the plans for each area, or any reviews on sustainable impacts after five years, as stipulated by the Land-Use Planning Act of 2006.

Extent and characteristics of urban agriculture

Crop farming

Ubungo Darajani, Chang'ombe 'A' and Goba smallholder farmers grow a wide range of crops (see Figure 8.5). These include African indigenous vegetables (AIVs) and exotic species. The primary AIVs are amaranth, sweet potatoes, cowpea leaves and pumpkins. In Goba settlement, smallholder farmers are involved in all types of crop farming, including pineapple, maize, beans,

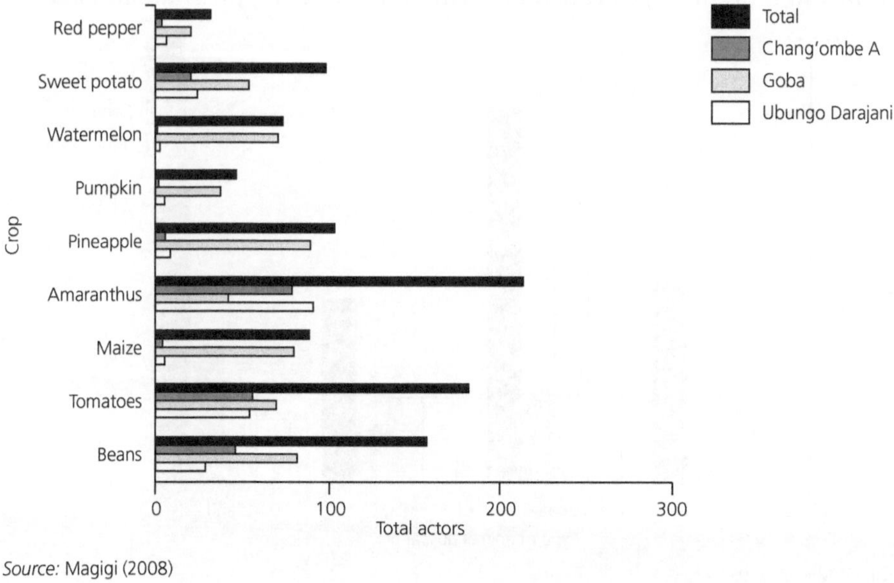

Source: Magigi (2008)

Figure 8.5 *Crops grown in the case study areas*

Table 8.2 *Location of market areas with respect to study settlements (km away)*

Market Case	*Kariakoo*	*Kisutu*	*Bugur-uni*	*Makum-busho*	*Tandale*	*Tegeta*	*Mwenge*	*Tandika*	*Temeke Sterior*
Goba	10.8	12	12	9	10	3.8	5	18	16
Chang'ombe	12	12.4	3	8	6	17	13	3	2.5
Ubungo Darajani	8	8.4	3	5.7	3.5	8	3	6	5.5

Source: Magigi (2008)

tomatoes and watermelon. Likewise, Figure 8.5 indicates that amaranth farming dominates in the urban zones compared to peri-urban settlement (i.e. Goba). Vegetables such as amaranth are perishable and require good packaging and proximity to the marketplace. Long distances to market (see Table 8.2) and the hot climate in the city affects leafy vegetables and other forms of crop farming. Maize and other crops dominate peri-urban settlements compared to urban areas. This is due to the availability of land and the presence of taps and deep water in these areas, which facilitates production. Generally, access to land and inadequate water for crop farming are primary constraints on urban agriculture.

Animal husbandry

Within Goba, Chang'ombe 'A' and Ubungo Darajani, both meat and dairy types of indoor and outdoor animal farming were observed. The large number of animals shows the importance of animal husbandry for food security for the producers themselves and for other people who purchase meat (see Figure 8.6). Milk and meat provide protein to smallholder farmers, their families and to

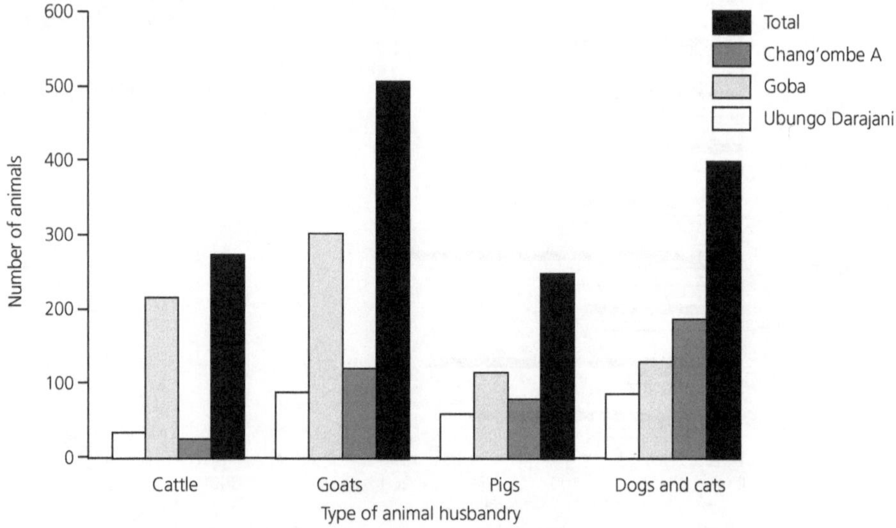

Source: Magigi (2008)

Figure 8.6 *Extent of animal farming in the case study areas*

other urban dwellers. The milk and the by-products produced by smallholder farmers in peri-urban areas of the city are mainly consumed in urban areas.

Poultry farming

Chicken farming is common in the case study areas. Free-range and indoor farming systems are practised, and both indigenous and exotic breeds are kept. This form of farming can be done on small plots. The cost involved includes the construction of a stall for keeping hens and treatment. For a production unit of 100 chickens kept to sell eggs alone, approximately US$600 in profit is gained per month, excluding other expenses.

Horticulture production

Vegetables grown in and around the case study settlements include *Amaranthus* spp. (known as African spinach), sweet potato leaves (*Ipomea batatas*) and pumpkin leaves (*Cucurbita moschata*)). Others include cowpea leaves, cauliflower, cabbage, carrot, spinach, mustard leaves, okra and tomato. In addition, a range of culinary herbs such as fenugreek and coriander are cultivated. These vegetables are perishable and must to be produced close to markets. Due to unregulated land use, the production is often done on fragile land that has been demarcated for conservation and on plots in residential premises (see Table 8.3).

Voices of different stakeholders regarding urban agriculture implementation in the city

Understanding the voices of different stakeholders helps to explore existing policy and practical problems. A total of 450 stakeholders were interviewed, including smallholder farmers (360), the private sector, donor community and government officials, policy-makers and ward councillors. The main problems indicated by the farmers were:

- distance to market areas, which increases their cost of production and lowers their profit;
- insecurity of land tenure, which was exacerbated by lack of designated areas for farming;
- lack of subsidies and credit schemes that can facilitate farming, due to lack of collateral;
- city harassment, which increases fear of eviction and restriction of farming in plot areas;
- theft of their crops;
- inadequate water facilities to enhance their farming activities;
- inadequate extension services or urban farming within the city; and
- poor understanding of policy and legal basis guiding urban land development.

Table 8.3 *Location of urban farming activities in the case study areas*

Form of urban agriculture	*Ubungo Darajani*	*Goba*	*Chang'ombe 'A'*	*Remarks*
Crop farming				
• Maize	Undeveloped plots and along the Kibangu River Valley	In plots and off plots	Undeveloped plots	Fertilizer is not used for farming; only manure to some smallholder farmers
• Pineapple	In plots	In plots and off plots	In plots	Organic fertilizer from cattle dung and decomposed waste used
• Bananas	In plots	In plots and off plots	In plots	Organic fertilizer from cattle dung and decomposed waste used
• Beans	In plots, undeveloped plots and along the valley	In plots and off plots	In plots and undeveloped plots	Organic fertilizer from cattle dung and decomposed waste used
Animal husbandry				
• Cattle	In plots	In plots	In plots	Create air pollution but the dung is beneficial for farming
• Domestic animals including dogs and cats	In plots	In plots	In plots	Can be accommodated since used for protection in homes Need medical care
• Pigs	In plots	In plots	In plots	Create noise and air pollution, and require more care in rearing Potential in peri-urban areas
Poultry keeping				
• Hens	In plots	In plots	In plots	Create noise and air pollution, and require more care in rearing in residential plots
• Pigeons	In plots	In plots	In plots	Most are domestic and are kept for aesthetic reasons
Horticulture (gardening)	In plots, undeveloped plots and along the river valley, institutional and industrial premises, road and railway strips	In plots and empty spaces, undeveloped plots and along the river valley	In plots, undeveloped plots, institutional and industrial premises, road and railway strips	Pollution of water affects the valley products Plots and undeveloped plots are preferred; tap water for irrigation
Floriculture	In plots, road and railway strips	In plots	In plots, road and railway strips	Mostly for selling within and outside the city; no information on international transportation
Seedlings/ tree planting	In plots, road and railway strips	In plots	In plots, road and railway strips	Both fruit and canopy trees are common

Source: Magigi (2008)

Urban planners, city managers and policy-makers, including councillors, admitted to the benefits of urban agriculture in improving and subsidizing the poor in the city. They were aware that the activity is not only practised by the poor, but even by rich individuals and elite groups. Some examples:

> *Dar es Salaam city has abundant land in peri-urban zones and its dormitory districts of Mkuranga, Bagamoyo and Kibaha. If land in these areas is put into farming and identified farmers to use the land, it will be easy for them to get organized, get credit and therefore influence decision-making processes in terms of improving their social well-being.* (Director of NEMC-Research and Publication)

> *We can tolerate gardening and floriculture farming in urban settings as they do not require much land for a person to get involved in. This is important in plots and off plots and not cropping of maize and other high raised crops. This can cause malaria, mice and hideouts for criminals. They also create fear, especially during harvesting, where raping is experienced in the city.* (Director of Human Settlement Development, Ministry of Lands and Human Settlements Developments)

> *We cannot allow urban farming such as animal husbandry, including pigs, cattle and goats, in a city setting. We need to accommodate only domestic animals that must be kept in a safe way [to prevent] endangering people, including dogs.* (Director of Ministry of Health)

Spatial planning barriers to urban agriculture in Dar es Salaam

Low urban agriculture rent price at market value as city investment expands

The guidelines on property taxes for urban farming activities are generalized and low compared with other urban land uses. For a person with high- and medium-density residential premises, the land rent for 2007/2008 is approximately 10,000 Tanzanian shillings (US$8); for commercial residential areas, it is 30,000 Tanzanian shillings (US$24); and for commercial and industrial premises, it is 450,000 Tanzanian shillings (US$360). This serves the government for revenue collection. A plot of equivalent size for urban agriculture requires the payment of 20 Tanzanian shillings only (US$0.02). While attractive to farmers, this serves to entrench the perception that it is not an important land use, and there is a constant motivation to replace the land with land uses that generate higher taxes per unit area for the city.

Legal and policy environment

The agriculture policy environment in Tanzania is marked by a common dichotomy between urban and rural development administration. It leaves little scope for acknowledging the specific characteristics and needs of agriculture in urban areas. Agricultural policies and programmes are primarily designed for rural areas and are therefore not always compatible with the needs of urban agriculture. Moreover, cross-sectoral policies do not recognize urban agriculture. For example, the 2005 National Strategy for Growth and Poverty Reduction does not cite urban agriculture as a strategy for poverty reduction, including employment creation and improved nutritious food security provision.

Inadequate water for meeting both city urban population and urban agriculture demand

The water demand for Dar es Salaam city is under pressure from the rapidly expanding urban and peri-urban population. The major water sources include the Ruvu chini and Juu plants, Mtoni plants and water wells in residential settlements. According to DAWASCO, the water consumption per capita is 187 litres per day in the city. The water infrastructure is plagued by breakdown of the water system, power interruptions and aged pipes. Only 18 per cent of households are connected to the tap water system, while the rest use deep and shallow wells for both domestic and agricultural uses. Thus, it is difficult to make policies and programmes for the provision of water for urban agriculture when the current systems cannot meet the requirements of the increasing urban population of approximately 3 million. Other sources of water need to be in place to supplement the city's water supply system, including use of deep wells and rainwater harvesting.

Housing density and mixed land uses

Table 8.4 shows that in all three sites housing density increased from 1970 to 1980. During this period, it was influenced by a 'villagization' and broad acre policy (1972), abolition of local governments (1972), the Uganda war (1978) and a food crisis (1980) in Tanzania. It was during this period that people began constructing their buildings without building permits from local authorities. In the years from 1981 onwards, the construction of buildings has been increasing beyond what urban planning officials can easily coordinate and control. Inadequate human resources and exclusion of community involvement in land-use planning and its implementation occurred throughout this time. Open spaces demarcated for public use were also encroached upon for other uses; as a result, housing density increased and agricultural land disappeared. Identifying available land close to particular settlements requires urgent attention.

Land-use zoning and conditions for change of use

In policy documents, including master plans for peri-urban areas, land currently under agricultural use is frequently designated as areas for future urban expansion, rather than protected for agricultural activities. Consequently, the local authorities place little emphasis on managing such land

Table 8.4 *Chronological documentation of housing density in case study areas*

Year	*Area*								
	Chang'ombe 'A'			*Ubungo Darajani*			*Goba*		
	Number of houses	*Housing density*	*Agricultural land (ha)*	*Number of houses*	*Housing density*	*Agricultural land (ha)*	*Number of houses*	*Housing density*	*Agricultural land (ha)*
1960–1965	240	1.2	165.0	12	0.4	19.0	20	0.3	56.0
1966–1970	320	1.7	162.8	78	3.0	16.2	55	0.9	54.0
1970–1975	820	4.3	138.0	165	6.3	10.4	146	2.4	51.0
1976–1980	1470	7.8	112.0	196	7.5	8.2	254	4.2	48.7
1981–1985	2045	10.9	98.0	240	9.2	6.1	340	5.6	46.2
1986–1990	3605	19.2	28.0	297	11.4	3.4	460	7.6	42.1
1991–1995	3820	20.4	19.2	310	11.9	3.1	570	9.5	41.5
1996–2000	4120	22.0	16.8	348	13.3	2.6	640	10.6	41.9
2001–2005	4196	22.3	17.3	436	16.7	2.8	702	11.7	41.8
2006–2007	4200	22.4	17.5	442	17.0	2.9	720	12.0	42.6

Note: Housing density = total housing per given unit area.
Source: Magigi (2008)

in order to meet proposed future needs. This permits the encroachment of residents and continued land subdivision, and later becomes difficult for the government to plan for peri-urban settlements. Awareness of land-use change procedures is low amongst residents when they want to use their plot for farming. This is an important aspect of urban planning when planners want to monitor and coordinate urban development in an organized fashion. Land-use planning often excludes farming in the land-use plans, and planning briefs rarely consider urban agriculture and its implementation modalities. These barriers enhance the exclusion of urban agriculture in spatial planning.

Land tenure

Within the three settlements, land tenure systems consist of a wide range of statutory, customary and non-formal categories. The statutory system refers to owning land through formal procedures where the title deeds are granted to the landholder. The customary system refers to the ownership that is inherited, and the non-formal categories include informal land transactions, such as buying from friends and the encroachment of public lands (see Figure 8.7).

Ubungo Darajani, Goba and Chang'ombe settlements were developed under traditional or customary tenure systems, which were at that time outside urban administrative boundaries. The land was later absorbed by the expanding city. During such times, various statutory tenure categories, such as formally granted rights of occupancy, were introduced to which later housing development had to conform. Alternatively, early developments have changed their tenure status as regularization programmes were introduced at Ubungo Darajani. This allowed informal status to become formal when local communities became involved in spatial land-use planning.

Tenure as a limitation to urban agriculture relates to access to resources. For example, 6 per cent of landholders in the case study had *short-term titles* of 2 to 15 years. This limits the ability of smallholder farmers to access financial

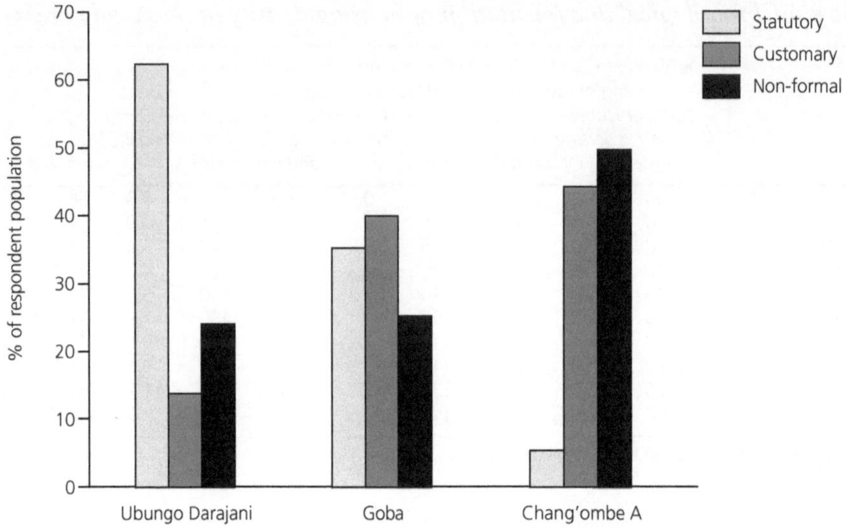

Source: Magigi (2008)

Figure 8.7 *Tenure modalities in the case study settlements*

credit that can help to improve their farming system and outputs. Therefore, there is a need to consider granting long-term titles to attract financial assistance to smallholder farmers who want to legalize their land. Providing a range of tenure options could be the most effective means of enabling the urban poor to improve their living conditions and livelihood opportunities and, therefore, to improve urban land governance.

Inadequate access to land-use planning information and language

Over three-quarters (78 per cent) of the respondents interviewed were not aware of the official requirements of land-use planning processes, as well as procedures to obtain land for urban agriculture. This is because they have little contact with government agencies. For others, a major constraint on their ability to conform to official requirements is the simple problem of understanding bureaucratic forms, procedures and even language. For example, planning regulations, standards and administrative procedures are published in English that the majority of landholders and tenants in Tanzania hardly use. There is a need for such documents and policies to be available in local languages (e.g. Swahili).

Adaptation of planning standards and building regulation

Formal land-use planning standards and regulations – for example, those used in the development of new formal settlements – would generally not be appropriate for upgrading many informal settlements. These will impose severe payment burdens on residents or the building of a particular type of housing that may be difficult for smallholder farmers to afford. Smallholder farmers

Table 8.5 *Summary of planning standards*

Site number	*Infrastructure facility*	*National planning standards*	*Community agreed standards approved by the Ministry of Lands and Human Settlement Development*		
			Ubungo Darajani	*Goba*	*Chang'ombe 'A'*
1	**Standard for residential areas**				
	High-density plots	400–800m²	12–800m²	400–800m²	20–800m²
	Medium-density plots	801–1600m²	900–1600m²	801–1600m²	801–1600m²
	Low-density plots	1600–4000m²	1601–4000m²	1600–4000m²	1600–4000m²
2	**Road right of way**				
	Access path (footpath)	3–6m	2–3m	3–6m	3–6m
	Access road (residential area)	10–20m	8–10m	10–20m	10–20m
	Local distributor (residential area)	10–20m	12m	10–20m	10–20m
	District road	20–30m	20–30m (existing)	20–30m	20–30m
	Primary road	20–30m	20–30m (existing)	20–30m	20–30m
	Trunk road	60–70m	60–70m (existing)	60–70m	60–70m

Source: Magigi (2008)

depending upon small plot farming cannot exist in a settlement upgraded according to conventional regulations. Besides, smallholder farmers are wary of situations where they cannot keep up with payments and do not have the means to begin building an approved permanent house because of their unsustainable income levels. This affects their living standards in urban areas; as a result, shifting to peri-urban land is common. The main planning standards used are summarized in Table 8.5.

Table 8.5 shows that the minimum plot standard area for Ubungo Darajani and Chang'ombe 'A' is 12 and 20 square metres, respectively. With plots this size it is difficult to consolidate urban agriculture due to minimal plot coverage ratio, which also accompanies building standards (see Table 8.6).

Tables 8.5 and 8.6 indicate that in promoting urban agriculture, the availability of space and density distribution needs to be considered (see Figure 8.8). For example, Table 8.6 shows that the minimal plot for Goba, a peri-urban settlement, ranges from 400 to 4000 square metres. Similarly, the availability

Table 8.6 *Minimum plot coverage*

Size	*Plot size*	*National standards used*	*Community agreed standards*	*Plot coverage in Ubungo Darajani*		*Plot coverage in Goba and Chang'ombe 'A'*	
				Range	*Number of plots*	*Range*	*Number of plots*
High density	400–800m²	40%	12–800m²	50–120%	234	50–120%	458
Medium density	801–1600m²	25%	900–1600m²	30–38%	35	30–38%	114
Low density	1600–4000m²	15%	1601–4000m²	10–15%	–	10–15%	67

Source: Magigi (2008)

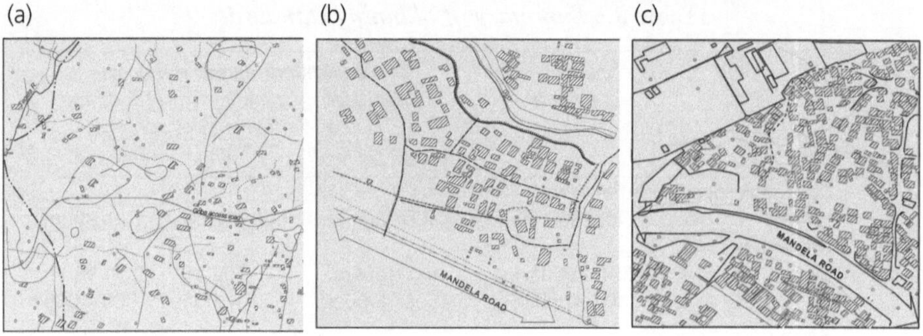

Source: Magigi (2008)

Figure 8.8 *Settlement growth processes and possibility of integrating farming: (a) Part of Goba neighbourhood showing initial stage where urban agriculture integration can be well planned in advance; (b) part of Ubungo Darajani neighbourhood showing consolidated stage where urban agriculture can be integrated; (c) part of Chang'ombe 'A' neighbourhood showing saturated stage where urban agriculture integration needs attention*

of unoccupied land within the peri-urban settlements becomes one of the criteria for consolidating the sector in urban areas. Likewise, it shows the potential of peri-urban land consolidation to accommodate urban agriculture in a city setting where mixed urban uses and density distribution are high.

Financial aspects

The annual budgets of the urban local authorities are approved by central government. Experience shows that these funds are inadequate to implement the land-use development programmes in the country. Increased cost of infrastructure provision and maintenance calls for heavy investment by the responsible ministries.

Policy and planning implications

Before an attempt is made to find the way forward, we need to raise the question: how do future urban land development policies and institutional structures of urban planning enable effective integration of urban and peri-urban agriculture within spatial land-use planning? This question becomes more significant in the context of current rapid urbanization processes, food insecurity, policy change and increased urban land market and investment.

Human settlements and building regulation policies and legislation

In Tanzania, the National Human Settlement Development Policy (NHSDP)-URT of 2000, Section 4.3.7, identifies urban agriculture as one of the urban development strategies with the potential to reduce urban poverty, therefore requiring appropriate management. However, although the policy advocates urban agriculture as an important component of sustainable development, it also recognizes that if not properly promoted and practised, it can lead to conflicts with other urban land uses, degradation and water pollution, which could threaten human health and safety. Similarly, increased urbanization and the consequent requirements for better water infrastructure are associated challenges, which demand trade-offs. Under consideration of these positive and negative consequences, the policy states that the government shall:

> *... designate special areas within planning areas whereby people will be granted legal rights to engage themselves in agricultural activities; and will continue to regulate and research on the conduct of urban agriculture and will ensure that it does not disrupt planned urban development, review existing laws to facilitate planned urban agriculture, and lastly will facilitate the construction of appropriate infrastructure to prevent land degradation, water pollution, health and safety hazards in areas where agriculture is permitted.*
>
> *... first ensure the need for administrative boundary expansion shall be substantiated scientifically in a participatory manner between the interest parties. Secondly, ensure villages engulfed in urban areas as a result of the expansion of township boundaries shall cease to be villages whether or not they are registered and their administration shall come under the respective urban local authority.*

These two government policy statements show an acceptance of urban agriculture and the need to accommodate it in an organized manner. City local authorities have even listed urban agriculture as one of the issues to be promoted in the city strategic plan since 1992. In 1993, a working group was formed to develop strategies for integrating urban agriculture within the city agenda. It designated special zones for agriculture within the city strategic plan. Whereas the master plan considered urban agriculture as a transitional land use, the strategic plan regards it as an important activity and affords it corresponding importance in development. But the practice shows that eviction of smallholder farmers remains a common occurrence. This is due to a lack of land-use plans or general planning schemes in the city that inform smallholder farmers of where they can, and cannot, cultivate. Even when a legal framework is given, there is a lack of implementation. This is common to other SSA cities, such as Kampala (Uganda) and Harare (Zimbabwe) (van Veenhuizen, 2006).

In line with adopting flexible and participatory planning approaches, the NHSDP shows that planning for urban development has been based on master plans as a tool for coordinating and controlling land development. The blueprint nature of master plans means that they are relatively inflexible and cannot be easily adapted to constantly changing social and economic circumstances. Master plans have not been as effective as intended, thus leading to haphazard growth of towns with inadequate infrastructure and services. The increasing demand for housing in cities beyond planning capacity triggers the haphazard growth of a city. In recognition of the NHSDP, the 1995 National Land Policy remarks that urban planning will aim to achieve the following:

> *First, identification of key planning issues in land and environmental management and in the provision of housing infrastructure services. Secondly, preparation of detailed land-use plans for land development. Thirdly, identification and mobilization of local and external resources for implementing urban development programmes; and, fourthly, promoting community participation in planning, integrating and coordinating the actions and resources of various sectoral implementing agencies, including those in the popular sector.*

In supporting the above, forms of participation in land-use planning processes explained in Figure 8.3 help to understand at which stage urban agriculture can be incorporated. It is at the outset, through interactive and consultative communication with the spatial land-use planning project, that such needs can be integrated. It is very difficult for city officials, including planners, to know what is happening at the local level in residential settlements. Communities need to engage with elected representatives and government officials and to report needs, concerns and problems.

2005 National Strategy for Growth and Poverty Reduction in Tanzania

Poverty reduction is a central tenet of achieving and strengthening urban land governance objectives and therefore meeting the Millennium Development Goals (MDGs). Poverty, here, is a relative term reflecting both income and non-income elements (IMF, 2002). Enabling landholders, including smallholder farmers, to access land and information, and to participate in policy and decision-making processes, is one contribution to empowerment and a step towards getting out of poverty.

Options to reduce barriers to integrating urban agriculture within spatial planning

In order to reduce existing barriers to the integration of urban and peri-urban agriculture within planning processes and outcomes, the following is needed:

- Adopt flexible land-use planning approaches, standards and values.
- Include smallholder farmers and their values in the design processes, outputs and implementation. To be successful in this respect, the following urban planning principles should be implemented:

  - pursue holistic-oriented values through the integration of community needs, and social, environmental and economic values in land-use planning;
  - develop and maintain cooperative working relationships during land-use planning initiation, decision-making and subsequent implementation;
  - rationally formulate land-use planning guidelines before the process starts, which helps to create awareness among actors, as well as during the review of outputs;
  - involve smallholder farmers to determine their needs through participatory planning methods and training; and
  - ensure access to resource and tenure security for the urban poor.
- Create community involvement and partnership among different stakeholders. This sounds easy but is the most challenging element because it requires consultation, dialogue and community organization, as well as the creation of new institutions, such as farmers' organizations or allotment associations (Drescher, 2001).
- Strengthen institutional arrangements.
- Harmonize policy and legislation arrangements.
- Enhance the city's economy and budget allocation.
- Develop and maintain information databases that can be accessed by different stakeholders and planning institutions.
- Raise awareness and exchange information among developers.
- Generate political commitment and support.
- Include urban agriculture in town, municipal, city and national land development programmes.
- Use technology in wastewater treatment and conduct environmental impact assessments for industries.

Conclusions

This chapter has investigated how land-use planning and urban agriculture can be integrated in order to improve urban land governance in African cities, using Dar es Salaam (Tanzania) as a case example.

The general situation in SSA shows that urban agriculture is often regarded as a misplaced rural enterprise that should not be located in the city; but awareness of its contribution to food security, employment, poverty reduction and sustainable city development is increasing. The efforts of local authorities can be observed in a number of African cities, including Dar es Salaam.

Rapid urbanization in SSA results in unguided land development, poor enforcement of policies and regulations, the inability to replace outdated policies inherited from the colonial period and inadequate community involvement in spatial land-use planning processes. This uncontrolled situation increases urban poverty, food insecurity and unemployment in African cities. The implications generated by these factors suggest that weak urban land governance is not the only cause; rather, the weakness of planning institutions to realize and to adapt to the new challenges has also had serious impacts.

It is likely that, for the foreseeable future, urban agriculture will grow at the same pace or faster than population growth in cities. It will thus remain an inevitable reality, attracting more participants as they seek to cope with the rapidly changing social, economic and environmental conditions of expanding cities. Recognition of urban agriculture-based livelihoods is required in all land-use planning processes and structures, including in decision-making, preparation, implementation and monitoring. This can be achieved through (but not limited to) settlement upgrading, institutional collaborations, adopting participatory urban planning approaches, decentralizing roles to the local level and creating platforms for dialogue. Moreover, strengthening smallholder organizations through institutionalization and giving them a voice in the political dialogue is crucial. Other key actions are to implement effective treatment of wastewater for farming, granting long-term tenure security, improving accessibility and adopting flexible planning standards.

Urban planners and policy-makers have little choice but to ensconce themselves and consolidate the emerging form of urban agriculture as a land-use practice for sustainable urban land development. If urban agriculture is not guided and properly institutionalized, then the negative effects, in the long run, will undermine the livelihoods of the urban poor and damage environmental quality. Thus, there is an urgent need to guide the processes and to strengthen institutional structures and linkages in land-use planning, as well as to adopt participatory approaches that involve affected stakeholders in improving urban land governance.

References

Arimah, B. C. (1999) 'User modifications in public housing estates: Some findings from Nigerian scenes', in Awatona, A. (ed) *Housing Provision and Bottom-Up Approaches: Family Case Studies from Africa, Asia and South America*, Ashgate, Aldershot, pp68–79

Armstrong, A. (1987) 'Master plans for Dar es Salaam – Tanzania', *Habitat International,* vol 11, pp133–145

Atukunda, G. (1998) 'An analysis of the impacts of IDRC funded research projects on urban agriculture in Uganda', Makerere Institute of Social Research, University of Kampala, Kampala, Paper presented at IDRC Cities Feeding People Workshop on Lessons Learned from Urban Agriculture Project in African Cities, Nairobi, Kenya

Bakker, M. D., Gundel, S. and Sabel-Koschella, U. (eds) (2000) *Growing Cities, Growing Food*, German Foundation for International Development (DSE), Feldafing, pp477–500

Dowall, D. E. and Clark, G. (1996) *A Framework for Reforming Urban Land Policies in Developing Countries*, Urban Management Programme Paper 7, World Bank, Washington, DC

Drescher, A. W. (2001) 'The German allotment gardens: A model for poverty alleviation and food security in Southern African cities?', in *Proceedings of the Sub-Regional Expert Meeting on Urban Horticulture, Stellenbosch, South Africa*, 15–19

January 2001, FAO/University of Stellenbosch, www.cityfarmer.org/germanAllot.html, pp159–167

Drescher, A. W. and Iaquinta, D. L. (1999) *Urban and Periurban Agriculture: A New Challenge for the UN Food and Agriculture Organisation (FAO)*, FAO, Rome

Faludi, A. (1970) 'The planning environment and the meaning of planning', *Regional Studies*, vol 1, pp1–9

Faludi, A. (1984) *Planning Theory*, Upplagan, London

Friedmann, J. (1987) *Planning in the Public Domain: From Knowledge to Action*, Princeton University Press, Princeton, NJ

Hall, P. (1988) *Cities of Tomorrow: An Intellectual History of Urban Planning and Design in the Twentieth Century*, Wiley and Sons, London

Halla, F. (2002) 'A century of urban development planning for Dar es Salaam City in Tanzania', *Journal of Building and Land Development*, vol 9, pp28–46

Healey, P. (1997) *Collaborative Planning: Shaping Places in Fragmented Societies*, McMillan, London

Iaquinta, D. L. and Drescher, A. W. (2000) 'Defining peri-urban: Towards guidelines for understanding rural–urban linkages and their connection to institutional contexts', *Land Reform 2000/2*, FAO, Rome

IMF (International Monetary Fund) (2002) *Comprehensive Review of the Poverty Reduction Strategy Paper (PRSP) Approach*, World Bank, New York, NY

Kombe, W. and Kreibich, V. (2000) 'Reconciling informal and formal land management: An agenda for improving tenure security and urban governance in poor countries', *Habitat International*, vol 24, pp231–240

Kyessi, A. (1997) 'City expansion and urban agriculture in Dar es Salaam: Lessons for planning', *Journal of Building and Land Development*, vol 4, pp28–37

Lynch, K. (1990) 'City and regional planning', in Banerjee, T. and Southworth, M. (eds) *City Sense and City Design: The Writings and Projects of Kelvin Lynch*, MIT Press, Cambridge, MA, pp535–562

Lupala, A. (2002) *Peri-Urban Land Management in the Rapidly Growing Cities: The Case of Dar es Salaam, SPRING Research Series* no 32, Dortmund, Germany

Madaleno, I. M. (2000) 'Urban agriculture in Brazil: A tale of two cities', *Trialog*, vol 65, pp24–27

Magigi, W. (2008) *Improving Urban Land Governance with Emphasis on Integrating Agriculture Based Livelihoods in Spatial Land Use Planning Practices in Tanzania*, PhD thesis, Freiburg University, Germany

Magigi, W. and Majani, B. B. K. (2006) 'Community involvement in land regularisation for informal settlements in Tanzania: A strategy for enhancing security of tenure in residential neighbourhoods', *Habitat International*, vol 30, pp1066–1081

Mougeot, L. J. A. (ed) (2005) *The Social, Political and Environmental Dimensions of Urban Agriculture*, Earthscan, London

Olima, W. H. A. and Kreibich, V. (eds) (2002) *Urban Land Management in Africa, Spring Research Series* no 40, Dortmund, Germany

Pretty, J. N. (1995) 'Participatory learning for sustainable agriculture', *World Development*, vol 23, pp1247–1263

Rakodi, C. and Lloyd-Jones, T. (eds) (2002) *Urban Livelihoods*, Earthscan, London

Santandreu, A. and Castro, C. (2006) *Management Models and Innovative Alliances for Influencing Public Policy Developed by Social Organisations of Urban and Peri-Urban Agricultural Producers of Latin America and Europe: A Comparative Analysis*, IDRC, Ottawa, Canada, pp18–20

Sawio, C. (1998) *Managing Urban Agriculture in Dar es Salaam Cities, Cities Feeding People Report 20*, IDRC, Ottawa, Canada

UNDP (United Nations Development Programme) (1998) *State of the World Population Report*, UNDP, New York, NY

UN-Habitat (2006) *State of the World's Cities 2006/2007: The Millennium Development Goals and Urban Sustainability*, UN-Habitat/Earthscan, London

URT (United Republic of Tanzania) (1979) *Dar es Salaam Master Plan*, Government Printer, Dar es Salaam

URT (2000) *National Human Settlement Development Policy (NHSDP)*, Government Printer, Dar es Salaam

URT (2003) *Population and Human Population Census Report*, Government Printer, Dar es Salaam, Tanzania

URT (2006) *Urban Planning Act*, Government Printer, Dar es Salaam

van Veenhuizen, R. (ed) (2006) *Cities Farming for the Future: Urban Agriculture for Green and Productive Cities*, RUAF, IDRC and IIRR, www.ruaf.org/node/961, accessed July 2006

World Bank (2006) *World Development Indicators 2006*, World Bank, Washington, DC

9

African Indigenous Vegetables in Urban Agriculture: Recurring Themes and Policy Lessons for the Future

Charlie M. Shackleton, Margaret W. Pasquini and Axel W. Drescher

Introduction

The preceding chapters have together presented a thorough coverage of the role of urban agriculture in livelihoods and local economies, and the benefits of integrating African indigenous vegetables (AIVs) within urban agriculture. Each chapter has dealt with a specific dimension; but when read as a whole and integrated, they make a compelling case that would be hard for any policy- or decision-maker to ignore. Integration across the different fields of nutrition, economics, urban planning, biodiversity and agriculture, supported by comprehensive case study material, allow us to gain a thorough understanding and insight regarding the production, consumption and trade of AIVs in urban spaces and places in sub-Saharan Africa (SSA). But what use is that insight if it does not influence actions and livelihoods on the ground?

Sub-Saharan Africa has the highest poverty rates in the world. It is also experiencing the highest rates of migration from rural to urban areas (UN-Habitat, 2006). These two dynamics alone demand innovative responses and programmes. These two have merged as the locus of poverty has slowly shifted from rural to urban areas, such that more than 56 per cent of the world's absolute or chronic poor are concentrated in cities and urban areas (WRI,

1996, cited in Drescher and Mackel, 2000). As much as 60 per cent of the income of the urban poor is spent on the purchase of food (Maxwell et al, 2000), and the issue of food supply, both its quantity and quality, is increasingly a central issue in poverty reduction and alleviation debates.

Since late 2006, the world has seen a sharp rise in the price of several internationally traded food commodities and staples (Mitchell, 2008), and there are growing concerns about global food and nutrition security. Already the livelihoods of millions are being affected. It is generally agreed that the current food price crisis has been triggered by a complex combination of factors (refer to Mitchell, 2008; UN High-Level Task Force, 2008; World Bank, 2008). The relative weighting of these different factors is the subject of much controversy, which has implications for the development of longer-term strategies to prevent similar crises from happening in the future. However, there is broad agreement that one course of action should be to boost smallholder farmer production. As the UN High-Level Task Force (2008) points out, the impacts will be most severe for households who spend over 60 per cent of their income on food, and will be more pronounced for those located in urban areas.

These recent escalations in food prices internationally have added urgency to poverty reduction targets and commitments as food security needs to be a central component of any poverty reduction strategies. Reducing poverty has been made a priority for governments at an international, continental and national level, as projected through the Millennium Development Goals (MDGs). Responses to ensuring that urbanization results in livelihood improvements rather than a decline have been highlighted by the United Nations Sustainable Cities programme. Synthesis of the work in the preceding chapters demonstrates that AIVs and urban agriculture can make a major contribution to meeting all of the MDGs (other than number 2, which relates to universal access to primary education) (see Table 9.1) and to improving the sustainability of cities and towns. The improvement in health and the reduction in the hunger and poverty goals of the MDGs are strongly advanced through activities and policies promoting urban agriculture and the use of AIVs. The environmental and equity dimensions are also covered, albeit to a lesser extent.

Various declarations advocating the development of sustainable cities all include commentary about the need to promote agriculture in urban spaces and places. This is largely for the numerous psychological and environmental benefits of green spaces within city limits (e.g. Fuller et al, 2008), contributions to food security and poverty reduction (Bryld, 2003; van Veenhuizen, 2006), as well as reducing the travel distance of fresh produce to market. However, we also need to bear in mind that cities themselves may offer, in some respects, more conducive environments for agriculture than their neighbouring rural areas. First, cities are typically 2° to 3°C warmer than surrounding rural areas (McGranahan et al, 2005), which serves to extend the growing season and growth rates of crops in those cities with marked temperature-induced seasonal growth periods. Second, there is greater access to inputs and at lower prices. Thus, urban farmers do not have to travel so far to access inputs such as seeds,

Table 9.1 *Contribution of AIVs to the Millennium Development Goals (shaded boxes represent direct contributions; darker shading represents greater contributions than lighter shading)*

Millennium Development Goals (MDGs)	*Benefits from African indigenous vegetables (AIVs)*		
	Home consumption of AIVs	*Local trade in AIVs*	*Domestication and cultivar selection*
1 Eradicate extreme hunger and poverty	■	■	■
2 Achieve universal primary education	□	□	□
3 Promote gender equality and empower women	□	▒	□
4 Reduce child mortality	■	■	□
5 Improve maternal health	■	■	□
6 Combat HIV/AIDS and other diseases	■	■	□
7 Ensure environmental sustainability	□	▒	□
8 Develop a global partnership for development	□	▒	▒

fertilizers and tools. Third, access to labour is usually much easier in cities than in rural areas (although usually more costly), where labour is frequently a key constraint to scaling up household agricultural production (Bah et al, 2003). Fourth, the greater spatial discontinuities in cities (Collins et al, 2000) serve to isolate areas of urban agriculture from one another, which may well minimize disease and pest transmission. Additionally, pest populations are substantially lower by virtue of the significant transformation of much of the urban landscape. Fifth, proximity to water is usually better, either from piped sources, household wastewater or runoff from hard surfaces (roads and paved areas). The importance of this better access is magnified in situations of drought, which may result in total crop failure in rural areas, but only crop reduction in urban ones. Agriculture does, of course, have high water demands. Many African cities are already struggling to meet domestic water use needs, and the predictions are that by 2025, 25 African countries will be in a situation of water scarcity or stress (UNEP, 1999). Activities to promote agriculture in urban and peri-urban areas should be underpinned by a careful stock-taking of available water supplies, population growth rates and a focus on water-saving technologies. Sixth, large-scale adverse phenomena, such as drought, floods and fires, are less frequent and less intense in urban areas (Collins et al, 2000). Last, the reduced availability of space for agriculture results in more intensive farming systems and higher productivity per unit area. All of these suggest that agriculture in urban and peri-urban areas should not be just an afterthought or an add-on in the quest for sustainable cities, but more of an argument that agriculture belongs in cities, and should be an integral component of urban planning.

Considering recurring themes and policy lessons

The preceding chapters have provided a comprehensive coverage of the myriad aspects of urban agriculture in sub-Saharan Africa, focusing on the previously neglected role and contribution of African indigenous vegetables. This review, synthesis and presentation of new information and data on AIVs within urban agriculture, all within a single text, is unique and represents a substantial contribution to knowledge. The integrated coverage will be useful to scholars in Africa and elsewhere dealing with similar issues. However, simply presenting new and comprehensive information and identifying links between different disciplines and ideas is insufficient. For science to have an impact, to change conditions, practices and livelihoods for the better, key lessons need to be summarized and communicated in a manner available to different audiences. The policy dimensions are particularly pertinent in this respect.

It would be an impossible task to develop a concise suite of policy lessons regarding AIVs in urban agriculture for the whole of SSA since the contexts and conditions are so widely different across the numerous SSA countries. With 39 countries in mainland SSA, it is hardly surprising to find large variation in key contextual attributes such as employment levels, gross domestic product (GDP), urbanization rates, biodiversity status, dietary preferences and bioclimatic zones. Varying combinations of these, and many other attributes too, present different contexts and, hence, opportunities and constraints to urban agriculture and the production, marketing and consumption of AIVs. The previous chapters have demonstrated that there is significant variation in urban agriculture and AIVs between the geographic regions of West, East and Southern Africa with respect to many aspects, such as the species used, complexity of market chains, profile of the different actors, contributions to livelihoods, agricultural practices adopted, and so on. While this diversity makes it an exciting area for research, it hinders identification of common themes and continent-wide policy lessons. Indeed, this multiplicity in itself demands further research attention as a foundation for appropriate policy options. Nonetheless, we do need to seek whatever generalities we can and consider appropriate responses to enhance the positive dimensions and limit or remove constraints.

Currently, relatively few SSA countries have policies specifically orientated at use, cultivation or marketing of AIVs. They are usually viewed as simply another crop or another product, and so are, by default, covered by general agricultural policies. The numerous advantages of AIVs relative to exotic vegetables are insufficiently communicated to, and appreciated by, policy- and decision-makers. This needs to change. In some instances, the advantages of AIVs, or specific species, are well substantiated by research – for example, the nutritional superiority of AIVs (see Chapter 4). In other instances, much more research is required to quantify the extent of a particular implied advantage; for example, the notion that AIVs are indigenous to the continent has resulted in oft-repeated claims that they can therefore better tolerate native pests and diseases than exotic vegetables (see Chapter 5). However, there is relatively little research on this, and it is probable that the reported relatively low pest and disease incidence can be linked to climate and typical AIV cropping practices (intercropping with a high diversity of species in a relatively small space; different crops grown in quick succession in the growing season). The evidence presented in Chapter 6 suggests that when AIVs are cultivated in greater densities or monocultures, in hot and humid climates, as happens in Benin, the dynamics of pest populations and outbreaks are vastly altered. Indeed, in a recent study in Benin, Adango et al (2006) reported heavy infestation by mites on *Solanum macrocarpon* and *Amaranthus cruentus* (the study did, however, indicate that *A. cruentus* was relatively less affected by *Tranychus ludeni* and suggested that this species could be used within an integrated pest management strategy).

In contrast to the weak or absent policies around AIVs, the policy dimensions of urban agriculture have received a lot more attention (see Chapters 1, 2 and 8). Implementation of these policies is, however, extremely variable from one country to the next, which is compounded by the different potential roles of urban agriculture as perceived by policy-makers or the actual urban households. For example, the policy interventions required to support commercial agricultural production in urban areas are different from those required to support food security of the poorest, or those seeking to promote food self-sufficiency and improve dietary diversity (see Chapter 1; Maxwell, 1995; Ellis and Sumberg, 1998). Identifying and removing constraints to policy implementation is as important as having a policy in the first place. This would be a fruitful area for future discussion and research.

Given the widespread absence of policies on AIVs, it is not surprising, therefore, that there are no examples of successful policies that integrate AIVs and urban agriculture. In this chapter we seek to identify some of the main themes emanating from the previous chapters and consider some potential policy options. These options are not prescriptive because, as mentioned above, the context within each country is very different. These are derived primarily from two sources. The first is from readily available literature, including the preceding chapters of this text. The second is the outcomes of a policy dialogue workshop held towards the end of the IndigenoVeg project.[1] At that workshop, the various research collaborators in the IndigenoVeg network developed a set on interim policy lessons. These were subsequently presented to an audience of approximately 30 policy- and decision-makers from seven different African countries. The audience included city mayors, government ministers, government bureaucrats, and heads of national or regional research agencies. This interaction and discussion of the draft policy options led to a robust set of issues. Moreover, many of the policy- and decision-makers undertook to table those policy recommendations relevant to their home situation back in their home countries in order to effect both mainstreaming of AIVs in urban agriculture and positive change on the ground.

Policy formulation must:

- have a clear objective or end result in mind;
- focus on achievable goals or steps; and
- be founded on what is known about the subject at hand.

The last is particularly crucial because policies formulated on conjecture or incorrect information and understanding will not be able to promote attainment of the desired objectives. The preceding chapters have reviewed existing knowledge pertaining to urban agriculture and the use of AIVs in SSA, but have also added to that knowledge base, largely, but not solely, through the results of the IndigenoVeg project. The key knowledge aspects bear repeating here as the context for distillation of subsequent summary lessons and recommendations. We do so under three headings – namely:

1 what we know;
2 what we suspect or think we know, but from an inadequate number of sites or studies; and
3 what we do not know (but the answers to which have a significant bearing on policy and intervention).

What we know

There is much knowledge and many case studies regarding urban agriculture in the cities of Africa and other continents (e.g. Mougeot, 2005; van Veenhuizen, 2006). However, relatively little of it has explored the significance and contribution of indigenous species, and the advantages that they may offer to local livelihoods and profitability in urban and peri-urban areas. It is in this dimension that the IndigenoVeg project and now this book have made a new contribution. The key dimensions of this knowledge, as substantiated in the preceding chapters, are as follows:

- AIVs and urban agriculture are an integral component of livelihoods throughout the continent.
- Hundreds of different species and varieties of AIVs are known and used.
- There is considerable diversity in the numbers of cultivated AIV species and varieties (which surpasses that of exotic vegetables), and an even greater diversity of wild-collected species.
- Select species collected from the wild can be locally very significant in terms of volumes consumed and traded, which raises conservation concerns.
- There is extensive trade in cultivated AIVs worth billions of US dollars per year, benefiting millions of producers and vendors.
- Women are currently playing strong roles in the production and/or marketing of AIVs in urban areas.
- Urban agriculture and the production of AIVs is not solely an activity of the urban poor.
- There are several popular AIVs with better nutrient and vitamin levels than exotic vegetables.
- There is considerable potential to increase the contribution of AIVs to local markets and livelihoods.
- Urban agriculture plays a significant contribution in supplying fresh and nutritious products to urban households and markets at low cost because

of proximity to the consumer.

- Current policies and planning processes in most countries and cities are insufficiently supportive of and proactive about urban agriculture and AIVs.

What we think we know

There are many agencies in SSA investigating aspects of AIVs or urban agriculture. It is our hope that the outputs of the IndigenoVeg project, including this book, will soon stimulate many of these agencies to look at AIVs and urban agriculture simultaneously and in a more networked fashion across agencies. Nonetheless, much excellent work is being done that could be used to inform policy if only we were sure it could be generalized. In other words, the results from one or two studies on a single species, or from a single region, or from a particular context need further supporting evidence for other species, or from other regions and contexts. This includes the following issues:

- The water-use efficiency or drought tolerance of AIVs is better than exotic vegetables.
- Yields of AIVs per unit area of land are equivalent to or are better than exotic vegetables.
- AIVs are more suitable crops than exotic vegetables for resource-poor farmers because reasonable yields can be obtained with lower inputs and because of their adaptability to local conditions.
- In the genetic and species diversity of AIVs, there is significant untapped potential in terms of climate change adaptation.
- The biggest constraint to urban farmers cultivating more AIVs is lack of seed.
- Farmers are willing to be involved in the domestication of wild AIV species if there is appropriate research and development support.
- Although AIVs have very good nutrient and vitamin levels in raw form, certain traditional preparation and cooking practices may result in the removal of the majority of these nutrients.
- Certain AIVs require lengthy preparation and cooking practices, or may be traditionally eaten with foodstuffs that are harder to prepare. This factor may explain why some, but not all, AIVs have declined in popularity amongst urban consumers with busy lifestyles.
- A high proportion of the AIVs found on urban markets are sourced from urban and peri-urban production areas; however, there can be significant flows from more rural locations for specific cultivated and wild AIVs.
- Consumers shopping in formal markets and supermarkets will buy AIVs.
- The cultural values of AIVs could be used to stimulate demand in the marketplace in West and Central Africa, and to a lesser extent in East Africa; but they would probably have limited effects in Southern Africa.
- In settings where women's access to land is traditionally very limited, land resources for cultivation are very scarce, and the production of AIVs is a highly profitable activity, men dominate the activity.

What we do not know

There are many claims in the literature about certain advantages or potential constraints of AIVs and of urban agriculture. While some can be substantiated from several case studies or trials, others currently remain more in the realm of conjecture and urgently require investigation. Amongst these we include the following:

- The pest tolerance of AIVs is better than exotic vegetables.
- Use of urban wastewater in urban agriculture is widespread and consequently poses hazards to the health of consumers.
- There is potential to add value to AIVs in many ways that could increase profitability.
- Niche markets exist for AIVs, such as organic production, which can be exploited to further increase profitability.

Areas for policy intervention

Having considered the state of knowledge around AIVs in urban agriculture in SSA, it then becomes possible to identify potential policy lessons. As stated previously, this is contingent upon the policy goals being identified. In this case, we argue that the preceding chapters have a made a very strong case for the advantages of both urban agriculture and AIVs individually and the merits of integrating the two. Consequently, the IndigenoVeg policy dialogue workshop considered policy requirements to prompt the substantiated merits of AIVs and urban agriculture while minimizing or removing constraints. The following six policy areas were identified.

Urban planning needs to accommodate urban agriculture

As described in Chapters 1 and 8, urban planning and regulation have only rarely given explicit recognition to urban agriculture. In many SSA cities, urban agriculture on public land is deemed illegal, although it is at times tolerated; but it is not explicitly planned for and accommodated (Ellis and Sumberg, 1998). This relates not only to the physical space for cultivation of crops, but also to access to the necessary infrastructure (such as water) and zonation so that it is located close to farmers' homes and markets. However, growing recognition of the consequences of increasing urbanization, the need for green and productive spaces in cities, coupled with concerns about the large distances that food from rural areas has to travel to reach urban consumers (which reduces its freshness, adds to final costs and has a significant carbon footprint) means that urban agriculture is to be promoted, both in public and private space. This requires policies and regulations that ensure that sufficient land is available (a minimum target should be stipulated), that tenure is secure, and that city officials support and foster urban agriculture in the same manner that they do other endeavours. The most pressing specific policy actions include:

- Designate areas (zoning) for agricultural production – high-quality agricultural land should not be zoned for infrastructure, housing or commercial businesses; residential allotments should be large enough to allow an area for food production.
- Promote high-value crops such as fruits and vegetables in urban areas, but facilitate mixed farming systems in peri-urban areas.
- Pass by-laws that legalize and promote urban agriculture and the conditions under, and place in which, it will be permitted.
- Introduce integrated land-use planning and practices.
- Develop sustainable water supply programmes for urban agriculture (harvesting rainwater, recycled wastewater, etc.).
- Secure user rights for producers.
- Facilitate stakeholders to organize into institutional governance structures for farmers, processors and traders, who can then interact with city officials.
- Regulate the interface between agriculture and waste management (use organic waste for compost and manage waste disposal to protect agricultural resources).

Urban administrations need to facilitate the marketing of produce

Chapter 7 has shown that markets for urban produce and, specifically, AIVs are extremely diverse. This is most likely an adaptation by role players to local conditions, regulations and consumer demands. In many areas, AIVs are marketed directly from farmer to consumer, mainly in close proximity to where the AIVs are grown, but at times further away. But in some cities there is a network of intermediaries, each fulfilling a specific function and each selling some of the produce to consumers, but also further up the market chain. In some cities there are designated physical spaces where marketing occurs; in others vendors move around in search of customers or vary their marketplace from day to day. Key issues for AIV farmers relate to insufficient designated market space, insufficient information regarding the volumes and prices of particular produce in the market at any time, and intolerance by city officials. On the other hand, officials are concerned about waste produce not being correctly dumped or removed, potentially unhygienic conditions for the selling of foodstuff, ensuring that access and thoroughfares are not congested, and ensuring that traders have the required permits to operate. Policy interventions to facilitate better market access, conditions and regulation include:

- Formally recognize AIV businesses, from farmers to traders.
- Support and promote urban and peri-urban farmer organizations, as well as trader organizations.
- Ensure that permit conditions are simple, well communicated and accessible, and that some or all of the funds collected via permits are reinvested in promoting the trade or improving physical market space.

- Designate market areas and provide appropriate infrastructure, which includes areas for dumping waste produce, and water for washing fresh produce and work surfaces.
- Include AIV produce in municipal awareness campaigns and support programmes on nutrition, health and local economic development.

Improved input systems need to be implemented to develop urban agriculture and AIVs to their full potential

It is clear that urban agriculture and AIV consumption and trade is extensive (see Chapter 7). However, in many areas this has been achieved in the face of, at best, neglect and intolerance, or, at worst, policies and regulations against such activities (see Chapters 1 and 2). Consequently, it seems reasonable to expect that the much-discussed benefits of urban agriculture and AIVs could be multiplied many times over if city, state and national policies provided a more supportive framework. That is at the policy level. But at a more tangible level, resources (human and financial) need to be allocated to scale up and enhance the input systems. Three areas in particular should be emphasized:

- Provide a steady supply of quality AIV seed to enhance profitability, sustainability and growth.
- Increase extension support about urban agriculture and AIVs to put them on an equal footing to other agricultural produce (i.e. remove the current bias to exotic species and export crops).
- Allocate funding in national agricultural research institutions for research on AIVs and development of information packages to farmers and consumers.

Gender programmes need to promote urban agriculture

In all of the cities examined by the IndigenoVeg programme (see Chapters 6 and 7), women played a significant role in either or both the production and marketing of AIVs and urban agriculture. Moreover, women are invariably responsible for preparation of the food for the household and thus are a primary target for the promotion of AIVs. Recognizing these two aspects means that policy interventions to promote urban agriculture amongst women can have beneficial outcomes both for women's employment and income generation, but simultaneously also for maternal and child health. This is especially pertinent if the cultivated plot is close to the homestead so that production activities can be easily accommodated within the other roles that women typically fulfil, such as child-minding, washing, cooking and so on (Bryld, 2003). This is highlighted as one of the key attributes of home gardens (see Chapter 5). Specific interventions include:

- Support and promote women's groups and organizations.
- Initiate maternal health programmes to promote AIVs and home gardening.

- Gender programmes should include urban agriculture in the portfolio of activities that they promote, investigating and understanding women's constraints in accessing land, and exploring how women could be included in land allocation programmes for urban agriculture and empowered to retain control over their plots in the longer term. This is likely to be important in those settings where AIVs have an established or growing commercial value.
- The higher nutritional benefits of AIVs relative to exotic vegetables need to be promoted, especially to women, who have the primary role of meal preparation. Good preparation and cooking techniques should be promoted.

AIVs need to be integrated within heritage and conservation programmes

The world's food systems rely on only a small proportion of known edible plants, which undermines the resilience of food systems, increases the costs of breeding, production and transport, and reduces cultural and dietary diversity. SSA has hundreds, if not thousands, of AIVs (see Chapter 3). Currently, little is known about the nutritional, production and growth characteristics of most of these species; some may be extremely nutritious, others may be drought tolerant, while yet others may be able to resist certain diseases or tolerate poor soil conditions. It is therefore crucial that efforts are made to document the local knowledge of useful species and to systematically examine the properties of these species. Specific actions required include:

- compilation of detailed national inventories of edible plants;
- systematic screening of such species;
- promotion of useful species in national and local heritage programmes and events.

Research policy shift towards diversification and 'traditional' systems

The research and development carried out by the National Agricultural Research and Extension Systems (NARES) in Africa has tended to focus on a restricted number of major crops, with particular support to export crops, crops with a locally high commercial value and important staples. Although most of the NARES have specific vegetable research programmes, they tend to be smaller and dispose of less funding. Even within these units research has undoubtedly been biased towards breeding improved varieties of a few select vegetables and improving agronomic practices in pure stand conditions. Although in recent years there has been increasing recognition of the importance of a number of AIVs, which is translating into some interesting projects, there is still a need for a more explicit policy change in terms of the research and development mandates of these institutions to:

- Fund research programmes on a much more diverse range of AIVs, including efforts to domesticate important wild species (following national inventory and screening of useful plant species, as detailed in the previous subsection).
- Develop agronomic practices packages for 'traditional' smallholder farming systems, incorporating AIVs (typically low input, with widespread intercropping).
- Develop specific programmes for urban and peri-urban areas that diversify the available range of AIVs and consider the entire value chain. It will not be sufficient to introduce new AIVs into production areas and provide farmers with technical support for their production; it will also be necessary to stimulate demand and support the stakeholders involved in the processing and marketing aspects.

Starting the policy change process

Triggering the process of change in these six policy areas will require active engagement by the research and development community of a diverse range of stakeholders. These will comprise the people who will be directly and indirectly affected by any changes in policy, as well as the people who are in a position to make and implement the changes.

There are several encouraging examples of how development and research organizations have catalysed the process of policy change. For example, the Resource Centres on Urban Agriculture and Food Security (RUAF) Foundation's Cities Farming for the Future Programme (see www.ruaf.org) has been facilitating participatory and multi-stakeholder policy formulation and action planning on urban agriculture in cities across Africa, Asia and Latin America. The programme has had impressive impacts at different levels of policy, developments plans and legal frameworks – for example, seeing the inclusion of urban agriculture in municipal planning documents in Bulawayo (Zimbabwe), Accra (Ghana) and Bobo Dioulasso (Burkina Faso), and supporting the development or revision of urban agriculture by-laws in Bulawayo and Accra.

A key message that has been emerging from this programme, as well as from events such as the IndigenoVeg policy dialogue, is that the development of forums where there can be a two-way conversation about what knowledge policy-makers are missing and what information scientists can provide to fill these gaps is essential to stimulate change. These forums foster the exchange of knowledge; but they are also important because they facilitate a discussion and compromise over competing priorities. This point was made very clearly during the IndigenoVeg policy dialogue when one policy-maker explained: 'I am here to support the research about the urban space. I am here to also protect the environment. I am supporting the research and we can together find way on how we can accommodate each' (IndigenoVeg, 2008, p56).

The dialogue between researchers and policy-makers should not be limited to a one-off event, however. One obvious reason why awareness-raising events need to be followed up with further dialogue is staff turnover. The process of policy change can be slow, and making one or two people aware of an issue does not guarantee that it will enter the institutional memory and be carried forward even after the individual has moved on. But a more fundamental reason is the fact that gaining an in-depth understanding of differing priorities to develop satisfactory compromise solutions for change requires trust, which can only be built over time and through social learning processes (Keen et al, 2005). The best results are therefore likely to be achieved when researchers make concerted efforts to build links with relevant policy-making institutions and policy-makers so that priority research areas are identified jointly and the results taken up and acted upon by a ready policy audience.

Conclusions

The evidence that has been collected so far, and presented in this book, makes a powerful case in favour of promoting AIVs in urban and peri-urban agriculture. However, developing further research on AIVs will simply not be enough to ensure that the production, marketing and consumption of these vegetables actually increase in cities across Africa. Researchers have been documenting and reporting on the immense advantages of urban and peri-urban agriculture for decades; but with rare exceptions, agriculture is not recognized as a legitimate urban activity and explicitly incorporated within municipal development plans. As this chapter shows, for AIVs to be successfully promoted in urban and peri-urban areas, several constraints need to be removed, which require policy change in a number of areas. Documenting and reporting the immense advantages of AIVs in urban and peri-urban areas in itself will not be enough; most vitally, the research and development community will also have to take *active* steps to communicate and discuss their findings to policy-makers in their cities.

Note

1 The policy dialogue was held at Rhodes University, Grahamstown, South Africa, 23–26 January 2008. For further information, refer to IndigenoVeg (2008) (download the report from www.indigenoveg.org/).

References

Adango, E., Onzo, A., Hanna, R., Atachi, P. and James, B. (2006) 'Mite pests of major importance on indigenous leafy vegetables in Benin: The search for appropriate control strategies. I – International conference on indigenous vegetables and legumes. Prospectus for fighting poverty, hunger and malnutrition', *Acta Horticulturae*, vol 752, pp311–318

Bah, M., Cissé, S., Diyamett, B., Diallo, G., Lerise, F., Okali, D., Okpara, E., Olawoye, J. and Tacoli, C. (2003) 'Changing rural–urban linkages in Mali, Nigeria and Tanzania', *Environment & Urbanization*, vol 15, no1, pp13–24, www.ucl.ac.uk/dpu-projects/drivers_urb_change/urb_economy/pdf_Urban_Rural/IIED_Bah_Mali.pdf

Bryld, E. (2003) 'Potentials, problems and policy implications for urban agriculture in developing countries', *Agriculture and Human Values*, vol 20, pp79–86

Collins, J. P., Kinzig, A., Grimm, N. B., Fagan, W. F., Hope, D., Wu, J. G. and Borer, E T. (2000) 'A new urban ecology', *American Scientist*, vol 88, pp416–425

Drescher, A. W. and Mackel, R. (2000) *Urban and Peri-Urban Food Security: The Example of Lusaka (Zambia), APT-Reports 11: Urbanization, Vulnerability and Resource Management in Developing Countries*, Freiberg, Germany, pp56–79

Ellis, F. and Sumberg, J. (1998) 'Food production, urban areas and policy responses', *World Development*, vol 26, pp213–225

Fuller, R. A., Irvine, K. N., Devine-Wright, P., Warren, P. H. and Gaston, K. J. (2008) 'Psychological benefits of green space increase with biodiversity', *Biology Letters*, vol 3, pp390–394

IndigenoVeg (2008) *African Indigenous Vegetables and Urban and Peri-Urban Agriculture Policy Dialogue*, Workshop Documentation, Rhodes University, Grahamstown, South Africa, 23–26 January 2008

Keen, M., Bruck, T. and Dyball, R. (2005) 'Social learning: A new approach to environmental management', in Keen, M., Brown, V. and Dyball, R. (eds) *Social Learning in Environmental Management: Towards a Sustainable Future*, Earthscan, London, pp3–21

Maxwell, D. G. (1995) 'Alternative food security strategy: A household analysis of urban agriculture in Kampala', *World Development*, vol 23, pp1669–1681

Maxwell, D., Levin, C., Amar-Klemesu, M., Morris, S., Ruel, M. and Ahiadekeke, C. (2000) 'Urban livelihoods and food and nutrition security in greater Ghana', Research Report no. 112, International Food Policy Research Institute, Washington, DC

McGranahan, G., Marcotillio, P., Bai, X., Balk, D., Braga, T., Douglas, I., Elmqvist, T., Rees, W., Satterthwaite, D., Songsore, J. and Zlotnik, H. (2005) 'Urban systems', in *Millennium Ecosystem Assessment: Ecosystems and Human Well-Being – Current State and Trends*, Island Press, Washington, DC, pp795–825

Mitchell, D. (2008) *A Note on Rising Food Prices*, Policy Research Working Paper 4682, World Bank, Development Prospects Group, Washington, DC

Mougeot, L. J. A. (ed) (2005) *Agropolis: The Social, Political and Environmental Dimensions of Urban Agriculture*, IDRC, Earthscan, London

UN-Habitat (2006) *State of the World's Cities 2006/2007: The Millennium Development Goals and Urban Sustainability*, UN-Habitat/Earthscan, London

UN High-Level Task Force (2008) *On the Global Food Crisis – Comprehensive Framework for Action*, www.un.org/issues/food/taskforce/Documentation/CFA%20Web.pdf

UNEP (United Nations Environment Programme) (1999) *Global Environment Outlook 2000*, www.unep.org/geo2000/english/0056.htm

van Veenhuizen, R. (2006) (ed) *Cities Farming for the Future: Urban Agriculture for Green and Productive Cities*, RUAF, Leusden, The Netherlands

World Bank (2008) *Double Jeopardy: Responding to High Food and Fuel Prices*, G8 Hokkaido–Tokyo Summit

Index